Living Without Mathematical Statistics

AF790032

Herbert Ruefer

Living Without Mathematical Statistics

Accurate Analysis, Diagnosis, and Prognosis Based on the Taguchi Method

Springer

Herbert Ruefer
Burghausen, Germany

ISBN 978-3-030-07617-7 ISBN 978-3-319-99632-5 (eBook)
https://doi.org/10.1007/978-3-319-99632-5

© Springer Nature Switzerland AG 2019
Softcover re-print of the Hardcover 1st edition 2019
This work is subject to copyright. All rights are reserved by the Publisher, whether the whole or part of the material is concerned, specifically the rights of translation, reprinting, reuse of illustrations, recitation, broadcasting, reproduction on microfilms or in any other physical way, and transmission or information storage and retrieval, electronic adaptation, computer software, or by similar or dissimilar methodology now known or hereafter developed.
The use of general descriptive names, registered names, trademarks, service marks, etc. in this publication does not imply, even in the absence of a specific statement, that such names are exempt from the relevant protective laws and regulations and therefore free for general use.
The publisher, the authors and the editors are safe to assume that the advice and information in this book are believed to be true and accurate at the date of publication. Neither the publisher nor the authors or the editors give a warranty, express or implied, with respect to the material contained herein or for any errors or omissions that may have been made. The publisher remains neutral with regard to jurisdictional claims in published maps and institutional affiliations.

This Springer imprint is published by the registered company Springer Nature Switzerland AG
The registered company address is: Gewerbestrasse 11, 6330 Cham, Switzerland

In memory of
Dr. Genichi Taguchi

Preface

New methods for dealing with multiparametric data are only accepted if they are relatively easy to learn and provide practical results. Both conditions are fulfilled with the method named after Dr. Genichi Taguchi (TM—Taguchi Method). The user does not need an advanced course in mathematics or special expert knowledge to interpret and exploit self-generated data in a meaningful way.

From a mathematical and statistical perspective, a close interweaving of the Taguchi method and statistics has emerged in numerous publications and books. However, since Taguchi's thoughts consist, among other things, in enabling reliable interpretation of results even with drastic reductions in effort, i.e., small amount of data, a treatise as a subchapter of statistics is not justified. This gave rise to the motivation to point out this approach as an alternative to the traditional point of view of statistics. For reasons of mathematical/physical consistency, some simplifications have been introduced which are in the sole responsibility of the author. The intent is to provide an easier understanding and thus, paving the way for applications and complete interpretability of experimental data or simulation calculations even for users confronted first time with these ideas.

The supporting pillars of the methodology named after Genichi Taguchi refer to the preparation of analyses, diagnoses, and prognoses from multiparametric data.

Accordingly, the structure of the book is divided into three parts. To make it easier to read, some mathematical derivations which are not relevant for the application of the method are spun off into Appendix A (book part—reference #*). The same applies to additional information to deepen some thoughts but is not part of the methodology. Tables are summarized in a separate Appendix B.

References are listed as far as they appear useful for the understanding of certain aspects, if they are available at least as an abstract in English language (book part—reference #). The gradual (historical) development of the methodology can be found in Japanese journals. The editorial is written by Genichi Taguchi in almost all issues of the listed volumes, understandably in Japanese language. If available, the English translation is quoted.

As far as possible, facts are explained using simple examples or geometric sketches instead of mathematical proofs and, if appropriate, are supplemented with easy-to-understand exercises. Due to the historical development, the experimental approaches are given more weight, whereby simulation calculations are becoming increasingly important today. For the first part of the Analysis of Variables, no special tools are needed apart from a calculator. For the following second part, Pattern Recognition and Diagnoses, a PC is advantageous, whereas the software Excel (or similar) is sufficient to follow up tasks or to work on own case studies. The same applies to Part III Prognoses, whereby in Part I and Part II, the mathematical prerequisites for its comprehension are developed.

Particularly noteworthy is the extraordinary support that the author found in all technical discussions in Genichi Taguchi himself. Even though not everything was easy to understand, Yuin Wu, Shin Taguchi, and Alan Wu were always helpful. James Kowalick, Teruo Mori, and Shoichi Teshima provided further decisive impulses for which the author is very grateful. Hideaki Kurosawa interpreted patiently Japanese texts and, finally, I would like to thank Regine Baumgärtel as a critical and constructive reviewer of the original manuscript.

Burghausen, Germany Herbert Ruefer
January 2018

Contents

Part II Pattern Recognition and Diagnosis

Chapter 1
Introduction

Contents

Abstract The origin and development of the introduced methodology is partly due to special circumstances of history. Burdened by the conditions in the 1940s and 50s, methods for practical problem solutions with relatively fast results were in demand. Further years passed until a solid methodical framework was available, which, however, was not easy to communicate. One of the main reasons for this was publications in the Japanese language, which only became accessible in translations in the proper sense at a later point in time. Ultimately, the method's persuasiveness was based on the low effort to obtain reliable results. Today, the consequent application of this method has found its way not only in manufacturing but into many other areas, such as biotechnology and diagnostics, to name a few, with even more opportunities any time.

1.1 History

In 1924, a highly gifted boy was born in Tohkamachi, Japan. It is called Genichi Taguchi. Already as an adolescent, he watched the silkworm pupating, which was cultivated in the family, until the silk thread could be won in laborious subsequent steps. From here to the finished kimono, complicated mechanical devices were used. Perhaps the overall impression of biological variability with not always predictable results, as well as the pitfalls of failure-prone technology with sensitive losses in textile production, shaped his later life. At the secondary school in Tohkamachi, which he graduated in 1941, his natural talent for mathematical contexts was already evident. At the age of 18, he obtained from libraries all publications of the Mathematical-Physical Society of Japan, which he was able to get hold of. His further education at Kiry Technical College was abruptly

© Springer Nature Switzerland AG 2019

H. Ruefer, *Living Without Mathematical Statistics*,
https://doi.org/10.1007/978-3-319-99632-5_1

interrupted by the war events. Even at this difficult time, he deepened himself, as best he could, in every free minute in mathematical literature.

He found his first job after World War II at the then Ministry of Health in Tokyo. Here, he conducted the first surveys to determine the health status of the population. His questionnaires have retained their validity and are still used in this form today. The Japanese mathematician Motosaburo Masuyama—who was urgently seeking reinforcement for a project in the pharmaceutical industry—had heard of Taguchi's talent. He succeeded in convincing the 23-year-old Taguchi of the urgency of this project, so that he quickly changed his job and joined the pharmaceutical company Morinaga. Masuyama introduced him to the problem situation and familiarized him with specific approaches that could provide valuable insights even with a very small database.

After the discovery of penicillin by Alexander Fleming (1881–1955), the most urgent task in the postwar era was the production of the fungal variant used as an antibiotic to treat severe inflammations. In addition to the visual evaluation of hundreds of vaccinated Petri dishes, strict mathematical validation of the results was also necessary before the antibiotic could be approved as a drug. In view of the postwar consequences, every day, every hour that could be used for development in the laboratories counted; the earlier a broad application was possible, the more lives could be saved. The usual statistical procedures are secured by the multitude of necessary data but at the expense of a disproportionately high effort. During the tedious statistical test calculations and under the external constraint of being able to produce reliable results quickly, Taguchi wondered whether a more efficient approach for such a project could not adopt Masuyama's ideas? Wouldn't it be much easier to create a model from the results with just a few systematic tests, to calculate the most suitable variant from them and only to confirm them experimentally? If such a solution were to emerge, statistics could be bypassed, at least in the run-up to technical research and development. Only the final result, if at all still necessary with the confirmed prediction of the underlying model, could be underpinned statistically.

Even though these considerations were not yet fully developed, they did not let him go. With his mathematical intuition, he ingeniously brought together methods that had already passed their test in various specialist areas. As a consequence, universal access to an objective evaluation of experimental results, regardless of their physical, chemical or biological nature, was possible.

On the occasion of his trip to India as visiting professor, Taguchi met the Indian mathematician Prasanta Chandra Mahalanobis (1893–1972) in 1954. He had developed a special algorithm for accurate discrimination of similar objects in a group. Its metric brought Taguchi together with his previously developed methodology. Extensive applications, including supporting medical diagnostics, opened up in this form.

The dissemination of his ideas outside Japan proved to be very difficult, not least because of the still existing language barrier to the western world. Thus, it was a lucky coincidence when Taguchi met the mathematician Yuin Wu from Taiwan in 1967. In addition to his mathematical professionalism, Yuin Wu was also didactically gifted and proficient in English. He was able to provide the first meaningful translations of Genichi Taguchi's thoughts into Chinese and English. From then on, they worked together intensively until Yuin Wu died in 2004. Genichi Taguchi continued to work tirelessly on developing reliable prognosis algorithms until he closed his eyes in Tokyo on June 2, 2012. Today, the methodical development is carried on and spread by the sons of both researchers, Shin Taguchi and Alan Wu.

1.2 Beginning

The function of technical systems is generally based on the use of machines, the reliability of which is of crucial importance for a wide range of applications. The energy transformations required for this, a substantial part of physical processes, are always expected to be highly efficient, i.e., undesirable forms of energy such as friction-related heating or noise emission must be kept to a minimum. If chemical reactions are considered, unwanted by-products must be avoided as far as possible, especially if they are toxic. When one thinks of pharmaceutical products, the decisive criterion is to minimize unintended, although unavoidable side effects. Despite the subject-specific nomenclature, it is easy to see that only the relationship between the intended effect and the undesired effect is the decisive factor in assessing the value of a system function. This is exactly what the following chapters deal with. In order to obtain the largest possible quotient—e.g., for reasons of competitiveness, for environmental protection or the therapeutic application of pharmaceutical products—it is essential to reliably determine the impact of numerous variables (=parameters). If their number is in the two- or three-digit range, the testing of all combinations that can occur is no longer justifiable.

Pioneering work in the field of multiparametric systems goes back to Dr. Genichi Taguchi, who found a way to reduce the effort for the determination of parameter effects enormously. In this case, the number of necessary experiments or simulation calculations—simply put—is proportional to the number of variables considered, i.e., it does not grow exponentially with them as a complete system description would require. For clarification, a simple numerical example: If any process is governed by 63 parameters with two setting options each, then its effects can basically be determined from 64 tests. By considering all existing parameter combinations, however, $\sim 10^{19}$ ($=2^{63}$) experiments or simulation calculations would be necessary. Supposed that a high-speed computer could perform one simulation calculation with 63 parameters in 0.1 μs, almost 32,000 years are required for all calculations. With the aforementioned approach—in contrast to the traditional practice—for the first-time, multiparameter systems do not have to be reduced in terms of the number of parameters but can even be extended for a more extensive description. The effort with 10, 50, 100 or more variables moves into the background in a controllable way, the experimental penetration into unknown territory comes to the fore.

With the relatively simplified data handling of multiparameter systems, a way is emerging enabling decision-making of great importance with high reliability. This usually refers to sensitive situations that require action. Whether, for example, a fire alarm has to be triggered to activate an automatic sprinkler system, immediate emergency braking has to be started due to a collision warning, immediate treatment has to be initiated in the event of an emergency, or—to put it more generic—a safe diagnosis with all its consequences has to be made. The pivotal point of such a decision-making process is always an error-free pattern recognition with the highest degree of reliability. For pattern recognition as such, the metric of P. C. Mahalanobis existed, as depicted in Part 2. Taguchi builds on this with its previously defined key figures and succeeds in improving discrimination power, sometimes even with reduced effort. The advantage of this combined procedure lies in the ability to exclude wrong decisions more reliably, i.e., to improve the diagnostic certainty decisively.

In a certain sense, the abovementioned situation can also be reversed. If one or more results, i.e., responses, from variable values of a multiparameter system are known, the system behavior can be predicted for other, i.e., unknown parameter values, as described in Part 3. This could gain far-reaching consequences not only for complicated technical processes but also in the event of imminent natural catastrophes since enormous damage can be avoided or at least mitigated by early and reliable forecasts.

The aim of this book is to arouse enthusiasm for unconventional ways of looking at things, which in the end turn out to be very successful. This encourages further action and to take on more demanding challenges. As experience grows, so too does the courage to transfer meaningful considerations to attractive future areas. With this in mind, the author wishes all readers using the Taguchi method that the corresponding tension of their own applications will arise, especially when it comes to realizing creative ideas, which may have previously failed because of too much effort involved.

Part I
Analysis of Variables

Chapter 2
Description of Systems

Contents

Abstract The central theme seen through the whole manuscript is the signal-to-noise ratio (*SNR*). It is introduced in the beginning to get the sense and the meaning as objective criteria for comparison and evaluation. For systems composed of numerous factors (or parameters, variables), special sets of equations guarantee the minimum effort to calculate the *SNR* index for analysis. Based on this index, the contribution of each parameter to the system response is derived and a model for prediction is deducted. Experiments or simulation calculations are designed in a way to gain the information of the interaction between parameters and sources of noise. Advantage can be taken from variables with a nonlinear transfer function to neutralize the impact of such noise factors. This results in the highest achievable numerical value of the *SNR* index with the consequence to hit a given target or goal most accurately and most reliably despite all sources of noise.

© Springer Nature Switzerland AG 2019
H. Ruefer, *Living Without Mathematical Statistics*,
https://doi.org/10.1007/978-3-319-99632-5_2

2.1 Concept and Usefulness of the Signal-to-Noise Ratio

If one takes a photo camera, a video camera, or any imaging system at hand, the expectation is that a generated image will provide a true-life image according to the scale of the optical components used. However, exact measurements with test patterns or examination of point-shaped images show that this assumption is sometimes the case, but only approximate. Lens defects and manufacturing tolerances affect contrast and resolution, which can go so far that the imaging function is no longer given. Based on the known physical relationships, the components involved for imaging can be calculated exactly. However, even with precise production, adjustment tolerances and material inhomogeneities are unavoidable, not to mention other sources of error. Taguchi's approach consists in assessing the real system behavior the better, the closer the image or a manufactured piece comes to its ideal dimensions or properties. In other words, the smaller the deviation appears to a given target to be hit, the better the intended function is fulfilled.

For the example of an optical image used here, a lens with a focal length of 50 mm is tested for fidelity. If one takes an object of a fixed size like the sun as the original and maps it (with an appropriate filter) to different spots of the existing image format on film or CCD sensor, e.g., to the diagonal end points and the center, then deviations from the practically circular object become obvious (Fig. 2.1).

From the circular original, some of the smallest and largest dimensions of the distorted images are recorded (Table 2.1).

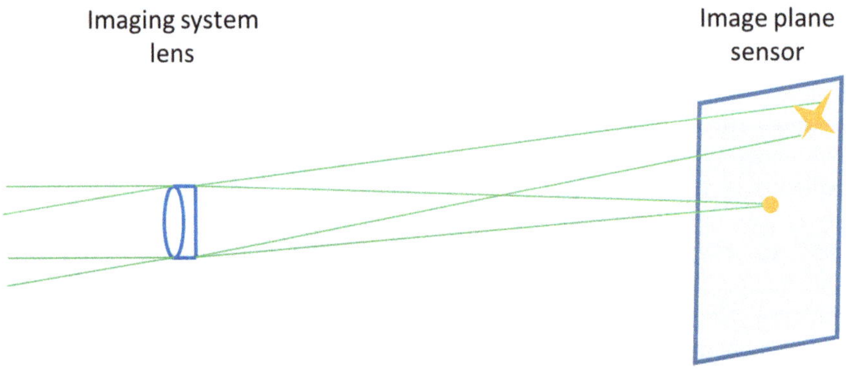

Fig. 2.1 Schematic diagram of the beam path of an image in the optical axis (connecting the center points) and outside from it

Table 2.1 Smallest and largest image diameter of a circular object at different points in the image plane of a 50 mm focal length lens

Diameter of the image (mm)	0.38	0.42	0.44	0.46	0.49	0.51

If the same test is performed with another 60 mm lens, the situation may look like this (Table 2.2).

How to find an objective comparison for an evaluation of the fidelity of the two lenses **(1-1*)**?

One could assume that the distances to the respective arithmetic average allow a conclusion **(1-2*)**. If they were simply added together, the result would be exactly zero because of the definition of the average. An information remains only if either the absolute numerical values are considered, or all signs are made positive by squaring. The latter approach, which can also be applied without precise knowledge of the distribution function of the measuring points, was used by Gauss to characterize dispersion or deviation **(1-3*)** (Table 2.3).

If one looks at the squares of the standard deviations of the two lenses, they are only slightly different with their units in square millimeters. However, such a comparison is too simple since the focal lengths of the lenses have changed the magnification. The longer focal length results in a larger image, so that with it also image defects increase (always at the same aperture ratio). From this follows that an objective evaluation is only meaningful if the scale is invariant. Scale invariance occurs when a quotient always returns the same numerical value, regardless of the originally selected unit. By squaring the average in relation to the squared standard deviation, the units **(1-4*)** are omitted. The result is a characteristically dimensionless number (Table 2.4).

Since the deviation corresponds to a distortion of the image, a higher fidelity is clearly in favor of the longer focal length lens (Table 2.5).

Table 2.2 Smallest and largest image diameter of a circular object at different points in the image plane of a 60 mm focal length lens

Diameter of the image (mm)	0.48	0.50	0.52	0.55	0.59	0.60

Table 2.3 Definitions of the arithmetic average and the squared standard deviation

Arithmetic average	$\bar{y} = \frac{y_1 + y_2 + \cdots y_n}{n}$
Squared standard deviation	$\sigma^2 = \frac{(y_1-\bar{y})^2 + (y_2-\bar{y})^2 + \cdots + (y_n-\bar{y})^2}{n}$

Table 2.4 Comparison of the measurement results of two imaging systems

	Lens with 50 mm focal length	Lens with 60 mm focal length
n: Number of measuring points	6	6
$\bar{y}$: Arithmetic average (mm)	0.45	0.54
σ^2: Squared standard deviation (mm^2)	0.001867	0.001967

Table 2.5 Scale-invariant evaluation of the image accuracy of optical systems

Characteristic value for lens with 50 mm focal length	$\frac{\bar{y}^2}{\sigma^2} = 0.45^2/0.001867 = 108.5$
Characteristic value for lens with 60 mm focal length	$\frac{\bar{y}^2}{\sigma^2} = 0.54^2/0.001967 = 148.3$

The given example is a comparison at two different fixed points (focal lengths of the lenses 50 and 60 mm). Therefore, in the case of target values that are given in this way due to the design, one speaks of static behavior or, more generally, of a static system. If one detaches oneself from the example and considers only the evaluation index, then the counter is defined as the target value of a function to be fulfilled, whereas the denominator is a numerical value that attenuates or disturbs the intended function. As in communication technology, this quotient is therefore referred to as the signal-to-noise ratio, abbreviated to *S/N* **(1-5*)**. The transfer of information is equivalent to a transformation from one form of energy to another, which is the more efficient, the less energy is diverted for malfunctions. This means that a single dimensionless number is sufficient for assessing the function of simple or complex devices or processes:

$$\frac{S}{N} = \frac{\text{Converted useful energy for system function (intended effect)}}{\text{Loss of energy (dissipation, harmful components, disturbances)}}$$

The Taguchi method is based on the concept of quantifying the performance of any given system in an objective way by applying this *S/N* ratio. With the goal

$$\frac{S}{N} = \text{Maximum}$$

the energy form involved is converted in such a way that the deviation from the ideal system function is minimized.

Let us return to the example of the optical image. Obviously, the evaluation index for each lens with a different focal length must be measured and calculated anew. What does this mean in the case of a zoom lens? Here, too, the expectation is that the original and the image correspond with a size according to the focal length. However, with a variable focal length, the fixed point of the previous representation is to be understood as an independent variable. The static state is extended by one dimension, with the task of finding an *S/N* evaluation index for the now called dynamic function. In order to involve this dimension, test measurements must be carried out at several setting levels of the variable, in this case the focal length (Fig. 2.2).

The sampling points are also referred to as signal values M, the ratio of output to input change as slope β equally to sensitivity coefficient, gain factor or sensitivity with the abbreviation *SEN*. In addition to the dispersion behavior at the associated signal values, deviation from linearity can also occur as a further type of disturbance. In case of an ideal optical imaging, there must be a linear relation, i.e.,

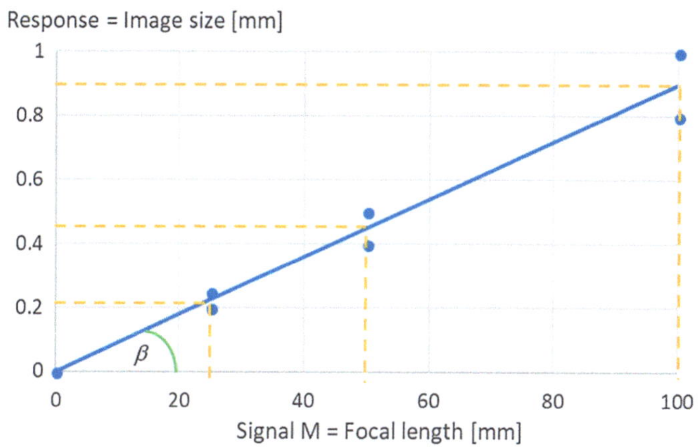

Fig. 2.2 Dimensions of two images at different spots in the image plane for several zoom settings

proportionality, to the image size for fixed object sizes, depending on the focal length. To describe the proportionality factor, a compensation line can be used for this purpose, which is arranged in such a way that the sum of all squared measuring point deviations from the line is minimized. This condition is followed by differentiating the sum of the squared distances after the slope and zero setting (extreme value = minimum) the straight-line slope through the origin with the associated definition of the squared standard deviation for dynamic systems **(1-6*)**.

If, in analogy to the static *S/N* ratio, the square of the arithmetic average is replaced by the squared slope, the numerical value of the quotient increases with both a steeper slope and a decrease in the squared standard deviation. The former is characterized by an increase in sensitivity, the latter by higher precision and reliability. On closer inspection, however, it should be noted that the ratio is dimensional in this form and cannot be scale-invariant under any circumstances. However, the strength of an *S/N* ratio lies precisely in its general validity as a nondimensional value. Therefore, an additional dimension factor is required for an equivalent dynamic *S/N* ratio. Since the slope β is applied in its squared form, the dimension factor must possess the squared unit of the applied signal. As such serves the second moment of the signal values **(1-2*)**. In the abbreviation S stands for the sum, the index 2 for the squared terms (Table 2.6).

The *S/N* ratio for dynamic behavior is the quotient of the squared slope and the squared standard deviation multiplied by the dimension factor. In Table 2.7, the characteristic quantities of static and dynamic systems are compared.

Table 2.6 Definition of the dimension factor of dynamic systems

Dimension factor S_2	
Second momentum of the signal values	$S_2 = \frac{1}{n}\left(M_1^2 + M_2^2 + \cdots + M_n^2\right)$
Number of signals	n

Table 2.7 Characteristic quantities of static and dynamic systems

Static system	Dynamic system
$\bar{y} = \frac{y_1 + y_2 + \cdots + y_n}{n}$	$\beta = \frac{y_1 M_1 + y_2 M_2 + \cdots + y_n M_n}{M_1^2 + M_2^2 + \cdots + M_n^2}$
$\sigma^2 = \frac{(y_1 - \bar{y})^2 + (y_2 - \bar{y})^2 + \cdots + (y_n - \bar{y})^2}{n}$	$\sigma^2 = \frac{(y_1 - \beta M_1)^2 + (y_2 - \beta M_2)^2 + \cdots + (y_n - \beta M_n)^2}{n}$
	$S_2 = \frac{1}{n}\left(M_1^2 + M_2^2 + \cdots + M_n^2\right)$
$\frac{S}{N} = \frac{\bar{y}^2}{\sigma^2}$	$\frac{S}{N} = \frac{\beta^2}{\sigma^2} S_2$

The static approach is included in the dynamic. If all signal values are reduced to one point in the dynamic *S/N* ratio, the static *S/N* ratio is displayed **(1-7*)**.

$$\frac{S}{N}_{\text{dyn}} = \frac{\beta^2}{\sigma^2} S_2 \rightarrow M_i \rightarrow M \rightarrow \frac{\bar{y}^2}{\sigma^2} = \frac{S}{N}_{\text{stat}}$$

If, for technical reasons, the signal values for experiments investigating the same system that are to be compared are different, or if the number of signals is different, S_2 balances the deviations and ensures objective evaluation. A more detailed examination is found in **(1-8*)**.

The logarithmic form is often required for further calculation steps in dealing with the *S/N* evaluation index. For this purpose, the usual definition in communication technology is adopted **(1-9*)**.

$$SNR = 10 \cdot \log\left(\frac{S}{N}\right)$$

The abbreviation *SNR* stands for the logarithmic signal-to-noise ratio. Although dimensionless, the designation Decibel [=dB] is added to distinguish the non-logarithmic quantity. Solving for the *S/N* ratio results in

$$S/N = 10^{\frac{SNR}{10}}$$

to base 2

$$S/N = 2^{\frac{SNR}{3}}$$

Thus, a change of 3 (dB) on the logarithmic scale corresponds to a multiple of factor 2 on the linear scale. Table 2.8 shows the equivalence of both approaches.

For example, if the *SNR* indices of two systems to be compared are 62 (dB) and 59 (dB), then the *S/N* ratio of one system is higher by a factor of 2. For comparative evaluations, the difference of the *SNR* indices alone is important, not their absolute numerical value. In this respect, the pre-factor of 10 is nothing else but a scaling factor. Since the calculations to be carried out include the transformation back into the *S/N* ratio, as described in the following chapters, the unit has no importance in this context. The main reasons for using a logarithmic scale are

Table 2.8 Equivalence of the logarithmic and the linear signal-to-noise ratio

SNR (dB)	*S/N*
…	…
9	8
6	4
3	2
0	1
−3	1/2
−6	1/4
−9	1/8
…	…

- a clear representation method for the illustration of large number ranges
- adaptation to physiological perception **(1-10*)**
- avoidance of over- or undervaluation of *S/N* ratios under the influence of numerous variables
- originally multiplicatively effective parameters are converted into additive relations.

Exercise

For different internal combustion engines, injection nozzles with a nominal diameter of 0.3, 0.4, and 0.6 mm are produced on two production units. Three samples are taken from each variety for optical measurements.

Signal	M_1	M_2	M_3
Nominal diameter (mm)	0.3	0.4	0.6
Production unit 1	0.278 0.296 0.312	0.384 0.409 0.413	0.585 0.590 0.605
Production unit 2	0.291 0.311 0.314	0.388 0.408 0.412	0.598 0.604 0.606

Which production unit is more precise?

Problem solution

The dynamic *SNR* key figure must be determined for each set of data. The larger numerical value stands for a more precise production in the entire specified diameter range.

	Calculation formulas	Unit 1	Unit 2
Slope	$\beta = \frac{y_1M_1+y_2M_2+\cdots+y_nM_n}{M_1^2+M_2^2+\cdots+M_n^2}$; $n=9$	0.992	1.007
Squared standard deviation$_{\text{dynamic}}$	$\sigma^2 = \frac{(y_1-\beta M_1)^2+(y_2-\beta M_2)^2+\cdots+(y_n-\beta M_n)^2}{n}$	0.000155	0.000104
Dimension factor	$S_2 = \frac{1}{n}\left(M_1^2+M_2^2+\cdots+M_n^2\right)$	0.203	0.203
SNR	$SNR = 10\log\left(\frac{\beta^2}{\sigma^2}S_2\right)$	31.11 (dB)	34.13 (dB)

Production unit 2 is more precise by 3 (dB).

2.2 Static Systems

For frequently occurring situations, test results can be assessed the easy way with the average value of the squared measurement data (=second moment) if scale invariance is not important. If, for example, the question arises to reduce undesired side effects or to achieve the smallest possible numerical values for

- the emission of by-products (NO_x, CO, other toxic compounds)
- noise generation (vibration, squeaking)
- defects in manufacturing processes
- side effects of medication
- the detection limit in chemical–physical analysis

then the only objective is to find the lowest possible value of **(1-11*)**

$$\frac{1}{n}\left(y_1^2+y_2^2+\ldots+y_n^2\right) = \bar{y}^2+\sigma^2$$

n: number of measurements.

Since only the smallest numerical value is important here but the measured values are afflicted with dispersion, the advantage of the second moment is that the average value and the deviation are taken into account simultaneously. Hence, the reciprocal form takes a maximum value which is equated to the signal-to-noise ratio, where y_0 stands for the respective unit of the measurement data with the numerical value 1.

$$\frac{S}{N} = \frac{n}{y_1^2+y_2^2+\cdots+y_n^2}y_0^2 = \frac{1}{\bar{y}^2+\sigma^2}y_0^2$$

The notation for the corresponding *S/N* ratio in logarithmic form is called *smaller-the-better (STB).*

$$STB = SNR = 10\log\left(\frac{n}{y_1^2 + y_2^2 + \cdots + y_n^2} y_0^2\right) = 10\log\left(\frac{1}{\bar{y}^2 + \sigma^2} y_0^2\right)$$

The logical counterpart, i.e., the largest possible numerical value, is also of great importance, as with the

- adhesion of a glue (adhesive strength)
- generation rate, for instance in biotechnological processes
- dielectric breakdown voltage (field strength) for insulators
- velocity of a gas stream for certain nozzle shapes
- hardness of an alloy for machining other materials.

For quantification, the reciprocal numerical values for the second moment are used **(1-11*)**.

$$\frac{1}{n}\left(\frac{1}{y_1^2} + \frac{1}{y_2^2} + \cdots + \frac{1}{y_n^2}\right) = \frac{1}{\bar{y}^2}\left(1 + \frac{3\sigma^2}{\bar{y}^2}\right)$$

The reciprocal value is maximal and describes again the signal-to-noise ratio for this situation.

$$\frac{S}{N} = \frac{n}{\left(\frac{1}{y_1^2} + \frac{1}{y_2^2} + \cdots + \frac{1}{y_n^2}\right)} \frac{1}{y_0^2} = \left(\frac{\bar{y}^2}{1 + \frac{3\sigma^2}{\bar{y}^2}}\right) \frac{1}{y_0^2}$$

In this case, the meaningful description of the *SNR* key figure is called *larger-the-better (LTB)*.

$$LTB = SNR = 10\log\left(\frac{n}{\left(\frac{1}{y_1^2} + \frac{1}{y_2^2} + \cdots + \frac{1}{y_n^2}\right)} \frac{1}{y_0^2}\right) = 10\log\left(\left(\frac{\bar{y}^2}{1 + \frac{3\sigma^2}{\bar{y}^2}}\right) \frac{1}{y_0^2}\right)$$

A special case occurs if a numerical value to be achieved is not only as small as possible, but the average value is zero, then the squared standard deviation remains for description in the equation of the type *STB* **(1-11*)**.

$$SNR = 10\log\left(\frac{n}{y_1^2 + y_2^2 + \cdots + y_n^2} y_0^2\right) = 10\log\left(\frac{y_0^2}{\sigma^2}\right)$$

If, on the other hand, the target value, abbreviated *m*, differs from the average value without any adjustable parameters being available to bring the average in line with the target, this difference must also be considered **(1-11*)**. Examples are mechanical fits such as the cylinder diameter and bore of a combustion engine or the chemical synthesis of a product with a certain molecular weight.

$$\frac{1}{n}\left((y_1 - m)^2 + (y_2 - m)^2 + \cdots + (y_n - m)^2\right) = (\bar{y} - m)^2 + \sigma^2$$

Here again, the *S/N* key figure follows from the reciprocal value with

$$\frac{S}{N} = \frac{n}{(y_1 - m)^2 + (y_2 - m)^2 + \cdots + (y_n - m)^2} y_0^2 = \frac{1}{(\bar{y} - m)^2 + \sigma^2} y_0^2$$

The corresponding *SNR* value has the additional notation *nominal-the-best I (NTB I)*

$$NTB\,I = SNR = 10\log\left(\frac{n}{(y_1 - m)^2 + (y_2 - m)^2 + \cdots + (y_n - m)^2} y_0^2\right)$$

or

$$NTB\,I = SNR = 10\log\left(\frac{1}{(\bar{y} - m)^2 + \sigma^2} y_0^2\right)$$

Whenever possible, the scale-invariant *S/N* definition introduced in this chapter should be preferred. Practically always suitable parameters can be found which allow an adjustment to variable target values. Figure 2.3 illustrates the procedure to hit a certain target, for example, with the process time at different deposition rates in a galvanic process.

The following table lists the *S/N* ratios for static systems. The transformation of the *S/N* ratios into the logarithmic scale is performed with.

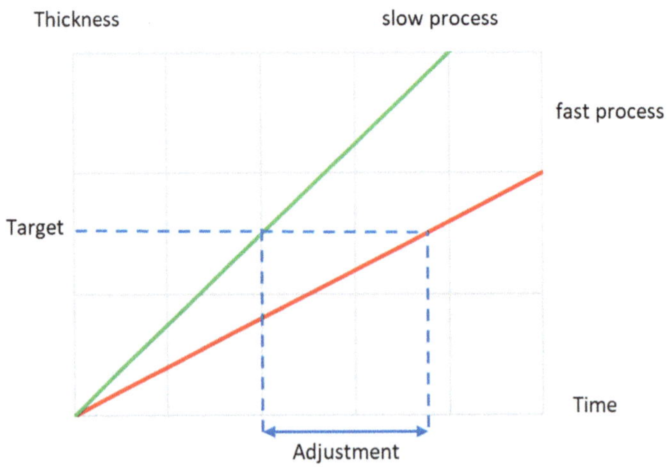

Fig. 2.3 Adjustment of a process on a certain target value

Table 2.9 Summary of the static *S/N* key figures

Application	*S/N* key figure	Properties
Target value as small as possible	$\frac{S}{N} = \frac{n}{y_1^2 + y_2^2 + \cdots + y_n^2} y_0^2$ or $\frac{S}{N} = \frac{1}{\bar{y}^2 + \sigma^2} y_0^2$	Not scale-invariant, the *SNR* key figure is defined as smaller-the-better, *STB*
Target value as large as possible	$\frac{S}{N} = \frac{n}{\frac{1}{y_1^2} + \frac{1}{y_2^2} + \cdots + \frac{1}{y_n^2}} \cdot \frac{1}{y_0^2}$ or $\frac{S}{N} = \frac{\bar{y}^2}{1 + \frac{3\sigma^2}{\bar{y}^2}} \cdot \frac{1}{y_0^2}$	Not scale-invariant, the *SNR* key figure is defined as larger-the-better, *LTB*
Average value zero, Nominal value = 0	$\frac{S}{N} = \frac{n}{y_1^2 + y_2^2 + \cdots + y_n^2} y_0^2$ or $\frac{S}{N} = \frac{y_0^2}{\sigma^2}$	Not scale-invariant, the *SNR* key figure is defined as *nominal-the-best type* II, *NTB* II, Nominal value = 0
Nominal value $\neq 0$	$\frac{S}{N} = \frac{n}{(y_1 - m)^2 + \cdots + (y_n - m)^2} y_0^2$ or $\frac{S}{N} = \frac{1}{(\bar{y} - m)^2 + \sigma^2} y_0^2$	Not scale-invariant, not adjustable, the *SNR* key figure is defined as *nominal-the-best type* I, *NTB* I
Variable target value	$\frac{S}{N} = \frac{\bar{y}^2}{\sigma^2}$	Scale-invariant, adjustment to target value: $\bar{y} = m$
n: number of measurements	y_0^2 = unit^2 numerical value = 1	*m*: Target

$$SNR = 10 \log \left(\frac{S}{N} \right)$$

Exercise

The adhesive strength (in Newton) of glues (for the same materials and adhesive surface) of two manufacturers is compared. Which glue has better properties?

Manufacturer 1 (N)	370	417	465	481	493
Manufacturer 2 (N)	381	399	413	427	465

Problem solution

The target value is of the larger-the-better type, i.e., the corresponding *SNR* key figure must be determined for each manufacturer. Since the units are eliminated by the additional constant y_0, they can be omitted during the calculation.

	Calculation formulas	*SNR (LTB)* (dB)
Manufacturer 1	$10\log\left(\frac{5}{\frac{1}{370^2}+\frac{1}{417^2}+\frac{1}{465^2}+\frac{1}{481^2}+\frac{1}{493^2}}\frac{[N^2]}{y_0^2}\right)$	52.8
Manufacturer 2	$10\log\left(\frac{5}{\frac{1}{381^2}+\frac{1}{399^2}+\frac{1}{413^2}+\frac{1}{427^2}+\frac{1}{465^2}}\right)$	52.3

The glue from manufacturer 1 shows slightly better properties.

2.3 Dynamic Systems

In contrast to the static view, there is an input signal in the dynamic view. This is imprinted on a system that performs an intended transformation according to its function that results in an output value or response. More abstract is the concept of a system function, i.e., the conversion of an input variable into an output variable (Fig. 2.4).

The more general approach lies in a dynamic view. The basic relationship between input and output variable is explained in Sect. 2.1 using the zoom lens as an example. Further systems are listed in the following table (Table 2.10).

At first, linear behavior is assumed for all such systems, i.e., the output variable y changes proportionally to the input variable, referred to as signal value M.

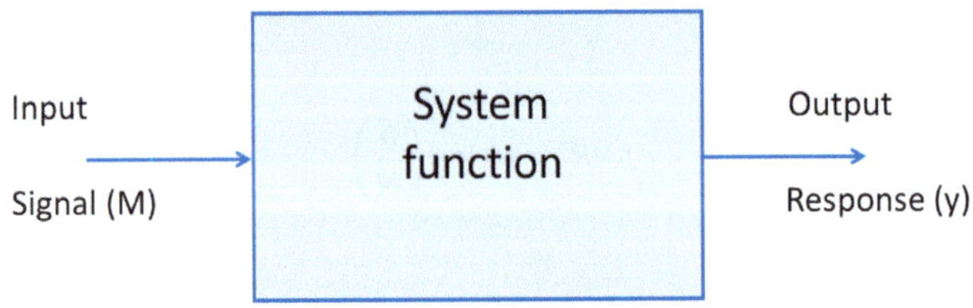

Fig. 2.4 Input variable related to an output variable through a system function

Table 2.10 Collection of some systems with dynamic behavior

System	Input	Output
Loudspeaker	Electric signal	Acoustic response
Car	Path of the accelerator pedal	Engine revolutions
Altimeter (Barometer)	Atmospheric pressure	Elevation above sea level
Balance	Weight	Needle deflection
Thermometer	Temperature	Height of mercury column
Grinding machine	Pressure	Removal rate
Radar pistol	Beat frequency	Speed

$$y \sim M$$

$$y = \beta M$$

The proportionality constant β stands for the slope, equivalent to an amplification factor or sensitivity coefficient.

Virtually all measuring instruments show such a proportionality due to the starting point at the origin of the coordinate system. This means that for a signal with the numerical value zero, the response is zero as well. In practice, the strict relation can be subject to disturbances that cause deviation from linearity, result in greater dispersion, and lower sensitivity. A typical real dynamic system characteristic is displayed in the sketch (Fig. 2.5).

No matter how all faults affect a system, exactly one measure, the *S/N* ratio, describes all properties together. The more the numerical value for *S/N* increases, the better the linear behavior between input and output variables is fulfilled, the smaller the deviation appears, and the higher the slope value **(1-12*)**. Regardless of the kind of system, whether more or less complicated, the *S/N* ratio always reacts unambiguously and describes the better system with the larger numerical value, provided that

- accurate linearity,
- minimum deviation, and
- highest slope

are considered. Even if measured values are far from the origin, the relationship through the coordinate origin is to be applied in case of physical, especially metrological measurement equipment. Only if the zero point is not part of the dynamic response and linear behavior is present, it can be shifted into another value range to compensate for an offset or a systematic error (distortion or bias). If a

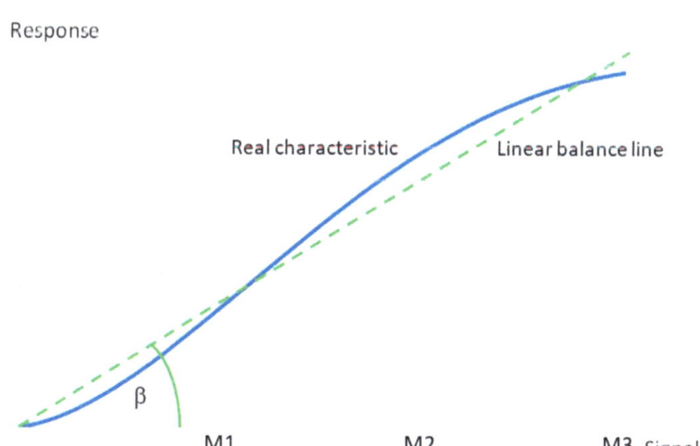

Fig. 2.5 Real system behavior

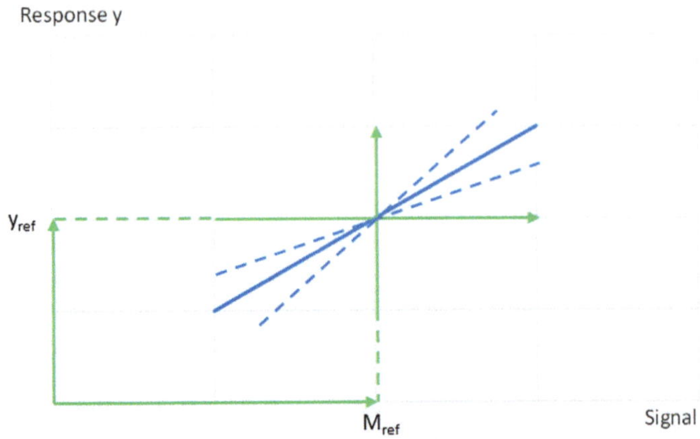

Fig. 2.6 Linear dynamic behavior with respect to a reference point

certain signal value with the corresponding response is given, the selection of the new origin is appropriate to such a reference point. Instead of a known reference point, the signal average and the averaged response can assume the role of the reference point (Fig. 2.6).

$$y_{\text{ref}} = a + \beta M_{\text{ref}}$$

a	Intercept
β	Sensitivity coefficient (slope)
$(M_{\text{ref}}, y_{\text{ref}})$	Reference point

The origin and the reference point are related through the mathematical equations (Table 2.11).

After moving to the new origin (subindices o and r represent the original and reference coordinates as introduced above), the slope β, the squared standard deviation σ^2, and the dimension factor S_2 are calculated with

$$\beta = \frac{(y_{o1} - y_r)(M_{o1} - M_r) + (y_{o2} - y_r)(M_{o2} - M_r) + \cdots + (y_{on} - y_r)(M_{on} - M_r)}{(M_{o1} - M_r)^2 + (M_{o2} - M_r)^2 + \cdots + (M_{on} - M_r)^2}$$

Table 2.11 Shift of the coordinate system with reference point as new origin

Linear function behavior with reference point	Linear function behavior with average value as reference point
$M = M_{\text{original}} - M_{\text{reference}}$	$M = M_{\text{original}} - M_{\text{average}}$
$y = y_{\text{original}} - y_{\text{reference}}$	$y = y_{\text{original}} - y_{\text{average}}$

$$\sigma^2 = \frac{1}{n}\left(((y_{o1} - y_r) - \beta(M_{o1} - M_r))^2 + \cdots + ((y_{on} - y_r) - \beta(M_{on} - M_r))^2\right)$$

$$S_2 = \frac{1}{n}\left((M_{o1} - M_r)^2 + (M_{o2} - M_r)^2 + \cdots + (M_{on} - M_r)^2\right)$$

The terms below have become established in literature to characterize the kind of linear relationship

- Zero-point-proportional equation (use of original data, no shift)
- Reference-point-proportional equation (shift of data to a fixed reference point as new origin)
- Linear equation (shift of data to newly defined origin)

After moving to the newly defined coordinate origin, this results in the familiar relationship for the *S/N* key figure (Sect. 2.1)

$$\frac{S}{N} = \frac{\beta^2}{\sigma^2} S_2$$

or the equivalent *SNR* value

$$SNR = 10 \cdot \log\left(\frac{\beta^2}{\sigma^2} S_2\right)$$

If the slope β is determined, the standard deviation σ can be derived. For this purpose, the representation of the slope β in logarithmic form is advantageous, where β_0 stands for the corresponding unit with the numerical value of one. Based on the additional designation for β as amplification coefficient, the logarithmic expression stands for sensitivity, in short form referred to as *SEN*.

$$SEN = 10 \cdot \log\left(\frac{\beta^2}{\beta_0^2}\right)$$

The squared term avoids problems in case of a negative slope.

Exercise

Two different HiFi amplifiers are subjected to a simple electrical test. For this purpose, the microphone input is fed from a frequency generator with 50 Hz, 1000 Hz, and 15 kHz at 3 signal amplitudes *M* each with 10, 100, and 1000 mV and the corresponding output current through the loudspeaker is measured. Are the amplifiers equivalent?

	Input signal	M_1 10 (mV)	M_2 100 (mV)	M_3 1000 (mV)
Amplifier 1 output (A)	Frequency: 50 Hz	0.04	0.45	3.44
	Frequency: 1000 Hz	0.05	0.42	3.08
	Frequency: 15 kHz	0.03	0.39	1.87
Amplifier 2 output (A)	Frequency: 50 Hz	0.03	0.36	3.21
	Frequency: 1000 Hz	0.04	0.35	3.12
	Frequency: 15 kHz	0.03	0.33	2.42

Problem solution

Without an input voltage, the output current is zero, i.e., a zero-point-proportional equation is the basis for data evaluation. To calculate the *SNR* indices, the slope, the squared standard deviation, and the dimension factor are figured out.

	Amplifier 1	Amplifier 2
Slope (A/mV)	0.0028	0.0029
Squared standard deviation$_{dynamic}$ (A^2)	0.157	0.0426
Dimension factor (mV^2)	336,700	336,700
SNR index (dB)	12.29	18.29

HiFi amplifier 2 is superior by 6 (dB) essentially due to the frequency response.

2.4 Relative Signal Values

Dynamic *S/N* indices can only be calculated if numerical signal values and their corresponding output responses are present. If only the signal exists, whose true value is also unknown, it is sometimes possible to generate further interpolation points by continuous extension or reduction. For clarification, the signal to be used is a substance concentration. If an unknown concentration of a substance is not known, the concentration can either be increased by adding the same substance or, conversely, reduced with a solvent. Depending on the range to be covered, it may be advisable to set certain intervals or ratios of the signal values. With both measures, relative signal values are generated for which corresponding *S/N* key figures can be determined, provided that an approximate linear relation can be assumed between the signal and output values.

The first step is the addition of the substance to generate certain intervals of the signals. At the beginning, there is a solution of a known substance with a low but unknown concentration, M_1^*. The addition of the same substance increases the concentration, whereby the intervals can be arbitrary; sufficient stock solution, simple units, and equidistant intervals are assumed only for reasons of clarity.

$M_1^* = 100\,\text{g Solution of concentration} \quad x$
$M_2^* = 90\,\text{g Solution} + 10\,\text{g Substance} \quad \sim x + 10\%$
$M_3^* = 80\,\text{g Solution} + 20\,\text{g Substance} \quad \sim x + 20\%$
$M_4^* = 70\,\text{g Solution} + 30\,\text{g Substance} \quad \sim x + 30\%$
and so forth

From this, the average concentration is calculated, by which the other values are grouped.

$$\overline{M} = \frac{x + (x + 10\%) + (x + 20\%) + (x + 30\%)}{4} = x + 15\%$$

The signals related to the average value result in

$M_1^* - \overline{M} = x - (x + 15\%) = -15\% = M_1$ (lowest concentration)
$M_2^* - \overline{M} = (x + 10\%) - (x + 15\%) = -5\% = M_2$
$M_3^* - \overline{M} = (x + 20\%) - (x + 15\%) = +5\% = M_3$
$M_4^* - \overline{M} = (x + 30\%) - (x + 15\%) = +15\% = M_4$ (highest concentration)
and so forth

Test results are objectively comparable, although the true signal value is not known.

Alternatively, the second example shows the formation of signal ratios. The addition of solvents creates additional measurement points. These can be adjusted as desired, whereby a continuous dilution by a factor ½ is selected for illustration purposes only.

$M_4^* = 100\,\text{ml Solution of concentration} \quad x$ (highest concentration)
$M_3^* = 100\,\text{ml Solution } M_1^* + 100\,\text{ml Solvent} \quad 1/2x$
$M_2^* = 100\,\text{ml Solution } M_1^* + 300\,\text{ml Solvent} \quad 1/4x$
$M_1^* = 100\,\text{ml Solution } M_1^* + 700\,\text{ml Solvent} \quad 1/8x$ (lowest concentration)
and so forth

The new signal values are compared with the ratios from the dilution stage and of the original solution.

$$M_4 = \frac{x}{x} = 1$$
$$M_3 = \frac{\frac{1}{2}x}{x} = \frac{1}{2}$$
$$M_2 = \frac{\frac{1}{4}x}{x} = \frac{1}{4}$$
$$M_1 = \frac{\frac{1}{8}x}{x} = \frac{1}{8}$$
and so forth

The described possibilities are completely equivalent. As the graphs show, the comparability of different tests is guaranteed due to equivalent signal values (Fig. 2.7).

If neither the true signal value is accessible, nor a relative description is possible, a comparative evaluation of experimental data can be carried out under the assumption of the proportional behavior of the output response to the input variable, provided that at least one measurement repetition exists. For this purpose, the average value is calculated from all test data representing the state of the first signal value and used as the signal value M_1. The next signal value M_2 is again averaged from the associated test data, similar M_3, etc. As a result, the slope is identical to one, so that for different tests, the deviation remains an essential criterion. The following example illustrates the situation: Two different procedures are investigated for measuring optical brilliance. Relative measurement values characterize the surface. Shiny etched brass, electropolished stainless steel, and gold foil serve as samples (Table 2.12).

Calculation of the respective signal values from the corresponding measurement data (Table 2.13)

$$SNR = 10 \cdot \log\left(\frac{\beta^2}{\sigma^2} S_2\right)$$

$$S_{2(1)} = \frac{1}{3}\left(71.5^2 + 83.5^2 + 95^2\right) = 7037$$
$$\beta = 1$$
$$\sigma_1^2 = 184.5$$
$$SNR_1 = 10 \log\left(\frac{1^2}{184.5} 7037\right) = 15.8\,(\text{dB})$$

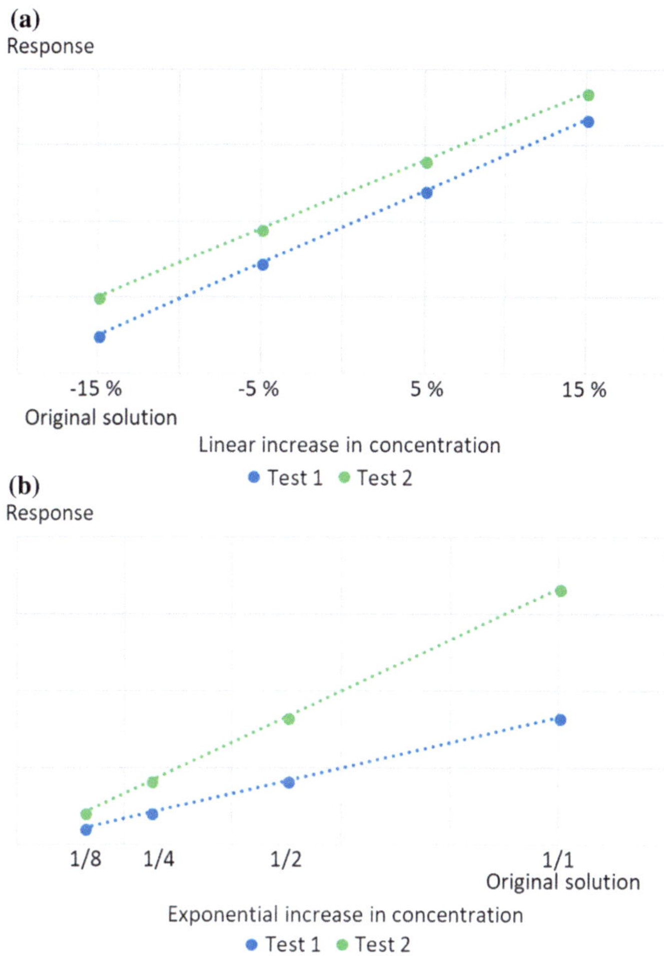

Fig. 2.7 Presentation of relative signal values for **a** intervals, **b** ratios

Table 2.12 Characterization of the surface property of different materials

Material	Brass		Stainless steel		Gold	
Procedure 1	62	81	65	102	84	106
Procedure 2	76	85	85	96	104	112

Table 2.13 Derived signal values from the measured data

Signal	71.5		83.5		95.0	
Procedure 1	62	81	65	102	84	106
Signal	80.5		90.5		108.0	
Procedure 2	76	85	85	96	104	112

$$S_{2(2)} = \frac{1}{3}\left(80.5^2 + 90.5^2 + 108^2\right) = 8778$$

$$\beta = 1$$

$$\sigma_2^2 = 22.2$$

$$SNR_2 = 10\log\left(\frac{1^2}{22.2}8778\right) = 26\,(\text{dB})$$

The second procedure appears to be more suitable due to lower deviation values.

Exercise

Two analytical methods are to be compared with respect to their linear response. An aqueous solution of low concentration of calcium chloride with unknown water content, i.e., unknown concentration, is used. To test the methods, the initial solution is diluted with distilled water to ½ and ¼ of the original concentration. Which method promises to deliver more reliable results?

Concentration	2. Dilution step	1. Dilution step	Original solution
Method 1 Measurement units	55	107	212
Method 2 Measurement units	265	524	1048

Problem solution

For the application of a zero-point-proportional equation, the concentration series is used for the signal values: $M_1 = 0.25$, $M_2 = 0.5$, $M_3 = 1.0$

	Method 1	Method 2
Slope	212.8	1057
Squared standard deviation	1.413	5.6
Dimension factor	0.4375	0.4375
SNR index (dB)	41.5	49.4

The linearity of the second method is superior by 8 (dB) to the first method.

2.5 Orthogonal Arrays and Their Design

Demanding technical processes such as multistage chemical reactions, successive steps in the production of integrated circuits, or design studies in aerospace are accompanied by a rapidly increasing number of variables. These affect the system function to varying degrees as can be seen in the output response. The extension of the system function with an input signal is illustrated. For static systems, the input signal is omitted (Fig. 2.8).

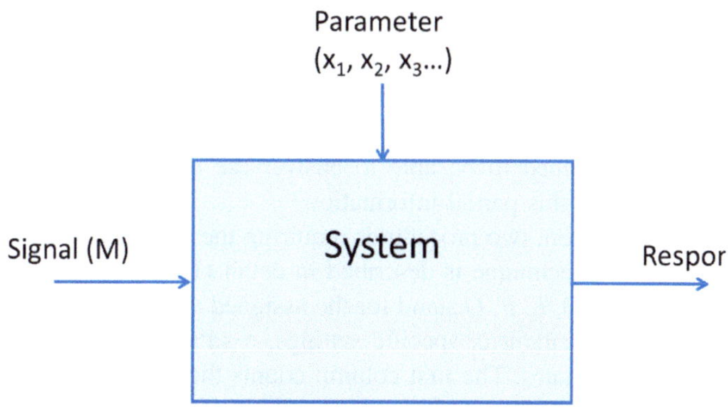

Fig. 2.8 Diagram of a dynamic system function

If all parameter effects are well known, they could be used in several ways

- to achieve maximum yields or efficiencies,
- to stabilize target values even under disturbing influences,
- to keep manufacturing costs in terms of material, energy, and time to a minimum.

In the standard method for the experimental determination of the parameter effects, only one parameter is changed at a time while keeping all other parameters constant. If in the simplest case, only two parameters, *A* and *B*, with two levels, 1 and 2 each, are described, four variants result to be tested (Table 2.14).

With each additional parameter, the total number doubles, i.e., the effort for testing all combinations increases exponentially with the number of variables *P*.

$$N = 2^P$$

Correspondingly for three levels

$$N = 3^P$$

and so forth

For example, this means 128 possible combinations for seven parameters with two levels each, although ultimately only the information regarding the seven

Table 2.14 Combination options for two parameters with two levels each

Number of combinations	Parameter *A*	Parameter *B*
1	1	1
2	1	2
3	2	1
4	2	2

parameters is searched for. This disproportionately large serial effort for the investigation of all variants was replaced at an early stage by a parallel approach. A part or all variables are changed at the same time according to a certain pattern. The consideration is to reduce the experimental effort to such an extent that exactly the information is obtained to be able to answer the question of the parameter effects. How reliable is this partial information?

To clarify this question, two procedures requiring the same effort are compared. First, the combination technique is described in detail **(1-13*)** (Table 2.15).

The letters *A*, *B*, *C*, *D*, *E*, *F*, *G* stand for the assigned variables, and the numbers 1 and 2 represent the parameter-specific settings, a smaller and larger numerical value or two different states. The first column counts the experimental effort, here eight tests. As can be seen, this procedure only changes one variable at a time. In a comparison of test #1 and test #2, changes in a result are therefore causally due to variable A, since only this setting was chosen differently. Up to test #8, the influence of all parameters *A*–*G* is thus continuously determined. This analysis, which appears obvious at first glance, presupposes, however, that a similar result would be achieved even if the setting values 1 and 2 were reversed. However, this cannot be taken for granted, since coupling between the variables or interactions in general cannot be ruled out. Interaction refers to a mutual influence or dependence, such as the change of a beneficial effect of a drug when another drug is administered at the same time. Thus, it cannot be excluded from the outset that the effect of a certain parameter setting depends on which setting value one of the other parameters takes. This means that the evaluation schema is not necessarily reliable, since only one certain configuration—or eight when all tests are carried out—of 128 possibilities is applicable. Conversely, there will be only a consistent result if all other variables with their respective setting values are taken into account for the evaluation of one variable. For two states, this means the already outlined four possible combinations for two variables. This arrangement considers the settings *A*1 and *A*2 for both *B*1 and *B*2, while the same applies for *B*1 and *B*2, i.e., all states of *A* are also included for *B*1 and *B*2. Can more variables be added according to this

Table 2.15 Assignment of variables with the combination technique

Test #	Parameter						
	A	*B*	*C*	*D*	*E*	*F*	*G*
1	1	1	1	1	1	1	1
2	2	1	1	1	1	1	1
3	2	2	1	1	1	1	1
4	2	2	2	1	1	1	1
5	2	2	2	2	1	1	1
6	2	2	2	2	2	1	1
7	2	2	2	2	2	2	1
8	2	2	2	2	2	2	2

principle? By trial and error, the above schema can be extended by one more column (Table 2.16).

However, there do not exist other columns with the same number of setting values in another combination, so that for a larger number of variables, in the example seven parameters A–G, the number of rows must be doubled to consider all variables equally.

With eight tests, the number of permutations with seven columns is exhausted. If there are more columns, i.e., variables to be analyzed, the number of rows must be doubled again, so that eight times level 1 and eight times level 2 occurs. There is a maximum of 15 columns for such an arrangement, all of which are equally represented. A further doubling of the number of lines to 32 offers space for up to 31 variables, similarly to 64 lines with up to 63 parameters, etc.

If all combinations of level settings in pairs of columns are equally frequent, such a balanced arrangement is referred to as an orthogonal array **(1-14*)**. Another look at Table 2.17 reveals now a pattern classified as an orthogonal array. The blueprint of this schema becomes obvious when the parameters are evaluated column by column. As already indicated, *A*1 is the average value from tests 1, 2, 3, 4 and *A*2 corresponding to 5, 6, 7, 8. For *B*1 1, 2, 5, 6 and 3, 4, 7, 8 are to be used and 3, 4, 7, 8 for *B*2. Continuously up to and including *G*, the average values from tests combined accordingly are to be used. For *G*1, these are the tests 1, 4, 6, 7 and for *G*2, the remaining 2, 3, 5, 8. The decisive consideration when dealing with orthogonal arrays is that only the averaged effect remains, since all other parameters are varied equally with all setting values. If a consistent effect is achieved for a

Table 2.16 Smallest possible test number to record three parameters with two levels each

Test #	*A*	*B*	*C*
1	1	1	1
2	1	2	2
3	2	1	2
4	2	2	1

Table 2.17 Balanced arrangement of seven parameters with two levels each

Test #	Parameter						
	A	*B*	*C*	*D*	*E*	*F*	*G*
1	1	1	1	1	1	1	1
2	1	1	1	2	2	2	2
3	1	2	2	1	1	2	2
4	1	2	2	2	2	1	1
5	2	1	2	1	2	1	2
6	2	1	2	2	1	2	1
7	2	2	1	1	2	2	1
8	2	2	1	2	1	1	2

parameter, even though all other variables change, this means high reliability. Conversely, an effect can be reduced or counteracted by changing the settings of the other variables, so that in this case, the parameter effect will be small, i.e., relatively close to the average value. The purpose and intention of the application of orthogonal arrays is to read out only the really reliable parameter effects from in this way designed experiments. In other words, an orthogonal array fulfills the function of a measuring instrument to detect the true effects in the parameter space. Only such effects are effective and can be used advantageously even for upscaling of laboratory results.

Orthogonal arrays can be constructed with three or more instead of two setting values. Three setting values have the advantage of being able to immediately recognize linear behavior, an extreme value (maximum/minimum also called peak/ valley) or a saturation characteristic (Fig. 2.9).

In addition, linear effect behavior at three (or more) levels allows settings to be interpolated or, more importantly, to be extrapolated to extend the corresponding system response beyond existing ranges of experience. The increase in information of orthogonal arrays with more than three values is relatively low. Orthogonal arrays with two setting values each are preferred for screening purposes because of the large number of variables to be accommodated. For all other applications, such with three or mixed orthogonal arrays, consisting of two and three setting values, are recommended.

Since the same frequency of occurrence of all pairwise combinations of levels limits the number of columns, the next larger numbering scheme must be used when the number of variables is exceeded. For mathematical reasons of symmetry, only certain orthogonal arrays exist, as some of them are represented (Table 2.18).

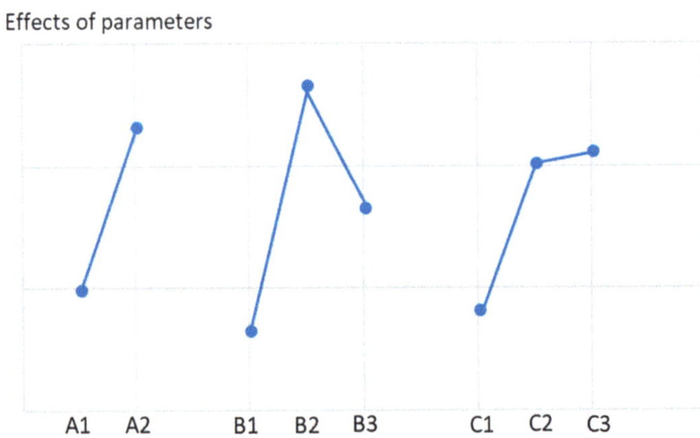

Fig. 2.9 Information increase for parameters with 2 and 3 levels

Table 2.18 Size and properties of some orthogonal arrays

Rows of the matrix = Number of tests to be performed	Columns of the matrix = Largest number of assignable parameters	Number of parameter levels = Number of sampling points
4	3	2
8	7	2
16	15	2
9	4	3
27	13	3
81	40	3

2.6 Types of Orthogonal Arrays

In addition to the purely orthogonal arrangements, there are also almost orthogonal arrays and mixed orthogonal arrays, i.e., parameter columns with two levels and three or more levels. The decisive difference lies in dealing with interactions between two arbitrary columns with the variables assigned to them. Interactions between three and more parameters are averaged out in practice, so that only the situation of low or strong coupling of parameter pairs needs to be considered in more detail.

If the coupling is so strong that the effect assumes similarly large values as the individual effect itself or even exceeds them, the stubborn consideration of only the individual effects leads to incorrect interpretations. If strong interactions as such are uninteresting and undesirable, they can be mitigated or even eliminated by

- replacing the used variable to find a better energy relation and
- decoupling of the parameter pair settings.

If the interaction itself is to be looked for, then it is treated as an additional independent parameter which, like all other parameters, can be determined using a suitable orthogonal array. This is where the decisive difference between orthogonal arrays with different properties comes into play. Only with purely orthogonal arrays, the coupling of any parameter pairs is uniquely assigned to a certain column of the orthogonal array. The accompanying interaction tables (Appendix B) are used to identify them. Partially, orthogonal or mixed orthogonal means that although these are balanced in the frequency of the occurrence of the variable's setting levels, possible interactions between parameter pairs are broken down and are partially redistributed to all available columns. Consequently, there exist no interaction tables for these orthogonal arrays and individual interactions cannot be determined. The decisive advantage of the application of partially orthogonal arrays is that existing, although unknown, not too strong interactions are rendered harmless by decomposition and redistribution to all columns. This means that partial orthogonal arrays are predestined for applications in production processes, which must be inherently free of interactions.

The smallest pure orthogonal array with variables on two levels requires four equations or tests for the unambiguous calculation of up to three parameters. The next larger one is the L8 with the possibility to consider up to seven parameters. This is followed by the L16 with up to 15 parameters, the L32 with up to 31 parameters, the L64 with up to 63 parameters, etc. Correspondingly, there are purely orthogonal arrays based on three level settings, starting with L9 to assign a maximum of four parameters, followed by L27 with 13 columns, then L81 with space for 40 variables, and so on. The modified L64 with up to 21 parameters is available with four level settings. With five levels, there is the corresponding L25 with space for a maximum of six parameters. In addition to the already mentioned purely orthogonal arrays, there exist the corresponding interaction tables suitable for determining the columns on which the interactions fall. Only purely orthogonal arrays may be modified according to certain rules and adapted to special experimental conditions. Table 2.19 shows a summary of pure and partially orthogonal arrays and their properties. The most frequently used orthogonal arrays with related interaction tables and modified orthogonal arrays are listed in Appendix B.

Table 2.19 Overview of orthogonal arrays and some characteristic features

Orthogonal Arrays and their acronyms	Maximum number of parameters	Levels
Pure orthogonal array L4	3	2
Pure orthogonal array L8	7	2
Pure orthogonal array L16	15	2
Pure orthogonal array L16	5	4
Pure orthogonal array L32	31	2
Pure orthogonal array L32	10	2 and 4
Pure orthogonal array L64	63	2
Pure orthogonal array L64	21	4
Pure orthogonal array L9	4	3
Pure orthogonal array L27	13	3
Pure orthogonal array L81	40	3
Pure orthogonal array L81	10	9
Pure orthogonal array L25	6	5
Partial orthogonal array L12	11	2
Partial orthogonal array L18	8	2 and 3
Partial orthogonal array L20	19	2
Partial orthogonal array L28	27	2
Partial orthogonal array L36	23	2 and 3
Partial orthogonal array L36	16	2 and 3
Partial orthogonal array L44	43	2
Partial orthogonal array L50	12	2 and 5
Partial orthogonal array L54	26	2 and 3
Partial orthogonal array L108	49	3

2.7 Modification of Orthogonal Arrays

The recognition and consideration of interactions is identical to the question of modifying a purely orthogonal array with respect to a column with more setting values. This is explained by the example of the already mentioned L8 (Table 2.20) with the accompanying interaction table (Table 2.21).

The interaction table is designed in a way that the number at the intersection point of the parameter pairs indicates the number of the column associated with the interaction of the pair. If the effect amplitude of the interaction is searched for, no other assignment of this column is permitted. If one looks at the parameter pair *A* (=column 1) and *B* (=column 2), the interaction $A \times B$ (symbolically represented by x) is defined to column 3. Columns 1, 2, and 3 therefore form a numerical triple. An overall parameter analysis discloses whether an effect is also visible on column 3, even though no variable is assigned. If the amplitude is relatively small in a cross comparison with the effects on column 1 (=parameter *A*) and column 2 (=parameter *B*), it can be neglected for practical applications. If it appears with a similar numerical value (=difference of the responses from setting value 1 to setting value 2) or greater, the consideration of only the preferential setting values of *A* and *B* leads to a misinterpretation, since the interaction is greater than the individual effects of *A* and *B*. In such a situation, the effect of all possible combinations of *A* and *B* must be revealed and evaluated individually. With one parameter pair of two levels, there are exactly four combinations.

The related number triple must be changed in such a way that a new parameter is created with these combinations, i.e., with four levels. An unambiguous assignment must be chosen for this purpose. The following table shows one possible definition (Table 2.22).

With this definition, columns *A*, *B*, and *C* of the orthogonal array L8 are to be replaced by the new variable *N* with four levels corresponding to the existing combinations of *A* and *B* (Table 2.23).

All possible combinations of the parameter pair (*A*, *B*) are available for evaluation, so that the most advantageous combination can be selected independently of

Table 2.20 Pure orthogonal array L8

Test #	Parameter						
	$A = 1$	$B = 2$	$C = 3$	$D = 4$	$E = 5$	$F = 6$	$G = 7$
1	1	1	1	1	1	1	1
2	1	1	1	2	2	2	2
3	1	2	2	1	1	2	2
4	1	2	2	2	2	1	1
5	2	1	2	1	2	1	2
6	2	1	2	2	1	2	1
7	2	2	1	1	2	2	1
8	2	2	1	2	1	1	2

Table 2.21 Interaction table L8

Parameter columns	Parameter columns						
	1	2	3	4	5	6	7
1		3	2	5	4	7	6
2			1	6	7	4	5
3				7	6	5	4
4					1	2	3
5						3	2
6							1

Table 2.22 Number of combinations of arbitrary column pairs with two levels each

Combination	Column_i; $i \neq k$	Column_k; $i \neq k$
1	1	1
2	1	2
3	2	1
4	2	2

Table 2.23 Resolution of an interaction for four levels

A	*B*	→	$A \times B$ or new variable *N* with 4 levels
1	1	→	1
1	2	→	2
2	1	→	3
2	2	→	4

the individual effects of variables *A* and *B*, Table 2.24. The modified L8 can also be regarded as an orthogonal array, which allows the assignment of one parameter with four levels, together with a maximum of four other parameters at two levels each (Table 2.25). In this way, purely orthogonal arrays can be modified for special experimental conditions. As experience has shown, an orthogonal array is rarely customized so that further examples with additional hints and information are listed in the appendix **(1-15*)**.

Table 2.24 Substitution of three parameters with two levels each by one parameter with four levels

Test #	Parameter						
	A, *B*		*N*	*D*	*E*	*F*	*G*
1	(1, 1)	→	1	1	1	1	1
2	(1, 1)	→	1	2	2	2	2
3	(1, 2)	→	2	1	1	2	2
4	(1, 2)	→	2	2	2	1	1
5	(2, 1)	→	3	1	2	1	2
6	(2, 1)	→	3	2	1	2	1
7	(2, 2)	→	4	1	2	2	1
8	(2, 2)	→	4	2	1	1	2

Table 2.25 Modified orthogonal array L8 with one variable with four levels

Test #	Parameter				
	A	*B*	*C*	*D*	*E*
1	1	1	1	1	1
2	1	2	2	2	2
3	2	1	1	2	2
4	2	2	2	1	1
5	3	1	2	1	2
6	3	2	1	2	1
7	4	1	2	2	1
8	4	2	1	1	2

2.8 Interpretation of Orthogonal Arrays

With the application of orthogonal arrays, as explained above, only those effects are retained that are maintained when all other parameters are changed at the same time. The expectation is that the numerical test results can be added, and the average value describes the effect of each variable. After the complete parameter analysis has been carried out, the desired or best possible functional characteristics can be determined by combining the settings considered to be advantageous in a configuration, i.e., adding their effects. To avoid over- or underestimations, which could lead to a negative absolute temperature or a negative number of defects or similar, all calculation steps are performed in logarithmic transformation, i.e., in the *SNR* scale. Only the end result is transformed back and is available for the product or process characteristic in the usual size and unit. Mathematically speaking, the execution of the calculations on a logarithmic scale corresponds to dealing with geometric average values. The reason for this is that the geometric average, with numbers of the same unit lying relatively far apart, represents a quantity that is better suited to an intuitively estimated center than the arithmetic average. Furthermore, when combining different parameter effects, their overall effect is usually multiplicative. On the logarithmic scale, this is transformed into a summation with the advantage to carry out all calculation steps without any exception as addition and subtraction.

To calculate the overall result, the effects of parameters *A*, *B*, *C*, etc., which are recognized as beneficial are added together as the values of the *SNR* indices in relation to the overall average and added together with the overall average. The graphic shows the geometric procedure (Fig. 2.10).

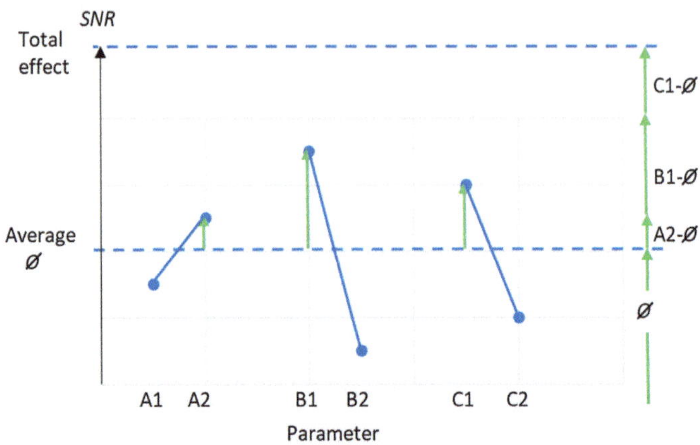

Fig. 2.10 Geometric illustration for the determination of the overall effect of an advantageous parameter configuration

Correspondingly, the numerical result is calculated as the sum of the effect quantities above the average $\varnothing$.

$$SNR_{\text{result}} = (SNR_A - \varnothing) + (SNR_B - \varnothing) + (SNR_C - \varnothing) + \cdots + \varnothing$$

Generally speaking, when adding n parameter effects, the $(n - 1)$-fold average value must be subtracted.

$$SNR_{\text{result}} = SNR_A + SNR_B + SNR_C + \cdots - (n - 1) \cdot \varnothing$$

The characteristic value previously defined as the target value is derived from the back-transformed numerical value of the *SNR* index. In principle, all parameters assigned to the columns of an orthogonal array could be used to calculate a maximum value of the *SNR* index. In practice, however, it can be assumed that, apart from possible interactions, tolerances of the setting values and of the measurements exist. To remain realistic in the prediction, the evaluation should be limited to half to two-thirds of the parameters with the greatest effects. This rule of thumb should be applied primarily to screening experiments.

If a parameter analysis is already carried out with a preselection of variables with expected larger effects at three setting values, then all parameter effects up to 1 or 2 with the comparatively smallest effects can be used for the calculation of the *SNR* key figure. The latter are suitable for estimating errors. In any case, the calculated effect of a new test configuration must be proofed, as this is usually not part of the orthogonal array and, for reasons of effort, there are no multiple test repetitions. The predicted result of the model has to be confirmed, i.e., must be compared with the experimental result of the identical parameter configuration.

Due to all tolerances, a hundred percent agreement will not occur, but the experimental result and the model calculation should lead to numerical values that are close to each other. It is purely based on expert knowledge whether the model of the analyzed parameter effects is confirmed (Sect. 2.10). An error analysis by one (or more) not assigned columns of the orthogonal array or parameters with the smallest effects that are not considered can be helpful for an interval limitation of the model calculations but carries the risk of a subjective influenceability. Such a calculation of errors is not a prerequisite and should not be regarded as conclusive proof of the model calculation if the experimental numerical value in the tolerance interval applies. By adding further parameters with similarly small effects, the error barriers could be successively widened, so that sooner or later the experimental value will appear in the tolerance range. Here, objectivity is to be maintained and the strict comparison of the average values from experiment and prediction is to be preferred in principle to all error calculations. If the values are too far apart according to professional estimates, this should be regarded as a warning. In such a situation, all steps of experimental preparation must be reviewed, and questions considered such as

- was the parameter set complete or were important variables not included,
- were test numbers confused and incorrectly assigned to the orthogonal array,
- have the external circumstances changed during the execution of confirmation experiments, e.g., by using a new, allegedly identical raw material, modifications of serial sequences, occurrence of defective parts, metrological problems or calibration, etc.?
- was coupling, i.e., interaction between the parameters ignored,
- can be assumed to deal essentially with a multiplicative parameter model **(1-16*)**?

The first three points are to be checked in detail, the fourth point leads to decoupling for instance between pressure and temperature, power and time, force and leverage. For this purpose, a table should be drawn up which varies the setting values of one parameter as a function of the other via its common coupling variable, the change of energy transformed in the system **(1-17*)**. Hidden, thus unknown interactions can be avoided by selecting other characteristic variables with a more direct energetic relationship. The associated thought is by adding the effects of the individual parameters to determine the total amount of energy as the end result. This is already expressed with the square of the measured quantity in the numerator of the *S/N* ratio. From a physical point of view, the measured variable is an amplitude and the square is directly proportional to the energy available for conversion in a process **(1-18*)**.

Now the meaning of orthogonal arrays becomes obvious. By balancing all possible combinations of the parameters used to determine always one parameter effect, the orthogonal array uncovers the existence of interactions between parameters and avoids misinterpretations by means of the confirmation experiment. The decisive advantage of using orthogonal arrays is not so much to increase the reliability of a process for generating certain product properties but to test the reliability as such. The improvement of an intended function is achieved by maximizing the *S/N* ratio.

Knowledge of elementary scientific contexts is a prerequisite for the successful application of orthogonal arrays. In most cases, the recognition of energy-related characteristics requires a good understanding of a technical process from a physical and chemical point of view. This poses a certain challenge to penetrate a system to be analyzed in advance up to a certain depth. Access is facilitated by recognizing the applied principles. From this, the type of the intended energy transformation becomes transparent and makes it possible to intervene directly in the process of energy conversion with the parameter analysis. As a result, the system function should be approached with a selected parameter configuration to achieve the best possible ideal behavior or function. Once again, physically speaking, technical processes are based on the principle of conservation of energy. By maximizing the *S/N* ratio, at least a portion of the energy previously consumed for malfunctions is attributed to the actual transformation process. If no other energy is available, i.e., for secondary paths, unintended and undesirable effects (noise emissions, wear, side reactions, side effects) can therefore no longer occur, or at least only to a much lesser extent.

Exercise

For improving the adhesion of an aluminum coating on plastic, parameters involved in the process are examined more closely. As a result, the abrasion force is measured for each test configuration, whereby one test repetition is carried out to ensure the measurement data. The table specifies the setting values for the parameters.

Parameter (*A*–*L*) and definition of the setting values

Parameter	Level 1	Level 2
A: Chemical plasticizer	Compound 1	Compound 2
B: Purity of the aluminum	Analysis grade (p.a.)	Chemical pure
C: Surface preparation	Sandblasting	Grinding
D: Mesh size	800	1000
E: Processing time	Short	Long
F: Cleaning procedure	Chemically	Ultrasound
G: Temperature of coating process	Low	High
H: Drying time	Short	Long
I: Drying temperature	Low	High
K: Posttreatment time	Short	Long
L: Post-drying temperature	Low	High

The number of parameters requires the use of an orthogonal array L12. Measurement values with one repetition have been recorded for each test configuration. The original measuring unit (Newton) is dimensionless by scaling. The greatest average adhesive force with the smallest deviation in the given parameter space is to be determined.

Orthogonal array L12 with measurement data

Test	Parameter											Data set	Data set
#	*A*	*B*	*C*	*D*	*E*	*F*	*G*	*H*	*I*	*K*	*L*	1	2
1	1	1	1	1	1	1	1	1	1	1	1	32.5	50
2	1	1	1	1	1	2	2	2	2	2	2	55	80
3	1	1	2	2	2	1	1	1	2	2	2	20	37.5
4	1	2	1	2	2	1	2	2	1	1	2	20	50
5	1	2	2	1	2	2	1	2	1	2	1	32.5	60
6	1	2	2	2	1	2	2	1	2	1	1	22.5	50
7	2	1	2	2	1	1	2	2	1	2	1	45	80
8	2	1	2	1	2	2	2	1	1	1	2	27.5	55
9	2	1	1	2	2	2	1	2	2	1	1	47.5	80
10	2	2	2	1	1	1	1	2	2	1	2	47.5	62.5
11	2	2	1	2	1	2	1	1	1	2	2	42.5	60
12	2	2	1	1	2	1	2	1	2	2	1	20	45

Problem solution
With the question of the largest average numerical value and smallest deviation, a test evaluation with the *SNR* index of the designation *LTB* (Larger-the-better) is to be carried out. For reasons of clarity, the *S/N* ratio is first calculated with the objective of the largest possible numerical value **(1-19*)**. The *SNR* key figure of the type *LTB* and the squared average value of the output data are displayed in the columns and extended by the equivalent logarithmic form.

Preparation of the measurement data

Test #	Data set 1	Data set 2	*S/N (LTB)*	*SNR (LTB)*	$\bar{y}^2$	*SEN*
1	32.5	50	1485	31.72	1702	32.31
2	55	80	4108	36.14	4556	36.59
3	20	37.5	623	27.94	827	29.17
4	20	50	690	28.39	1225	30.88
5	32.5	60	1633	32.13	2139	33.3
6	22.5	50	842	29.25	1314	31.19
7	45	80	3077	34.88	3906	35.92
8	27.5	55	1210	30.83	1702	32.31
9	47.5	80	3336	35.23	4064	36.09
10	47.5	62.5	2860	34.56	3025	34.81
11	42.5	60	2406	33.81	2627	34.19
12	20	45	668	28.25	1056	30.24

The first and most important step is the evaluation of the parameters regarding the *SNR* key figure. For the respective columns, the parameter analysis is performed, i.e., the effects of parameters *A*–*L* are summarized and displayed in tabular form and graphically for easier interpretation.Effects of parameters *A–L* as *SNR* key figures of the type *LTB*

	A	*B*	*C*	*D*	*E*	*F*	*G*	*H*	*I*	*K*	*L*
Level 1	30.93	32.79	32.26	32.27	33.39	30.96	32.57	30.3	31.96	31.66	31.91
Level 2	32.93	31.07	31.6	31.58	30.46	32.9	31.29	33.56	31.9	32.19	31.95

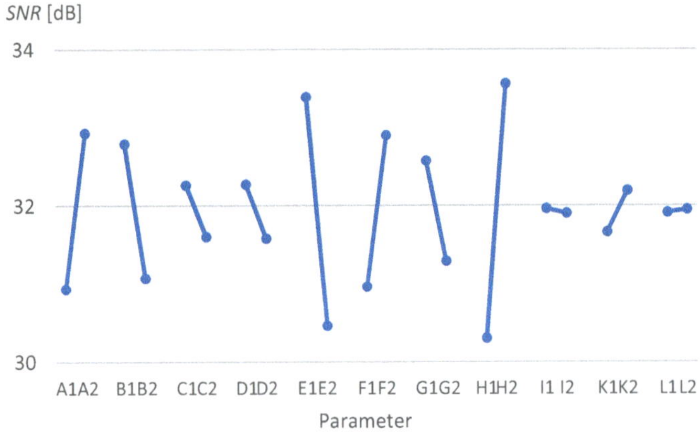

SNR (*LTB*) key figures graphics of parameter effects

For model prediction, about half of the strongest parameters should be used. This is represented by the variables *A*, *B*, *E*, *F*, and *H*. The greatest expected numerical value is obtained with the configuration:

$$A2 + B1 + E1 + F2 + H2$$

The remaining parameters can also be used advantageously, i.e., with their greatest effect or regarding process simplification, e.g., *K*1 instead of *K*2, to shorten process time. To return to the measured values, the parameter effects are calculated as well from the *SEN* (Sensitivity) column.

Effects of parameters *A–L* as *SEN* key figures

	A	*B*	*C*	*D*	*E*	*F*	*G*	*H*	*I*	*K*	*L*
Level 1	32.24	33.73	33.38	33.26	34.17	32.22	33.31	31.57	33.15	32.93	33.18
Level 2	33.93	32.44	32.78	32.91	32	33.95	32.86	34.6	33.02	33.24	32.99

From the model above, an expected average value for the abrasion force is calculated.

$$SEN = A2 + B1 + E1 + F2 + H2 - 4\varnothing$$
$$SEN = 10\log\left(\bar{y}^2\right) = 38.04$$
$$\bar{y} = 79.8 \quad \text{(Newton)}$$

2.9 Percentage of Effect Contribution

Apart from the effect size (=amplitude) of the parameters and the direction related to the setting values, applying the percentage effects of all variables allows an easy estimation for their advantageous use. As a rule of thumb to have an impact on the system response if the contribution, referred to as ρ (rho), exceeds 5%. Smaller contributions are lost in unavoidable tolerances and residual dispersion. Variables of this magnitude are not usable for technical improvements but should preferably be considered in terms of cost or time savings. For the calculation, the amplitudes of the effects are to be squared and compared to an error amplitude or the smallest parameter effect. The conversion in percent based on the sum of all parameter effects results in the respective effect contribution ρ.

The relative contribution to affect a process or change the product properties is given by

$$\text{Relative effect contribution} = \frac{(\Delta\ \text{parameter effect})^2}{(\text{error})^2\ \text{or}\ (\Delta\ \text{smallest parameter effect})^2}$$

$$\text{Percentage effect contribution}\ \rho = \frac{\text{Relative effect contribution}}{\text{Sum of total effect contributions}}\,100\%$$

As a numerical example, the percentage effects ρ of the 11 parameters from the previous exercise are calculated (Table 2.26).

Table 2.26 Calculation of the percentage contributions from the effect table of the parameters A–L of the exercise

	A	*B*	*C*	*D*	*E*	*F*	*G*	*H*	*I*	*K*	*L*
Level 1	30.93	32.79	32.26	32.27	33.39	30.96	32.57	30.3	31.96	31.66	31.91
Level 2	32.93	31.07	31.6	31.58	30.46	32.9	31.29	33.56	31.9	32.19	31.95
Delta^2	3.999	2.973	0.430	0.471	8.599	3.772	1.633	10.594	0.004	0.279	0.001
Quotient	3318	2467	357	391	7137	3130	1356	8792	3	232	1
Sum	27,182										
ρ (%)	12.21	9.08	1.31	1.44	26.25	11.52	4.99	32.34	0.01	0.85	0.00

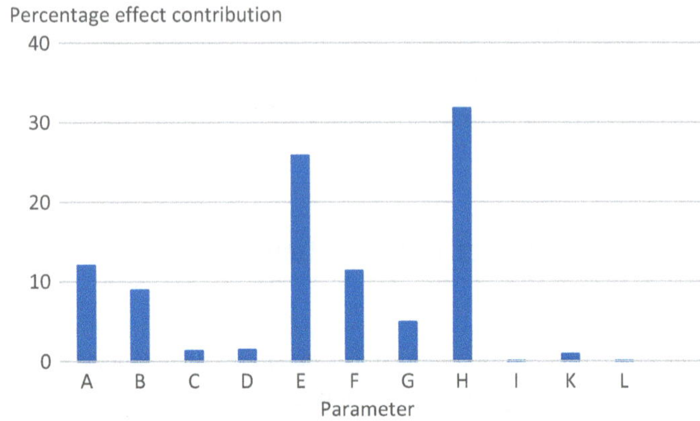

Fig. 2.11 Graphic view of the percentage effect contribution of the parameters A–L

The graphical presentation allows a cross comparison to distinguish between process effective and ineffective parameters with respect to a 5% threshold (Fig. 2.11).

2.10 Scope of Validity of a Model Prediction

In the case of several variables with different contribution to the system response, those with the largest numerical values of *SNR* indices are combined to improve technical targets and reduce dispersion. Preferably, orthogonal arrays are used for the analysis of the parameter effects, whose task is to detect interactions between the parameters. If the variables are in an energy relation, they will have an additive effect on the system response in terms of their overall effect, as predicted by the model of the summed effects on a logarithmic scale. If, on the other hand, there are interactions between pairs of variables that have not been decoupled, a model prediction cannot be confirmed experimentally. This information is to be understood as a kind of warning that an important fact has been overlooked or ignored, but in no way to be interpreted as a bad experimental art. Since model verification has an important role to play, the statement of a confirmation or non-confirmation must be clear and unambiguous.

In static systems, a target value can usually be hit or even improved. The initial values of the *SNR* and *SEN* key figures are derived by means of a back transformation from the model of the selected parameter configuration, from which the average target value and the standard deviation are obtained for proofing.

$$SNR = 10\log\left(\frac{\bar{y}^2}{\sigma^2}\right)$$

$$SEN = 10\log\left(\frac{\bar{y}^2}{y_0^2}\right)$$

The same applies to fix target value for the *SNR* indices of type *NTB*, *LTB*, and *STB*. The closer the test results of the underlying configuration are to this value, the more perfect is the agreement with the prediction, i.e., the model appears to be confirmed. If the numerical values differ markedly from the prediction, it must be distinguished which distance is still to be considered as model confirmation or non-confirmation. For reasons of practicality, a range of plus/minus the threefold standard deviation within which the model is still considered valid is permitted. If the experimental results exceed this range, the model is not suitable for an approximate description of the true facts (Table 2.27).

In dynamic systems, the slope for system characterization replaces a specific target value. The introduced dimension factor S_2 guarantees the scale invariance. The model calculation of a selected parameter configuration provides the numerical values for the *SNR* and *SEN* indices.

$$SNR = 10\log\left(\frac{\bar{\beta}^2}{\sigma^2}S_2\right)$$

$$SEN = 10\log\left(\frac{\bar{\beta}^2}{\beta_0^2}\right)$$

From the back transformation, the average value of the slope and associated standard deviation follows, whereby the latter must always be applied together with the dimension factor. Since the slope and not its square is taken, the arithmetic square root of the dimension factor must be considered. The validity area of the model is defined in the same way (Table 2.28).

The straight lines through the zero point with the slopes β_{min} and β_{max} represent the boundary lines within which the experimental test results fall for all signal values in the case of model confirmation. (Fig. 2.12)

Table 2.27 Verification of model prediction for static systems

Range	$y_{min} = \bar{y}_{prediction} - 3\sigma$ (for *STB* : $y_{min} \geq 0$) $y_{max} = \bar{y}_{prediction} + 3\sigma$
Model confirmed	$y_{min} < y_{experiment} < y_{max}$
Model not confirmed	$y_{min} \geq y_{experiment} \geq y_{max}$

Table 2.28 Verification of model prediction for dynamic systems

Range	$\beta_{\min} = \bar{\beta}_{\text{prediction}} - 3\sigma/\sqrt{S_2}$ $\beta_{\max} = \bar{\beta}_{\text{prediction}} + 3\sigma/\sqrt{S_2}$
Modell confirmed	$\beta_{\min} < \beta_{\text{experiment}} < \beta_{\max}$
Modell not confirmed	$\beta_{\min} \geq \beta_{\text{experiment}} \geq \beta_{\max}$

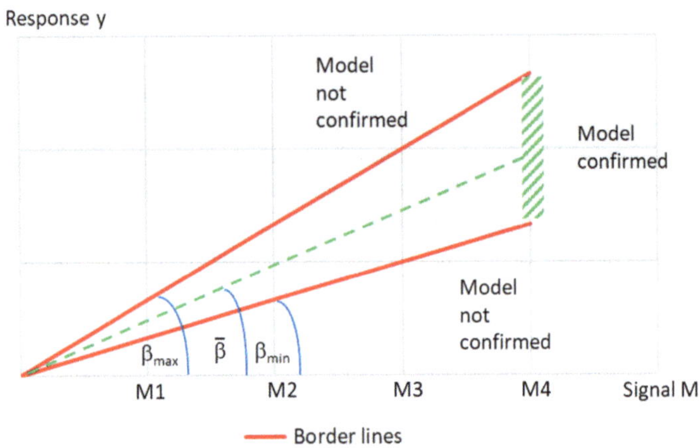

Fig. 2.12 Boundary lines for model confirmation of dynamic systems

2.11 Nonlinear Transfer Functions of Variables

The previous remarks were accompanied by the idea that the setting values of the variables themselves are constant and precise. In practice, there are also tolerances here, so that unintentional minor setting errors according to the transfer function of the considered system influence the initial response. The characteristic curve, i.e., the change in an output value as a function of the change in an input value, determines to what extent deviations are transferred unchanged, amplified, or attenuated. If there is an approximate linear relationship, a transmission of an input-side deviation is determined by the ratio of the slope, i.e., it cannot be changed in any way by choosing a certain setting value (Fig. 2.13).

If, on the other hand, the transfer function is curved, input deviations will have a correspondingly strong effect on the output variable in the case of a steep slope but become almost negligible as the slope decreases (Fig. 2.14).

If a change in the setting value shows no or only a minor influence on the result, it is not possible to intervene in the response as such or affect the output deviation. Parameters with consistently linear response are used for adjustment to one or more target values. Curved transfer functions lead to insensitivity of the system response to deviations when appropriate setting values are selected. Parameters that have no

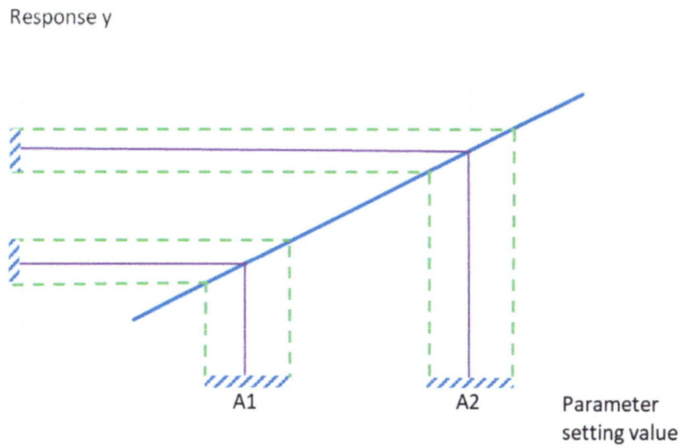

Fig. 2.13 Linear transfer function of a variable

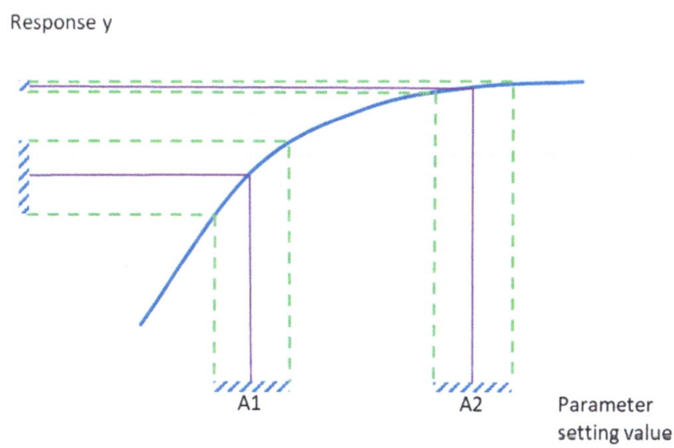

Fig. 2.14 Nonlinear transfer function of a variable

effect at all are freely selectable in their setting values, and are therefore available as the most advantageous, profitable or fastest option. After a parameter analysis, the *S/N* ratio is therefore always maximized in three steps with

Priority 1: Selection of parameters and setting values, so that the deviation of the system response is as small as possible. This is achieved with the merger of the largest *S/N* numerical values of the type $\frac{\bar{y}^2}{\sigma^2}$ (static system) or $\frac{\beta^2}{\sigma^2} S_2$ (dynamic system). In this context, the stability of the output value is also referred to as insensitivity to input deviations or robustness.
Priority 2: Utilization of the parameters and setting values which negligibly change of the abovementioned largest possible *S/N* ratio for adjusting the response exactly

to the target value. This ensures that the best properties of the process/product features even with input deviations are maintained.
Priority 3: Parameters and setting values that neither affect deviation nor adjustment are chosen to realize the most economical condition.

The more parameters are available for analysis, the better the chances of being able to use all possibilities. It is therefore necessary to rethink and not to save on variables because of superficially less effort but to create a set of parameters in a varied and extensive way. In addition to the variables based on theoretical and empirical considerations, this also includes those that influence the cost side and processes flow, as well as some that take intuition and creativity into account. Only if one goes beyond existing views or theories, there is a chance to break with behavioral patterns and investigate unknown territory with unexpected development capabilities.

2.12 Interaction with Noise Factors

The only way to integrate interactions with noise variables in an analysis is to repeat an experiment. Since the purpose of a repetition is to impose additional information, in particular, extreme situations that can occur in practice. For example, a source of noise can be the ambient temperature (in a disturbing environment it is also referred to as "outer noise"), so that a first series of tests can be carried out at noon during the day, and the repetition may be done at night when all devices or components are working at a slightly lower temperature. Dispersion can also be caused by a process drift (this material fatigue in the broadest sense is also referred to as "inner noise"), so it is advisable to plan a series of tests on machines running from Monday to Friday on Thursday/Friday, the repetition on Monday, 2 days after all instruments or systems have been shut down. A further possibility for deviations can be present in material modifications, also chemical reactions (here one speaks of batch dispersion and connects this with the term "product noise"). In this situation, both test series are to be carried out with the same substance, but from two different manufacturers. The most diverse noise factors extend the concept of a dynamic system function again, as schematically shown in the illustration (Fig. 2.15).

In the case of static behavior, only the signal, i.e., the input variable, is not considered. To detect each source of noise, a repetition of the experiment for each noise factor would have to be carried out to figure out parameter settings that make the system function robust or insensitive to the sources of noise. Due to all kind of thinkable tolerances, however, the effort would become unfeasible, so that the following procedures have developed from a practical point of view (Fig. 2.16):

1. reduction of all potential noise variables to approximately 2–3, which, based on the knowledge of the situation, play a substantial role in the process to be analyzed,

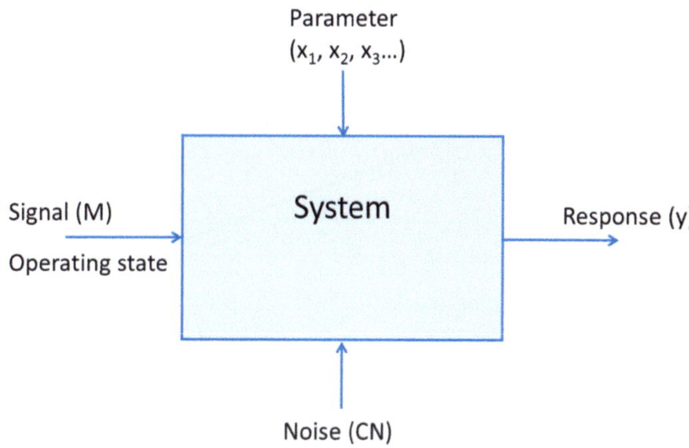

Fig. 2.15 Dynamic system function with control and noise parameters

2. compression of the most important noise sources by means of an orthogonal array, if three or more factors are to be considered,
3. merging of all noise parameters into one state of the largest "+" tolerance and the largest "−" tolerance; this leads back to exactly one combined source of noise, in which different boundary conditions add up to larger or smaller numerical values. The effort is reduced to exactly one repetition of each test of the applied orthogonal array, known as "compound noise" (CN) technique.

It should be emphasized that sources of noise by definition are inevitable, or rather their elimination fails due to excessive technical effort, i.e., costs. The temporary additional expenditure must be accepted once, for carrying out a unique parameter analysis under separate noise levels to gain the information and implement it in a useful way. The claim is precisely to avoid the elimination of the sources of noise, since the achieved robustness results in the adherence to one or several target values, even with deviations acting on it. Only if a parameter analysis is exhausted and the achieved insensitivity has to be further improved, a so-called tolerance analysis must be followed up, which is dealt with in Sects. 5.4 and 5.6.

2.13 Neutralizations of Noise Factors

The fundamental methodological concern is to neutralize noise factors. A parameter effect (*A*1, *A*2) under the influence of a source of noise with the states *N*1 and *N*2 is shown for illustration purposes in Fig. 2.16.

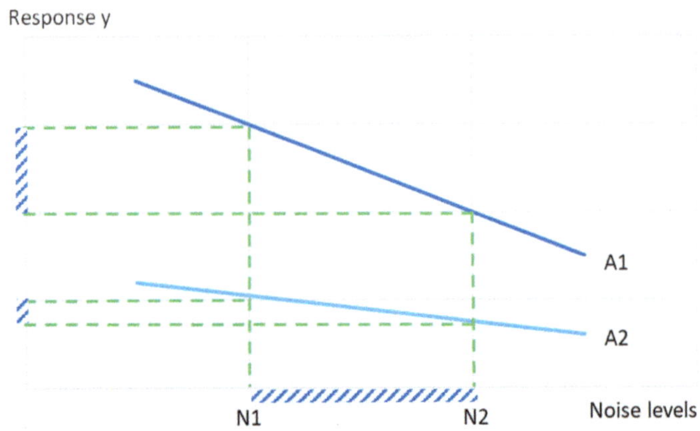

Fig. 2.16 Dispersion range of the system response under the influence of a source of noise

The advantageous setting value of parameter *A* consists, as immediately apparent, in *A*2, as it approximately compensates for the influence of the source of noise *N*. Therefore, an experimental test design is preferably concerned with the analysis of the interactions of the parameters with the existing noise sources and not with possible interactions of the parameters with each other. As already explained, the latter are to be eliminated by decoupling, for example, but the former are to be disclosed in a targeted manner so that the greatest possible benefit can be derived from test repetitions. Depending on the effort involved, a combination of noise factors is also justified. In the case of unknown tolerances, it will be necessary to apply an orthogonal array with two levels to determine all positive and negative deviations. This initially looks like an additional effort but pays off immediately with only one repetition of the test series, compared to using all repetitions according to the size of the orthogonal array with all assigned sources of noise for each test to be performed. To determine the sources of noise with their setting values to merge them into one combined noise variable, average parameter values must be assumed. With three monotonously increasing or decreasing numerical values, *A*2, *B*2, *C*2, etc. would stand for it. When viewed dynamically, an average signal value is the starting point. For the sources of noise, internal, external, and product noise, abbreviated as noise parameters NA, NB, NC etc., two levels each are completely sufficient for this consideration. The noise parameters are assigned to the smallest possible orthogonal array at two levels each, so that the following scheme is used, using the example of an orthogonal array L8 with seven noise factors (Table 2.29).

For an evaluation of the parameter trends under noise effects, an evaluation of the raw data is preferable to an *S/N* analysis. Applied graphically, smaller sources of

Table 2.29 Noise source analysis NA–NG for average parameter settings

Test #	NA	NB	NC	ND	NE	NF	NG	Measurement value
1	1	1	1	1	1	1	1	11.86
2	1	1	1	2	2	2	2	9.27
3	1	2	2	1	1	2	2	13.13
4	1	2	2	2	2	1	1	10.52
5	2	1	2	1	2	1	2	12.59
6	2	1	2	2	1	2	1	14.09
7	2	2	1	1	2	2	1	13.73
8	2	2	1	2	1	1	2	15.05

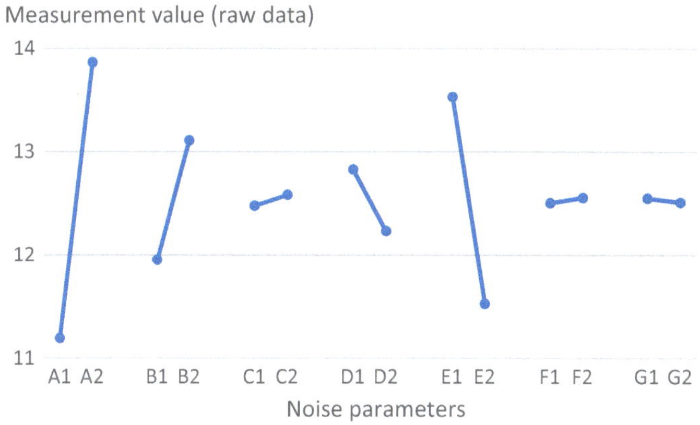

Fig. 2.17 Trend analysis of noise sources NA–NG

noise can be identified which can be neglected. According to their directional dependency, the remaining ones are combined to form a compound noise factor, as shown in the illustration (Fig. 2.17).

This results in the setting values for the compound noise parameter CN as follows:

$$CN1 = NA2 + NB2 + ND1 + NE1 \rightarrow \text{largest positive tolerance}$$

$$CN2 = NA1 + NB1 + ND2 + NE2 \rightarrow \text{largest negative tolerance}$$

For the execution of the parameter analysis with robust requirements, the orthogonal array corresponding to the problem and the technical possibilities is used, e.g., L18, then with only one compound noise source CN, i.e., exactly one repetition of the test series (Table 2.30).

Table 2.30 Layout for the implementation of an orthogonal array with one compound noise source

Test #	*A*	*B*	*C*	*D*	*E*	*F*	*G*	*H*	CN1	CN2	*S/N*	*SNR*	*SEN*
1	1	1	1	1	1	1	1	1	y1	y19			
2	1	1	2	2	2	2	2	2	y2	y20			
3	1	1	3	3	3	3	3	3	y3	y21			
4	1	2	1	1	2	2	3	3	y4	y22			
5	1	2	2	2	3	3	1	1	y5	y23			
6	1	2	3	3	1	1	2	2	y6	y24			
7	1	3	1	2	1	3	2	3	y7	y25			
8	1	3	2	3	2	1	3	1	y8	y26			
9	1	3	3	1	3	2	1	2	y9	y27			
10	2	1	1	3	3	2	2	1	y10	y28			
11	2	1	2	1	1	3	3	2	y11	y29			
12	2	1	3	2	2	1	1	3	y12	y30			
13	2	2	1	2	3	1	3	2	y13	y31			
14	2	2	2	3	1	2	1	3	y14	y32			
15	2	2	3	1	2	3	2	1	y15	y33			
16	2	3	1	3	2	3	1	2	y16	y34			
17	2	3	2	1	3	1	2	3	y17	y35			
18	2	3	3	2	1	2	3	1	y18	y36			

Without compound noise technique the possibility remains, as mentioned under point 2, to compress the noise parameters also with the smallest possible orthogonal array and to repeat the orthogonal array of the technically oriented parameters according to the required test number. As the table shows, the parameter effects result from multiplying the applications of both orthogonal arrays (Table 2.31).

The orthogonal array for analyzing the effects of technical parameters is also known as an internal orthogonal array, while that with the noise parameters is the external one. The advantage of such an arrangement is that it is possible to distinguish the effect of each source of noise on each variable. If a further group of parameters with all individual effects had to be distinguished, a third orthogonal array can be multiplied with the outer orthogonal array according to the same principle; with such a "number cube", the sources of noise of the first outer orthogonal array as well as, for example, special tolerances assigned to a second outer orthogonal array could be investigated. Since the logistical effort becomes relatively large, such a more detailed analysis is usually limited to simulation calculations with model character.

Table 2.31 Application of an orthogonal array with assigned sources of noise to an orthogonal array for effect analysis of technical parameters

								Nr.	**1**	**2**	**3**	**4**	**5**	**6**	**7**	**8**			
								NA	1	1	1	1	2	2	2	2			
								NB	1	1	2	2	1	1	2	2			
								NC	1	1	2	2	2	2	1	1			
								ND	1	2	1	2	1	2	1	2			
								NE	1	2	1	2	2	1	2	1			
								NF	1	2	2	1	1	2	2	1			
								NG	1	2	2	1	2	1	1	2			
Test Nr.	**A**	**B**	**C**	**D**	**E**	**F**	**G**	**H**				**Messwerte**					**S/N**	**SNR**	**SEN**
1	1	1	1	1	1	1	1	1	y1	y19	y37	y55	y73	y91	y109	y127			
2	1	1	2	2	2	2	2	2	y2	...					...	y128			
3	1	1	3	3	3	3	3	3	y3							y129			
4	1	2	1	1	2	2	3	3	y4							y130			
5	1	2	2	2	3	3	1	1	y5							y131			
6	1	2	3	3	1	1	2	2	y6							y132			
7	1	3	1	2	1	3	2	3	y7							y133			
8	1	3	2	3	2	1	3	1	y8							y134			
9	1	3	3	1	3	2	1	2	y9							y135			
10	2	1	1	3	3	2	2	1	y10							y136			
11	2	1	2	1	1	3	3	2	y11							y137			
12	2	1	3	2	2	1	1	3	y12							y138			
13	2	2	1	2	3	1	3	2	y13							y139			
14	2	2	2	3	1	2	1	3	y14							y140			
15	2	2	3	1	2	3	2	1	y15							y141			
16	2	3	1	3	2	3	1	2	y16							y142			
17	2	3	2	1	3	1	2	3	y17	...					...	y143			
18	2	3	3	2	1	2	3	1	y18	y36	y54	y72	y90	y108	y126	y144			

To come back to the first point, a limitation to 2, at most 3 of the most important sources of noise is equivalent to a four or eight times application of the inner orthogonal array. Obviously, it must be considered whether even with a small number of noise sources their merging considerably simplifies the overall experimental effort.

As already explained above, a minimum requirement for the performance of a parameter analysis is exactly one test repetition. This also requires a doubling of the testing effort when analyzing dynamic systems. Since three or more different signal values are used to describe dynamic systems, except that the zero point can be accepted as a reference point, the consequent thought for generating robustness already leads to a relatively high-test effort. Contrary to the previous remarks, a repetition can be dispensed for dynamic systems, if the deviation of linear behavior is considered as sufficient noise imposed to the system. If it is possible to minimize linearity deviations by means of suitable parameter selection and their setting values, this is equivalent to the neutralization of noise factors from "test repetitions", although set as signal values at different points.

Using a robust parameter analysis with an assumed orthogonal array L18, the different procedures for dealing with static systems with noise sources and a dynamic system result in the following total effort and expense:

Table 2.32 Test effort in comparison with different ways of dealing with sources of noise

Analysis of parameters with an orthogonal array L18	Repetitions	Total test effort
18 Tests	2^5 (5 noise parameter)	576
18 Tests	8 (L8 with up to 7 noise parameter)	144
18 Tests	2 (1 CN + 8 preliminary tests L8)	44
18 Tests	3 Signal values, dynamic system	54

When viewed with a sound sense of proportion, the effort under minimum requirements in a static system with noise parameters is similar to the dynamic situation, so that the dynamic system should be preferred already for this reason (Table 2.32).

Important items to memorize A dynamic system is preferable to a static system when performing a parameter analysis.

Pure orthogonal arrays have to be applied for the investigation or testing of interactions.

The analytical approach of a robust design is to disclose the interactions of parameters with existing noise sources and not between the parameters.

Chapter 3
Input–Output Relations

Contents

Abstract The mathematical tools introduced to analyze systems based on *SNR* key figures are expanded to any kind of data especially digital ones. For obvious reasons, the greater part of modern technology we deal with is digital. Less well known is the fact that this counts also for medical application of pharmaceuticals. Every combination, i.e., continuous or digital data fed to the input with a continuous or digital response as output is considered and outlined with practical examples. The intent is not to stick with area-specific nomenclature but to recognize the applied principles. In case of critical medical treatment, an objective scientific way is presented to extract an effective dosage to be on the safer side for a patient. The extension of a functional range to improve the reliability of digital functions by continuous data turns out to be one of the most efficient formulations for optimization.

3.1 Classification of Systems

The previous approach with continuously changing input variables (parameter, signal value), which continuously change the system output (response) according to their transfer function, must be extended by the important aspect of binary variables **(1-20*)**. Possible existing relations between system input and output are listed in the table (Table 3.1).

© Springer Nature Switzerland AG 2019

H. Ruefer, *Living Without Mathematical Statistics*,
https://doi.org/10.1007/978-3-319-99632-5_3

Table 3.1 Classes of systems with different data relationships

Input (variable, signal)	Output (response)
Continuous	Continuous
Continuous	Digital
Digital	Continuous
Digital	Digital

Table 3.2 Dynamic systems with continuous input/output relationship

System	Input	Output
Photographic camera, copying machine, TV, camcorder	Object, source image	Image
Radar	True position	Indicated position
Wind turbine	Revolutions per minute (rpm)	Velocity of wind
Steering	Turning angle	Turning radius
Sonar	Signal propagation delay	Water depth
Measurement instrument (analog construction)	Voltage, electrical current	Needle deflection
Electric engine	Power	Torque
Galvanization	Electric charge	Coating thickness

3.2 Systems with Continuous Data

Analog systems process continuous input and output data. When more energy is pumped into the system input, the energy content increases and is expressed as a system response in the form of higher speed, higher temperature, increasing pressure or similar. Some typical examples are listed in the table (Table 3.2).

All systems are characterized by a linear relation

$$y = \beta M$$

(Section 2.3) with applicability of the mathematical procedures listed so far.

3.3 Systems with Continuous Data Input and Digital Data Output

This combination of continuous data as system input and response in binary or digital form is also frequently encountered. All warning systems are based on this principle of triggering an alarm as soon as a critical value of one or more sensors is exceeded. Further examples are listed in the table (Table 3.3).

Table 3.3 Dynamic systems with continuous input and digital output relationship

System	Input	Output
Electrical fuse	Electrical current	Switch off when a limit value is exceeded
Authentication, identification	Analog characteristics	Accepted/rejected
Collision warning system	Speed, direction, distance	avoidance maneuver, emergency stop: yes/no
Filling level sensor	Rise height	Switch on/off
Air condition	Heating, cooling	Switch on/off
Analog–Digital-Converter (ADC)	Analog signal	Digital signal
Seismometer	Mechanical displacement, speed/ acceleration of the sea floor	Tsunami warning yes/no
Automatic transmission	Revolutions per minute (rpm)	Gear shifting

Table 3.4 Calculation of the *SNR* index from two states to be optimized

Determination of the parameter effects for the ON state	$SNR_{\mathrm{ON}} = 10 \log\left(\frac{\bar{y}^2_{\mathrm{ON}}}{\sigma^2_{\mathrm{ON}}}\right)$
Determination of the parameter effects for the OFF state	$SNR_{\mathrm{OFF}} = 10 \log\left(\frac{\bar{y}^2_{\mathrm{OFF}}}{\sigma^2_{\mathrm{OFF}}}\right)$
Highest achievable numerical value for maximum functional reliability	$SNR = SNR_{\mathrm{ON}} + SNR_{\mathrm{OFF}} = \mathrm{Maximum}$

The listed systems of type ADC (analog–digital converter) are characterized by the fact that switching thresholds or trigger points must function reliably and reproducibly according to the input variable even under noise conditions. The trigger points must be adapted to the external situation such as the ambient temperature or special requirements, i.e., with adjustable target values for a lower and upper threshold. This means that there are at least two target values, which are evaluated with the *S/N* ratio

$$S/N = \frac{\bar{y}^2}{\sigma^2}$$

The decisive factor for trouble-free operation is that the output states do not overlap, i.e., even with extremely close trigger points, the range of deviation is sufficiently small. An analysis of the variables involved under noise conditions must be carried out separately for each ON and OFF state to improve the system properties. If the target value is the temperature, the absolute temperature scale has to be used (Table 3.4).

For air-conditioning systems with heating and cooling, the maximum value must be determined in the same way for four *S/N* ratios, as shown in the sketch (Fig. 3.1).

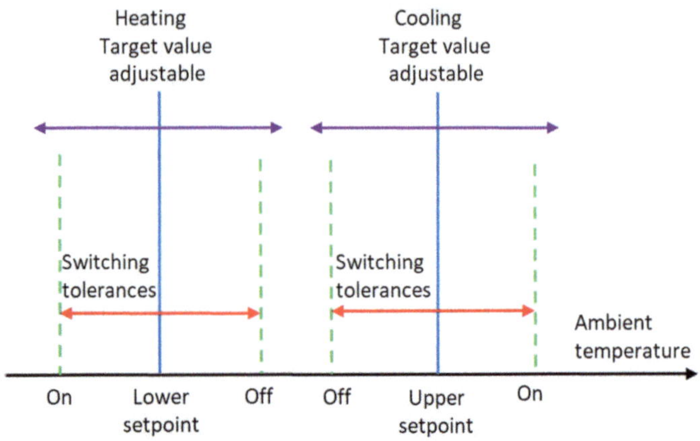

Fig. 3.1 Temperature control under noise conditions with four trigger points

3.4 Systems with Digital Data Input and Continuous Data Output

Systems that convert a binary input signal into an analog output signal are digital–analog converters (DAC) of modern communication technology. In this way, the digitally read signal from a CD or DVD is converted back to make the information audible/visible in analog form. Further examples are listed in the table (Table 3.5).

When converting digital data, the ideal characteristic consists of the maximum possible linearity of the output signal according to the digital signal on the input. For parameter analysis, a precise digital signal of the corresponding Bit width is to be fed in and linearity is evaluated with the *S/N* ratio

$$S/N = \frac{\beta^2}{\sigma^2} S_2$$

From the principle, an ADC (for measuring purposes) and a DAC (for system analysis) are coupled, so that the overall behavior is traced back to the processing of continuous (analog) data.

Table 3.5 Dynamic systems with digital input and continuous output relationship

System	Input	Output
Step motor	Pulses	Turning angle, revolutions per minute (rpm)
Pulse-width modulation	Rectangular signal of variable frequency	Sinusoidal wave
Counting systems	Units, pieces	Current, amplitude
Digital–Analog Converter (DAC)	Digital signal	Analog signal

3.5 Systems with Digital Data

Besides electronic modules for digital data processing, systems with binary data input and output appear in many different ways. All separation systems or mixing systems of physical, chemical, and biological nature are based on the binary principle. The table shows some examples (Table 3.6).

The processes listed above are always carried out in fixed assigned units for quantities that cannot be further divided. Seen in this way, all examples represent a binary data transfer. Access to a continuous description of such systems can be achieved by introducing a relative frequency p with a numerical value of 0 to 1 based on the observed or measured number of events, certain appearances or even electronic signals, etc., in standardized form (Table 3.7).

The traceability to a continuous relation raises the question of an equivalent transformation of the *S/N* ratio for digital systems. The amplitude square y^2 of analog systems can be replaced by the square of the observed or measured frequency p^2. Correspondingly, the square of the deviation coefficient σ^2 must be replaced by the measure of mixing $p(1 - p)$, which assumes the highest value at a frequency of 50% ($p = 0.5$) **(1-21*)**.

$$\frac{S}{N} = \frac{\bar{y}^2}{\sigma^2} \rightarrow \frac{S}{N} = \frac{p^2}{p(1-p)} = \frac{p}{1-p}$$

Table 3.6 Dynamic systems with digital input/output relationship

System	Input	Output
Cyclone, air classifier	Mixture of materials	Particle separation in fractions
Safety system air bag	Sensor signal in case of impending collision	Trigger: yes/no
Desalination plant reverse osmosis	Sea water	Drinking water and wastewater
Smelting	Mineral bearing rock	Ore and rock
Separation of isotopes (e.g., centrifuge)	Isotopes of an element	Sorting by atomic weights
Stirring device	Discrete components	Uniform mixture
Electronic amplifier	Digital signal from transmitter	Digital signal from receiver
Sorting system	Products with scrap	Separation into good/defect products
Counting systems	True number of pieces	Measured number of pieces
Separation of soluble liquids with different boiling points distillation/rectification	Mixture	Separation of alcohol/water

Table 3.7 Example for introducing relative frequency

	Number of pieces true to form	Number of pieces with defective form	Total number
Products from hydraulic press	$n_0 = 4985$	$n_1 = 15$	$n = 5000$
Relative frequency	$1 - p = \frac{n_0}{n} = \frac{4985}{5000}$ $1 - p = 0.997$	$p = \frac{n_1}{n} = \frac{15}{5000}$ $p = 0.003$ in percent $p = 0.3\%$	1

Relative error rate: p

This definition leads to a higher rating if the frequency p is to be as high as possible. If an improvement of the system behavior demands that the frequency should be as low as possible, the reciprocal value is to be used. The table shows the appropriate signal-to-noise ratios on the logarithmic scale (Table 3.8).

The dimensionless quotient $p/(1 - p)$ is identical to the Omega transformation for basic calculation with percentages. This ensures that back-transformed results from an effect analysis of system improvement parameters do not exceed or fall below the existing range of 0–1, i.e., 0–100%.

If there are two types or forms of appearance, the frequency of errors can be different for both. Correspondingly, the definition of the error frequency is extended and referred to as error type 1 and error type 2 (Table 3.9).

If the appearance of an error does not come across in the observation or measurement time frame, the definition in the table is made (Table 3.10).

Dealing with more than two different appearances (substance classes) is treated in connection with qualitative characteristics (Sect. 6.4).

If two components are processed simultaneously, communication technology also refers to a two-channel system. As schematically illustrated, the frequency p (error type 1) represents a transfer error of the state 0 of a digital signal and the frequency q (error type 2) stands for non-transmitting the state 1. Since p, q can only assume values between zero and one, the rate of error-free transmission is $1 - p$ or rather $1 - q$ (Fig. 3.2).

For an easy comprehension of the error rates p and q, examples from a wide variety of application areas are given.

In inspection systems, sorting machines or functional tests, sensors as well as the human eye are not free of defects. However, it is crucial that the automated test procedure corresponds as closely as possible to the visual inspection or the actual evaluation. For this purpose, the threshold value of the sensor is adjusted until the result reproduces a previously known appearance of test products. Setting a suitable threshold minimizes p, at the expense of q or vice versa (Fig. 3.3).

Table 3.8 Signal-to-noise ratios of relative frequency of type *LTB* or *STB*

Evaluation in case of $p \rightarrow 1$	$SNR_{p(LTB)} = 10\log\left(\frac{p}{1-p}\right)$
Evaluation in case of $p \rightarrow 0$	$SNR_{p(STB)} = 10\log\left(\frac{1-p}{p}\right)$

Table 3.9 Example for error type 1 and 2 of an automatic inspection system

Homogeneous anodized products	Products identified as good	Products classified as defective although good	Total number
	$n_{00} = 3852$	$n_{01} = 148$ Error type 1	$n_0 = 4000$
Anodized products with stains	Products classified as good although defective	Products identified as defective	
	$n_{10} = 7$ Error type 2	$n_{11} = 35$	$n_1 = 42$
Relative frequency of the error type 1	$1 - p = \frac{n_{00}}{n_0}$	$p = \frac{n_{01}}{n_0}$ $= \frac{148}{4000} = 0.037$ $p = 3.7\%$	1
Relative frequency of the error type 2	$q = \frac{n_{10}}{n_1}$ $= \frac{7}{42} = 0.166$ $q = 16.6\%$	$1 - q = \frac{n_{11}}{n_1}$	1

Relative frequency of the error type 1: p
Relative frequency of the error type 2: q

Table 3.10 Definitions of errors type 1 and 2 if no observations are made or measured values exist

For $n_{01} = 0$	Error type 1: $p = \frac{1}{2n_{00}}$
For $n_{10} = 0$	Error type 2: $q = \frac{1}{2n_{11}}$

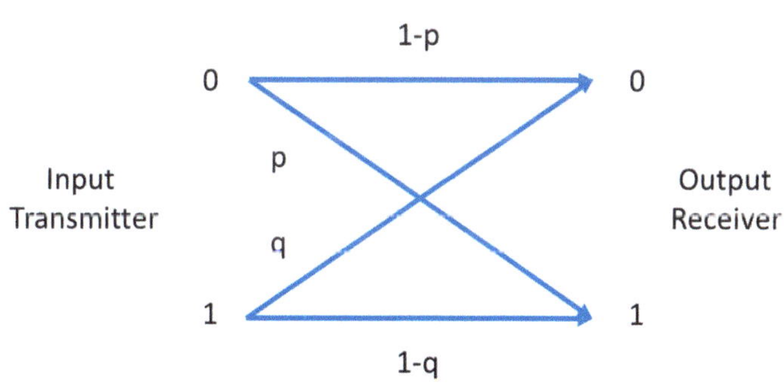

		Output receiver	
		0	1
Input transmitter	0	$1-p$	p
	1	q	$1-q$

Fig. 3.2 Channel scheme with corresponding table of a digital data transmission susceptible to faults

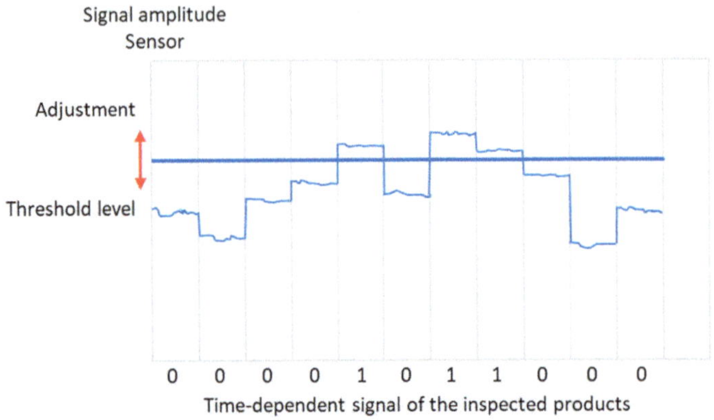

Fig. 3.3 Adjustment of a threshold setting for matching of sorting/inspection machines

The separation of miscible liquids with different vapor pressures is usually done by distillation. It is known, for example, that alcohol with a higher percentage is distilled of must or wine. Apart from discarding the pre-/post-distillate to get rid of fusel alcohols or impurities that are not digestible, the purity, i.e., the concentration of the separated product in a range determined by the hardware (separating power) and the given chemical properties, can be adjusted via the return flow (cooling device). In the case of a drinkable high-strength final product, the water content is set to approximately $p = 0.6$, i.e., 40% alcohol, as the defect value. A smaller defect size can be achieved for taste reasons but the higher energy input for separation does not pay off if the end-product must be reprocessed by adding water for consumption.

Reverse osmosis is a widespread process to produce drinking water from seawater **(1-22*)**. For desalination, the side of a membrane exposed to salt water is pressurized, which allows the passage of fresh water and retains salts. The desalinated water is collected and the seawater with an increased salt concentration is discharged. Since seawater is almost always available in abundance, it is sufficient to only slightly shift the defect ratio from $q = 0.965$ (=3.5% salt content) to $q = 0.96$ and return it to the sea. On the freshwater side, the error rate must be much smaller, but by no means zero. A small percentage of remaining salts (minerals) is advantageous for taste and to prevent corrosion, which can be controlled by the pressure difference and the flow rate depending on the membrane.

As the examples show, the respective operating point settings vary from system to system for reasons of practicality. For the objective evaluation of the function for comparison or optimization, it is necessary to balance the transmission error rates. The point at which the two error rates p and q are equal is also referred to as the crossover with the property that the error sum assumes a minimum. The symmetrical error of the same size occurs with the largest error mixing (Table 3.11).

Table 3.11 Characteristics of the error rates at the crossover (=symmetrical error rates of equal size)

Highest mixing ratio $p : q = 1 : 1$	$p = q = p_0$
Minimum error sum	$2\ p_0$

Only the error rate at the crossover characterizes a binary system and only at this point the *S/N* ratio and its maximization, derived from a parameter analysis, is meaningful. The operating point with identical error rates does not have to be set up experimentally but is calculated from the measured error rates.

$$2p_0 = p + q$$

$$2 \cdot 10\,\log\left(\frac{p_0}{1-p_0}\right) = 10\,\log\left(\frac{p}{1-p}\right) + 10\,\log\left(\frac{q}{1-q}\right)$$

$$\left(\frac{p_0}{1-p_0}\right)^2 = \left(\frac{p}{1-p}\right)\left(\frac{q}{1-q}\right)$$

$$\frac{p_0}{1-p_0} = \sqrt{\frac{p}{1-p}\frac{q}{1-q}}$$

$$p_0 = \frac{1}{1+\sqrt{\left(\frac{1}{p}-1\right)\left(\frac{1}{q}-1\right)}}$$

The effect of an error-symmetrical observation on the *SNR* index suitable for this purpose is shown in the illustration (Fig. 3.4).

If the error rates are different from zero, the slope β is less than 1.

$$\beta = \frac{\frac{1}{2}-p_0}{\frac{1}{2}} = 1 - 2p_0$$

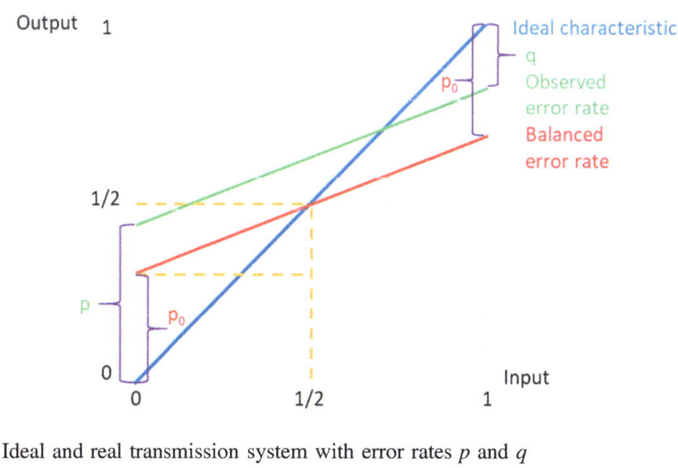

Fig. 3.4 Ideal and real transmission system with error rates p and q

The importance of the slope lies in the error-free transmission rate. The squared slope value is defined as the contribution rate ρ_0 (rho$_{zero}$) since the signals 0 and 1 are or are not distinguishable depending on the size of the symmetrical error (Fig. 3.5).

$$\beta^2 = (1 - 2p_0)^2 = \rho_0$$

For the *SNR* key figure for the evaluation of binary systems with two transmission channels (as well as substance classes, components, etc.) the Omega transformation is to be used as for percentage numbers or their fractions.

$$SNR = 10 \log\left(\frac{\rho_0}{1 - \rho_0}\right)$$

$$SNR = 10 \log\left(\frac{(1 - 2p_0)^2}{4p_0(1 - p_0)}\right)$$

With the derivation of the *SNR* index of a binary system, it can be stated that the smallest error sum is always included in the calculation since only this is related to the system evaluation. If it is not the error rates as such that are decisive, but the economic consequences, the error proportions can be adjusted according to their loss in value in such a way that the previously symmetrical error is shifted in favor of the more valuable component. Under no circumstances does this mean that an operation mode with the smallest error sum must be searched for. Instead, the more advantageous situation must always be set with system-specific variables that change the ratio of the error frequency p and q.

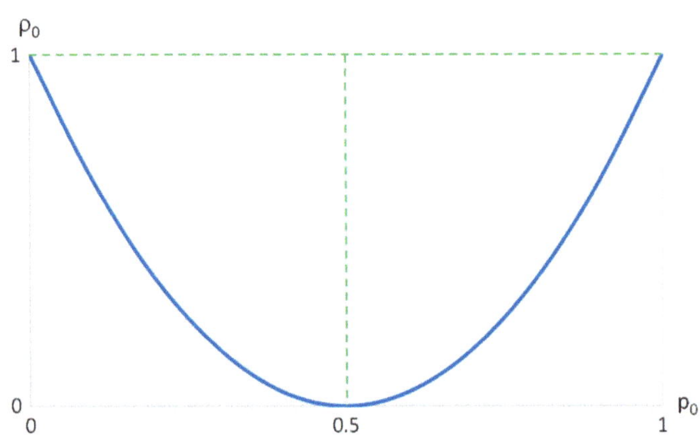

Fig. 3.5 Distinguishability of digital signals depending on the error mixing

3.6 Numerical Examples of Systems with Binary Data

The most frequently asked questions arise in dealing with system functions which are characterized by a binary data input and output. Since such systems can also occur in a wide variety of different manifestations, it seems more appropriate to look at practical examples in more depth than further theoretical explanations for special cases. The user is incumbent upon generalization or adaptation to other conditions.

3.6.1 *Comparison of Two Sorting Machines*

After counting a larger number of products, it can be determined which machine delivers the better results, i.e., with a lower error rate (Table 3.12).

The relative error frequencies p and q are calculated (Tables 3.13 and 3.14).

The precision or accuracy of sorting machine A is 3.7 (dB) better. If the sensor threshold is adjusted to the smallest error sum, the balanced error frequency must be used (Tables 3.15 and 3.16).

The setting of the minimum error sum is generally detrimental from a commercial point of view. Assuming that the products are cameras or other high-cost devices, an unsold product that is incorrectly classified as defective results in a greater loss than the complaint of a delivered product that is defective but has not been recognized as

Table 3.12 Defect rates of two sorting machines

Machine A		Output		Total
		Good	Defect	
Input	Good	5433	67	5500
	Defect	25	135	160
Machine B		Output		Total
		Good	Defect	
Input	Good	3948	52	4000
	Defect	50	65	115

Table 3.13 Relative error frequencies of sorting machines A and B

Machine A		Output		Total
		Good	Defect	
Input	Good	$1 - p = 0.9878$	$p = 0.01218$	1
	Defect	$q = 0.1563$	$1 - q = 0.8438$	1
Machine B		Output		Total
		Good	Defect	
Input	Good	$1 - p = 0.9870$	$p = 0.013$	1
	Defect	$q = 0.4348$	$1 - q = 0.5652$	1

Table 3.14 Error balancing and evaluation of machines A and B with the *SNR* index

	p_0	ρ_0	*SNR* (dB)
Machine A	0.04561	0.8259	6.76
Machine B	0.09145	0.6676	3.03

Table 3.15 Triggering threshold for error-symmetrical appearance (Crossover)

Machine A		Output		Total
		Good	Defect	
Input	Good	$1 - p_0 = 0.9544$	$p_0 = 0.04561$	1
	Defect	$p_0 = 0.04561$	$1 - p_0 = 0.9544$	1
Machine B		Output		Total
		Good	Defect	
Input	Good	$1 - p_0 = 0.9086$	$p_0 = 0.09145$	1
	Defect	$p_0 = 0.09145$	$1 - p_0 = 0.9086$	1

Table 3.16 Number of incorrectly identified products when using the error-symmetrical operating point

Machine A		Output		Total
		Good	Defect	
Input	Good	5250	250	5500
	Defect	7	153	160
Machine B		Output		Total
		Good	Defect	
Input	Good	3634	366	4000
	Defect	11	104	115

such. In other words, while maintaining the precision of the sorting machine, the response of the sensors is intentionally adjusted in such a way that the error of the second type is increased by a factor of 100, in comparison with the error of the first type. The minimum error sum equation can be used to adjust the error type 1 as a function of the error type 2 and to determine the error frequency to be set.

$$\left(\frac{1-p_0}{p_0}\right)^2 = \frac{1-p}{p}\frac{1-q}{q}$$

with use of $q = 100p$

$$\left(\frac{1-p_0}{p_0}\right)^2 = \frac{1-p}{p}\frac{1-100p}{100p}$$

$$\left(1 - \left(\frac{1-p_0}{p_0}\right)^2\right)p^2 - \frac{101}{100}p + \frac{1}{100} = 0$$

Solution of a second-degree equation **(1-23*)**

$$p = \frac{\frac{101}{100} \pm \sqrt{\left(\frac{101}{100}\right)^2 - \frac{4}{100}\left(1 - \left(\frac{1-p_0}{p_0}\right)^2\right)}}{2\left(1 - \left(\frac{1-p_0}{p_0}\right)^2\right)}$$

The following applies to the better sorting machine $p_0 = 0.04561$ (Table 3.17)

$$p = \frac{\frac{101}{100} - \sqrt{\left(\frac{101}{100}\right)^2 - \frac{4}{100}(-437)}}{2(-437)}$$

$$p = 0.003757$$
$$q = 0.3757$$

If the lost profit of a product, which is wrongly classified as defective, is 100$, while the complaint and shipping costs of defective goods are 30$/piece, the loss streams can be balanced in this way (without considering the loss of reputation).

3.6.2 *Magnetic Separator*

The visible indicator of any separation process (sorting machines, isotope separation, etc.) consists of one material stream that is fed into the system and at least two material streams that are discharged. In the latter, a material flow is enriched regarding a more valuable component, while the other is appropriately depleted. The perfection of a separation system is characterized by the fact that the undesired admixtures in both output currents are as low as possible. In the example of a magnetic separator, ferromagnetic minerals are separated from litter rock so that the pure ore content can be further processed. The broken material is transported by a conveyor belt into a strong magnetic field and ejected. If, for example, neodymium-iron-boron magnets are located underneath the material stream in the form of the deflection roller of the conveyor belt, the ferromagnetic components are more strongly deflected to the vertical, whereas the iron ore free material will continue to fly unaffected. By means of a splitter, the separation into the enriched or depleted fractions takes place in this fanned out material flow. If the magnets are

Table 3.17 Adjustment of the triggering threshold of machine A for a similar economical loss

Machine A		Output		Total
		Good	Defect	
Input	good	5479	21 $p = 0.003757$	5500
	Defect	60 $q = 0.3757$	100	160

mounted at the discharge point above the conveyor belt, the opposite applies, i.e., the ore-containing rock will be spread further from the discharge point in a stretched trajectory than non-ferromagnetic material. To improve selectivity, the parameter effects are determined by collecting and chemically analyzing both, a container of the waste stream and a container of the material enriched with ore, for each test carried out. The position of the splitter serves as an adjustment factor, which is to be set after the parameter analysis in such a way that, in case of model confirmation, the requirements are met. The assigned parameters are listed in the Table 3.18.

The chemical composition of the partial flows is described in the following diagram (Table 3.19).

An orthogonal array L4 is used for testing (Table 3.20).

Since the collected substance quantities are all different, the first step is the conversion into relative defect proportions p and q followed by the symmetric defect rate p_0 at the crossover, the contribution rate ρ_0, and the *SNR* key figures (Table 3.21).

The parameter analysis is displayed in tabular and graphical form (Table 3.22, Fig. 3.6).

The best possible configuration has already been proven with test #2

$$A1 + B2 + C2$$

Table 3.18 Parameters and settings for a magnetic separator

Parameter	Level	
	1	2
A: Particle size	Fine Up to 5 mm	Coarse Up to 10 mm
B: Kind of magnet	Nd–Fe-B anisotropic sintered	Nd–Fe-B isotropic resin-bonded
C: Magnetic array	Polarity pattern 1	Polarity pattern 2
Adjustment factor	Position of the splitter	

Table 3.19 Designation of the material flows for chemical analysis

	Ore	Rock
Valuable product, pure iron ore	n_{00}	n_{01} Error type 1
By-product, pure rock	n_{10} Error type 2	n_{11}

Table 3.20 Experimental results of testing a magnetic separator

Test	*A*	*B*	*C*	n_{00} (g)	n_{01} (g)	n_{10} (g)	n_{11} (g)
1	1	1	1	455	55	215	6440
2	1	2	2	510	36	125	6010
3	2	1	2	390	59	118	5890
4	2	2	1	405	77	195	6570

With the idea of keeping the sum of the incorrectly classified material flows as low as possible, the splitter position must be adjusted in such a way that the defect proportions become similarly large. Since the mineral concentration in the rock is close to 10%, the error ratio must be shifted by the same order of magnitude **(1-24*)**.

For this case error type 1 is calculated with

$$p = 10q$$

in the same way as in the previous example from the definition of the balanced error frequency.

$$\left(\frac{1-p_0}{p_0}\right)^2 = \frac{1-p}{p}\frac{1-q}{q}$$

$$q = \frac{\frac{11}{10} - \sqrt{\left(\frac{11}{10}\right)^2 - \frac{4}{10}\left(1 - \left(\frac{1-p_0}{p_0}\right)^2\right)}}{2\left(1 - \left(\frac{1-p_0}{p_0}\right)^2\right)}$$

This results in the individual error portions with the respective symmetry being offset, as an example defined with 1:10. Other ratios are calculated accordingly. For the configuration

$$A1 + B2 + C2$$

follows

$$p_0 = 0.03690$$

with the individual error rates (Table 3.23)

$$q = 0.011327$$
$$p = 0.11327$$

3.6.3 *Satellite Receiver*

For a satellite dish with receiver, the arrangement between receiver, amplifier, and frequency converter (LNB = low noise block converter) is varied with four parameters (i.e., the design of the preamplifier, the arrangement of the components, the cable routing and the shielding) to three setting values each. Since the error rate is already extremely small, the antenna area is covered (attenuation with wire mesh or similar) until a pixilation becomes visible, or, to put it more technically, the error

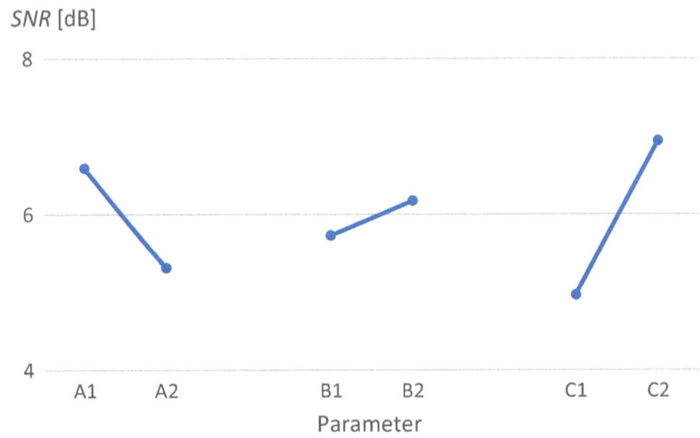

Fig. 3.6 Graphics of the parameter effects

Table 3.21 Processing of the test results

Test	A	B	C	p	q	p_0	ρ_0	SNR
1	1	1	1	0.1078	0.03231	0.05973	0.7753	5.3798
2	1	2	2	0.06593	0.02038	0.03690	0.8578	7.8062
3	2	1	2	0.1314	0.01964	0.05218	0.8022	6.0798
4	2	2	1	0.1598	0.02883	0.06987	0.7400	4.5436

Table 3.22 Effects of parameters

Parameter	Levels	
	1	2
A	6.5930	5.3117
B	5.7298	6.1749
C	4.9617	6.9430

Table 3.23 Smallest error proportions of both error types with nonsymmetrical adjustment

Magnetic separator		Output		Total
		Ore	Rock	
Input	Ore	484 g	62 g $p = 0.11327$	546 g
	Rock	69 g $q = 0.011327$	6066 g	6135 g

rate of the digital output signal increases dramatically. Thus, for each test configuration, the error frequencies can be measured at short signal lengths. The following applies to the error type 1

$$p = \frac{\text{Number of states 0 interpreted as 1}}{\text{All states 0 in signal of a certain length}}$$

and according to the definition of error type 2

$$q = \frac{\text{Number of states 1 interpreted as 0}}{\text{All states 1 in signal of a certain length}}$$

The signals and related error rates are shown in the table schematically (Table 3.24).

For calibration, a receiver system is characterized according to the state-of-the-art with a known error rate of $1{:}10^6$ for an unshielded antenna dish area. The table shows the experimental layout with the recorded error frequencies and calculation of the *SNR* indices (Table 3.25).

The evaluation of the parameter effects is displayed in the graphic (Fig. 3.7).

The largest value of the *SNR* index follows from the configuration

$$SNR(B3 + C1 + D3) = 8.287\ (\text{dB})$$

Table 3.24 Schematic diagram of the signals and error frequencies

Signal state		Output (signal from amplifier)	
		0	1
Input (signal from satellite)	0	$1 - p$	p
	1	q	$1 - q$

Table 3.25 Orthogonal array L9 with associated measured values and processed data

Test	A	B	C	D	p	q	p_0	ρ_0	*SNR* (dB)
1	1	1	1	1	0.27	0.1	0.169	0.439	−1.057
2	1	2	2	2	0.07	0.12	0.092	0.666	2.995
3	1	3	3	3	0.05	0.03	0.039	0.851	7.564
4	2	1	2	3	0.08	0.23	0.139	0.522	0.380
5	2	2	3	1	0.33	0.04	0.125	0.562	1.075
6	2	3	1	2	0.03	0.06	0.043	0.837	7.108
7	3	1	3	2	0.16	0.11	0.133	0.539	0.673
8	3	2	1	3	0.02	0.15	0.057	0.786	5.659
9	3	3	2	1	0.03	0.22	0.085	0.688	3.424
	Satellite receiver as benchmark				0.10	0.10	0.100	0.640	2.499

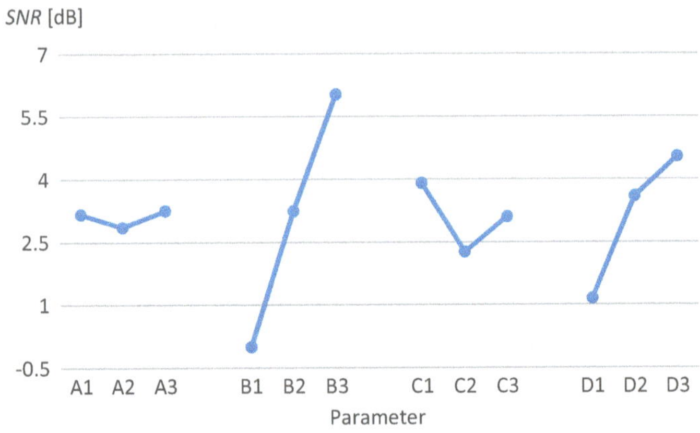

Fig. 3.7 Graphic of parameter effects

Table 3.26 True error transmission rates and *SNR* index

p	q	p_0	ρ_0	*SNR* (dB)
10^{-6}	10^{-6}	10^{-6}	0.999996	53.979

The improvement regarding the benchmark device is

$$\Delta SNR = 8.287\,(\text{dB}) - 2.499\,(\text{dB}) = 5.788\,(\text{dB})$$

If the true error transmission rate of the existing device is taken as a basis, the actual *SNR* index is obtained (Table 3.26).

Without shielding the antenna dish, the true value of the *SNR* index is obtained for the modified receiver design.

$$SNR = 53.979\,(\text{dB}) + 5.788\,(\text{dB}) = 59.767\,(\text{dB})$$

The new error rate is thus reduced to $1/2.6 \times 10^7$, i.e., a quarter of the initial one.

3.7 Functional Range with Two Signal-to-Noise Key Figures

Due to their robust trigger thresholds, systems with digital output possess such a high reliability that error messages, false alarms, etc., can be almost excluded. As shown previously, the highest stability despite noise factors is achieved by the sum of the *SNR* indices of the existing trigger points. From an analog viewpoint, the range between the trigger points can be defined as working area or functional range.

In general, it is advantageous to extend the functional range, i.e., the distance between the lower (smaller) and the upper (larger) threshold, till the function is no longer given. At these points the meaning of the trigger threshold is replaced by the functional or specification limit used in technical language. How are these limits defined?

As a guideline, the experience gained in the pharmaceutical and medical fields are valuable, particularly where typical limits are defined by the lethal dose of drugs used for therapeutic medication. In this case, the critical limit is determined by a survival rate of 50% as a function of an applied dosage (usually in mg/kg weight), in short LD50. The extension of such critical limits in static systems—which was invented by Don P. Clausing **(1-1)**—leads to an improved reliability of the function (in this context the survival rate). Don P. Clausing then coined the term *Operating Window* for the working area with the maximum possibility of use. This concept turned out to be of great importance as outlined in an example of medical investigation for the treatment of cancerous cells.

When drugs are used in therapy, the dosage is of the utmost importance. The traditional use of empirical findings to determine the appropriate dosage based on physiological factors is still widespread. Weight, age, sex, to name just a few, are considered to estimate the dosage. A compromise must be reached whereby a "trade-off" takes place between the intended effect on cancer cells and the toxic side effects on the patient.

A completely different approach can be realized with the application of signal-to-noise ratios. The first step consists of a robust independent therapy, which is followed by the second step to arrive at the most effective dose. For achieving this robust therapy, healthy, and pathological cells are taken for cultivation from the object of investigation, i.e., from plants, animals or the human body. Different cells of both types represent noise factors. The therapy will only be successful if all types of pathological cells are affected adversely to healthy cells which possess a far greater survival rate. For testing promising drugs, more parameters such as mixing ratios, added minerals, as well as different chemical groups are considered to improve the chance of intensifying the intended effect and to mitigate toxic side effects. The variables are assigned to an orthogonal array where the dosage is gradually increased until the biological function ceases. The concentration at this point (=LD50) is interpreted for all healthy cells with the signal-to-noise Ratio larger-the-better (*LTB*), and for all pathological cells with smaller-the-better (*STB*). The analysis of the parameters results in the most efficient concentration when

$$SNR = LTB + STB = \text{Maximum}$$

For $SNR > 0$ the survival rate of healthy cells is still higher even when a higher dose is administered, compared to a low survival rate of pathological cells even at a lower dose. In this way, the width of the functional range stands for an increased robustness or reliability of a drug-based therapy. The experimental approach to determine the lethal dose and the interpretation of data is outlined in Tables 3.27 and 3.28.

Table 3.27 Determination of the lethal dose LD50 for healthy and pathological cells

Test		Dosage in μg/ml									
		2	4	6	8	10	12	14	16	18	20
Healthy cells of type N	N_1	1	1	1	1	0					
	N_2	1	1	1	0						
	N_3	1	1	1	1	1	1	1	0		
	N_4	1	1	1	1	1	1	1	1	1	0
	N_5	1	1	1	1	1	0				
	…										
	N_n	1	1	1	1	1	1	0			
Pathological cells of type M	M_1	1	1	1	0						
	M_2	1	0								
	M_3	1	1	0							
	M_4	1	1	1	1	0					
	M_5	1	1	1	0						
	…										
	M_n	1	1	1	0						

Legend 1 = living
0 = dead
LD50 = Threshold 1/0
Noise factors: $N_1, N_2, \ldots, N_n, M_1, M_2, \ldots, M_n$

Table 3.28 Assignment of factors to an orthogonal array to increase the width of the therapeutic function

Test #	Factors A B C D E … Orthogonal Array	LD50 Healthy cells				*SNR* *LTB*	LD50 Pathological cells				*SNR* *STB*	SNR_{total} $LTB + STB$
		N_1	N_2	⋯	N_n		M_1	M_2	⋯	M_n		
1		x_1	x_2	⋯	x_n		y_1	y_2	⋯	y_n		
2 …												

Legend x_i = LD50 concentration (μg/ml) for healthy cells, $i = 1, \ldots, n$
y_i = LD50 concentration (μg/ml) for pathological cells, $i = 1, \ldots, n$

$$LTB = SNR = 10 \log\left(\frac{n}{\left(\frac{1}{x_1^2}+\frac{1}{x_2^2}+\cdots+\frac{1}{x_n^2}\right)}\frac{1}{x_0^2}\right)$$

$$STB = SNR = 10 \log\left(\frac{n}{y_1^2+y_2^2+\cdots+y_n^2}y_0^2\right)$$

$$LTB + STB = 10 \log \left(\frac{1}{\left(\frac{1}{x_1^2} + \frac{1}{x_2^2} + \cdots + \frac{1}{x_n^2}\right)} \cdot \frac{1}{y_1^2 + y_2^2 + \cdots + y_n^2} \cdot \frac{n^2 y_0^2}{x_0^2} \right)$$

In case of an identical survival rate for healthy as well as pathological cells, based on an identical dose, one arrives at the following:

$$SNR = 0$$

i.e., the drug is unsuccessful.

The dosage is no longer a decisive factor if the LD50 concentration limit for healthy cells can be shifted to higher values whereas pathological cells die already at far lower concentrations. The LD50 figure for all noise conditions (types of healthy as well as pathological cells) for the parameter configuration resulting in the widest possible functional range must be confirmed through experiments.

The adequate dose is derived after confirmation of the corresponding reaction of the cells for the largest characteristic *SNR* index. In this case, the average figures of the experimental cell-specific LD50 concentrations are calculated.

$$\bar{x} = \sqrt{\frac{n}{\left(\frac{1}{x_1^2} + \frac{1}{x_2^2} + \ldots + \frac{1}{x_n^2}\right)}}$$

$$\bar{y} = \sqrt{\frac{y_1^2 + y_2^2 + \ldots + y_n^2}{n}}$$

$$\text{Dosage}_{\text{center}} = \frac{\bar{x} + \bar{y}}{2}$$

The drug concentration to be applied is safe within the therapeutic window, that means between $\bar{x}$ and $\bar{y}$. For the medical treatment of a patient, the individual situation could be decisive for using a dosage in the center or closer to the limits of the functional range. The development of such a robust approach is principally obtained for any therapy, such as chemotherapy, gamma radiation, treatment with proton rays, as well as emerging new medical procedures **(1-25*)**. The extension of a functional range—to increase the reliability of digital functions (on/off, living/dead) by means of continuous data—proves to be one of the most efficient methods in both, the technical as well as the medical field.

3.8 Ideal Function

The function of a dynamic system is to relate the range of an input signal to a corresponding range of the output, referred to as response. The slope represents the gain. The steeper the slope, the more the output amplitude changes due to an applied signal. If input and output stand for an energy transformation, then a slope of 45° means an efficiency of 1 (=100%) when energy units of the same scale are applied. Due to unavoidable losses such as friction, damping, absorption or other dissipative paths, conversion efficiencies are always less than 100%. Based on understanding natural laws, almost all energy and material transformations (chemical reactions) can be ideally calculated. The results are perfect efficiencies or yields, hereinafter referred to as ideal function. This means a system performance without limiting boundary conditions. However, the real function is not the same as the ideal function. To avoid unwanted energy pathways or side reactions, ideal functional behavior is to be achieved as good as possible. With its numerical value, the *SNR* index stands for the degree of approximation to ideal behavior (Fig. 3.8).

Before any parameter analysis, the question arises as to what the ideal transfer function of the system to be analyzed looks like, even if the answer is not necessarily easy. However, the successful implementation of a parameter analysis depends on this knowledge since deviations from the ideal function can only be reduced by means of the *S/N* ratio. This is the decisive advantage of the dynamic system approach compared to the simple static analysis. In a static system analysis, an improvement of system parameters can be achieved, i.e., existing deficits can be compensated in the intended way, but other harmful symptoms can occur instead. The reason for this is lack of affecting the actual energy conversion process, i.e., the slope β of the operating characteristic, equivalent to the efficiency, remains unchanged. Energy losses continue to exist and are looking for other paths that may give rise to new problems. Although quick results are achieved due to the simplicity of static approaches (*NTB*, *LTB*, *STB*), their application is only recommended in well-considered cases.

First, a good preparation for multiparameter experiments consists of theoretical considerations of the maximum achievable numerical value of the slope. For

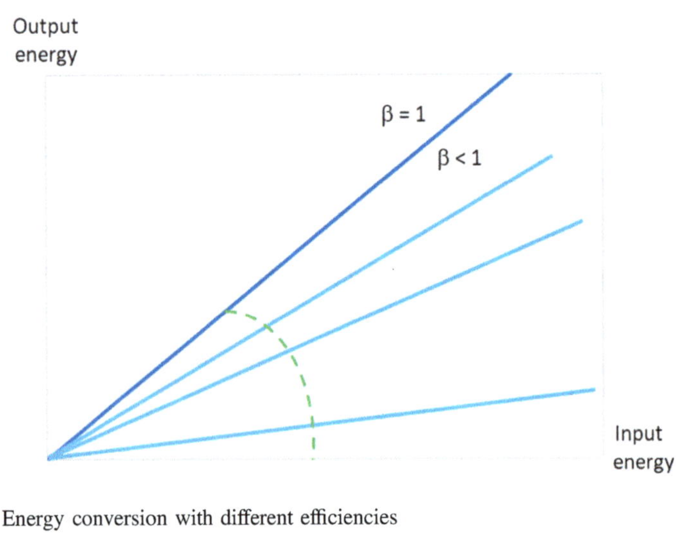

Fig. 3.8 Energy conversion with different efficiencies

example, if a system to be investigated is based on Carnot's loop process, the efficiency will be limited to about 40% for thermodynamic reasons. Approaching this value already corresponds to the best possible situation to be achieved. In addition, there are systems that change considerably as a function of time. For example, the ratio of the final velocity (v_t) achieved by a rocket to the jet velocity (v_g) at which the thrust gases leave the nozzle is logarithmically dependent on the ratio of the rocket mass at launch (m_0) to the mass at firing end (m_t) **(1-26*)** (Fig. 3.9).

An evaluation of linear behavior would be completely misleading. However, since the explicit relation is known, the transmission function can be linearized by applying the inverse mathematical operation. Exponential scaling of the ordinate transforms the relation into a straight line (Fig. 3.10).

The reverse behavior can be observed in biological growth or propagation processes. If there is food oversupply, the number of individuals grows exponentially. Linearization is achieved by transforming the ordinate on a logarithmic scale. In more general terms, transfer functions that can be represented explicitly can be linearized by the inverse mathematical operation and suitable substitution. The *S/N* ratio is generated from the identically transformed measurement data and evaluated only in this form. Commonly used functional relations, the associated inverse operation and the substitution of the measured variable are listed (Table 3.29).

Dealing with transfer characteristics of any form is discussed in Sect. 5.1.

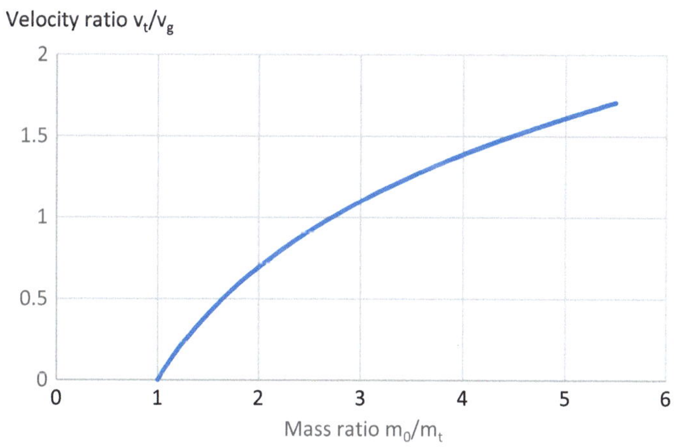

Fig. 3.9 Ratio of the rocket velocity to the thrust velocity of the reaction gases as a function of the mass ratio

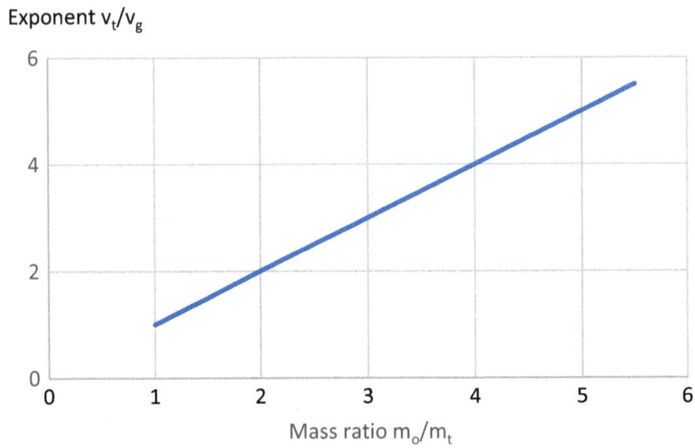

Fig. 3.10 Linearized transfer function of the ratio of rocket velocity to thrust velocity in relation to the mass ratio

Table 3.29 Frequent transformations to linearize the input/output relationship

Function	Inverse mathematical operation	Substitution for zero-point-proportional equation $y = \beta M$		
		y	β	M
$y = e^x$	$\ln(y) = x$	$\ln(y)$	1	x
$y = 10^x$	$\lg(y) = x$	$\lg(y)$	1	x
$y = \frac{1}{x}$	$\frac{1}{y} = x$	$\frac{1}{y}$	1	x
$y = \sqrt[2]{x}$	$y^2 = x$	y^2	1	x
$y = ax^b$	$\sqrt[b]{y} = \sqrt[b]{a}x$	$\sqrt[b]{y}$	$\sqrt[b]{a}$	x
$y = \frac{x}{a+bx}$	$\frac{1}{y} = \frac{a}{x} + b$	$\frac{1}{\frac{1}{y}-b}$	$\frac{1}{a}$	x
$y = a + \frac{b}{x}$	$\frac{1}{y-a} = \frac{x}{b}$	$\frac{1}{y-a}$	$\frac{1}{b}$	x
$y = a + b\ln(x)$	$\frac{y-a}{b} = \ln(x)$	$e^{\frac{y-a}{b}}$	1	x
$y = ae^{bx}$	$\ln\left(\frac{y}{a}\right) = bx$	$\ln\left(\frac{y}{a}\right)$	b	x
$y = 1 - e^{-bx}$	$\ln(1-y) = -bx$	$\ln\left(\frac{1}{1-y}\right)$	b	x

Important items to memorize The operating window concept represents a very powerful approach for extension of a functional range.

The evaluation with dynamic *SNR* indices requires approximately linear behavior; linearization methods may be used if necessary.

Chapter 4
Chemical Reactions and Special Systems

Contents

Abstract In almost all chemical reactions occur side reactions with side products. These lower the yield, are unwanted, and could be toxic. On an industrial scale, this is an issue of cost incurred because of the necessary effort to convert toxic material into safe material. In general, it cannot be avoided but at least minimized if the reactions are understood in detail. Although on the atomic or molecular scale, a chemical reaction is digital in nature, a macroscopic outcome is what can be observed and measured. This necessitates more experiments to extract some characteristic features—e.g., rate constants—for an overall kinetic description. These are the key figures to calculate the *SNR* index and thus determine the best conditions for a chemical reaction when applied to the desired intermediate product, either in a batch or a continuous process. This approach is not confined to chemistry alone but applicable to other time-dependent systems as well. Electrical engineering is based on a solid theoretical background. In contemporary technology, each design is optimized through software simulation before it is manufactured. Nevertheless, the *SNR* index in its complex form can be applied to reduce unnecessary loss of electrical power.

© Springer Nature Switzerland AG 2019

H. Ruefer, *Living Without Mathematical Statistics*,
https://doi.org/10.1007/978-3-319-99632-5_4

4.1 Macroscopic Description

Molecular reactions or the formation of molecules from atoms can be described in a way that a molecule or an atom reacts with the valence electrons of the impact partners in the intended way or not. In this sense, the event as such is of digital character. Usually, the individual molecular reaction is not directly observed, i.e., only the initial and final state of the molecule ensembles or substance concentrations are accessible for quantitative measurements. However, since common substance concentrations consist of a huge number of particles (Avogadro constant), such a system behaves quasi-continuously and can therefore be described in analog instead of digital form. In a simple macroscopic approach, the reaction takes place in a black box with characterization of the substance quantities/concentrations in the reaction input (reactants) and reaction output (products). For practical reasons, a chemical reaction is described by the following standard parameters:

1. Conversion (U) = Fraction of consumed reactants

$$U = \frac{c_0 - c}{c_0}$$

c_0 Initial concentration of reactants.
c Final concentration of reactants.

2. Selectivity (S) = Ratio of the specific product obtained to the consumed reactants

$$S = \frac{c_{\text{product}}}{c_0 - c}$$

3. Yield (A) = Ratio of the specific product obtained to the initial concentration of reactants. This is identical to the product of conversion (U) and selectivity (S)

$$A = \frac{c_{\text{product}}}{c_0} = U \cdot S$$

Since the latter definition follows from the aforementioned equations, the evaluation of a chemical reaction on the basis of the achieved conversion and selectivity of the target product is sufficient. For comparison of similar chemical reactions, the representation $S = f(U)$ appears to be reasonable since the selectivity can be derived at the same conversion in each case. The setting of different conversions is adjusted by the reaction temperature or the residence time as an independent variable (=signal parameter). Parameters affecting a reaction process such as concentration, pressure, composition of the components, catalysts, grain sizes, morphology, etc. will change conversion and selectivity to varying degrees. Due to the influence of such reaction parameters with the temperature, identical conversions can be adjusted only approximately. Therefore, it is recommended to use at least two, better three, different temperature settings for a comparative evaluation of selectivity and conversion.

The residence time of the reaction partners, i.e., the period from the beginning of a chemical reaction to sampling or termination of the reaction, must be kept constant in this simple procedure. The measurement of reaction products as a function of time, on the other hand, is of crucial importance in the context of reaction kinetics, as described in the following chapter.

Evaluation criteria of test series result from

1. Slope of the characteristic curves $S = f(U)$ (or of the inverse representation $U = f(S)$),
2. Selectivity for U = constant, and
3. Reaction temperatures for U = constant.

From the course of the slope $S = f(U)$, the sensitivity of selectivity to the conversion becomes apparent; the numerical value of selectivity stands for the purity of the target product formed and, in addition, the temperature for an adjustable starting point of the reaction.

In the case of multiparameter analyses of chemical reactions, model calculations depend on the additivity of the effects. If the measured values of the concentration of the components are taken as percentages or fractions, then the transformed value is to be used to restrict the response to 0–100%

$$\frac{S}{N} = \frac{x}{1-x}$$

with the characteristic *SNR* key figures

$$\begin{array}{ll} SNR(U) = 10\log\left(\frac{U}{1-U}\right) & \text{reaction temperature } T = \text{constant} \\ SNR(S) = 10\log\left(\frac{S}{1-S}\right) & \text{conversion}(U) = \text{constant} \\ SNR(T) = 10\log\left(\frac{T^2}{1K^2}\right) & \text{with constant conversion in the form } LTB \end{array}$$

For the dimension of the temperature, the absolute temperature scale (Kelvin) is used. The slope of the function $S = f(U)$ can be considered as additional information to set the slope as flat as possible ($\beta = 0$), i.e., conversion independent with maximum selectivity at the same time. Using model calculations, the standard parameters listed here allow to figure out the best trade-off for U, S, and T. Likewise, there is no problem to maximizing individual parameters at the expense of others. However, strict reaction optimization in favor of one component can only be achieved approximately, as the molecular processes involved in a chemical reaction are ignored as a black box. Without a black box, a wealth of details of chemical reactions can come to light, such as numerous intermediate reactions with parallel and consecutive reactions paths that lead to the target product. This is accompanied to varying degrees by inverse reactions, which cannot be neglected in case of approaching equilibrium conditions. In such a mixture of simultaneously existing reaction paths, the respective individual reaction speeds are responsible for the amount and purity of the target product generated. Even the “right” reaction speed determines the function of a system. For example, the explosion velocity of hydrocarbons with air at the combustion temperature in combustion engines must be within a range that triggers the mechanical piston movement without destroying it. For the blasting of hard rock, on the other hand, a much higher explosion velocity is required. In addition, the question of how much product can be produced per time comes up in industrial processes. Thus, the consideration of velocity determining processes in a chemical reaction can no longer be ignored. This extensive field, also known as reaction kinetics, requires a greater amount of experimental effort for a more comprehensive analytical interpretation but provides a better understanding of the kinetic processes at the molecular level.

4.2 Signal-to-Noise Ratios in Reaction Kinetics

For most chemical reactions, the reaction speed increases with increasing concentration due to increased collision frequency of the reacting components. In this relation, the proportionality constant is called the rate constant (k).

$$r = k[A]^{\alpha}$$

r Reaction speed.
k Rate constant.
$[A]$ Concentration of chem. components.
α Order of chemical reaction.

Applies $\alpha = 1$ to the observed regularity of the above function, a chemical reaction of first order exists. Analogously if $\alpha = 2$ is valid, this corresponds to a reaction of second order, etc. Half-number and fractional exponents can also occur if, for example, different reaction paths run simultaneously. Since the rate constant generally depends only on the absolute temperature, a major task of reaction kinetics is to determine k **(1-27*)**. If the k's of the involved intermediate reaction products that lead to the target product are known, all quantities, i.e., U, S, and T, and the generated product quantity/time can be calculated. Consequently, the easiest way would be to find out the rate constants in literature and to make use of them to avoid any experiments. However, as a matter of fact, the number of chemical reactions is extremely large, and the rate constants are often not easy to determine due to experimental shortcomings, and even values that can be found are not necessarily reliable. This means that nothing else is left to do if the necessary detailed knowledge is required for product optimization but to accept additional effort and to determine the k's experimentally. Due to the large number of kinetic reaction properties, these are classified according to the exponent (α = order of chemical reaction). In order not to exceed the scope and purpose of the book, the explanations are limited to frequently occurring first- and second-order reactions. The selection made is shown in the table (Table 4.1).

For higher reaction orders and special cases, refer to subject literature **(1-2)**.

The basic procedure for optimization of the different types of reactions is outlined.

Table 4.1 Chemical reactions with different reaction kinetic properties

Chemical reactions	
First order	Second order
(A) irreversible	(B) irreversible
(C) reversible	–
(D) parallel reaction	–
(E) consecutive reaction	–
(F) double consecutive reaction	–

4.2.1 *Irreversible Chemical Reaction of First Order*

When the reaction starts, the concentration of the reactants $[A]$ is converted into the product $[C]$ as a function of time. The integration of the corresponding differential equation results in an ideal function of the temporal concentration progression.

$$\ln\left(\frac{[A]}{[A]_0}\right) = -kt$$

$[A]$ = Concentration of the reactant at the time t
$[A]_0$= Concentration of the reactant at the time $t = 0$

or in exponential notation

$$[A] = [A]_0 \cdot \mathrm{e}^{-kt}$$

The concentration of the product $[C]$ increases simultaneously with

$$[C] = 1 - \mathrm{e}^{-kt}$$

If the initial concentration of the reactant is expediently set to 1 ($[A]_0$ = 100%) and the substitutions are introduced,

$$y = \ln\left(\frac{1}{[A]}\right)$$
$$\beta = k$$
$$M = t$$

the time-dependent concentration of an irreversible chemical reaction of first order follows the zero-point-proportional equation.

$$y = \beta M$$

This means that the reaction behavior can be understood as a dynamic system that is to be evaluated with the key figures:

$$SNR = 10\ \log\left(\frac{\beta^2}{\sigma^2} S_2\right)$$

Table 4.2 Exemplary layout of parameters and setting values for the investigation of reaction kinetics

Parameter	Levels		
	1	2	3
A: Components	Chloride	Nitrate	Sulfate
B: Concentration of the components [%]	5	10	20
C: Reactor pressure [bar]	1	2	3
D: Stirring speed [rpm]	10	20	30
M: Sampling intervals [min]	10	30	60

Reaction temperature T = constant (for example 100 °C)

$$SEN = 10 \log\left(\frac{\beta^2}{\beta_0^2}\right)$$

To carry out a multiparameter analysis, reaction-related variables such as components of different chemical compositions, concentrations, form and distribution of substances, geometric boundary conditions, etc. are assigned to the columns of an orthogonal array of suitable size. In practice, deviations from the ideal function can occur, e.g., due to diffusion limitation or additional reaction paths, so that preferably two or three samples can be taken at different residence times (or also at the same time at different flow velocities) and the concentrations of the components are measured. Only by comparing the experimental results with the postulated kinetic equation, it becomes obvious whether the approach is justified. If the assumption applies in good approximation, a parameter analysis can achieve the best possible agreement with the assumed reaction order by appropriate selection of the setting values. An experimental procedure can be based on the scheme outlined in the table (Table 4.2).

The independent variable for determining the kinetic approach is the residence time. There are four reaction parameters assigned to the columns of the orthogonal array L9 (Table 4.3).

In the first step of the evaluation, the chemical reaction is kinetically stabilized with the largest achievable *SNR* index, which equates to a robust process; in the second step, unless the *SNR* index is significantly impaired, the slope, i.e., the rate constant, can be adapted to specific requirements. In addition, it can be aggravating that in a limited temperature interval, the exponential relation of the rate constants with the temperature does not have to be exactly valid **(1-28*)**. To clarify this, it appears appropriate to test the preferred parameter configuration with sampling at a constant reaction time for three different reaction temperatures, e.g., with $T1$ = 80 °C, $T2$ = 100 °C, and $T3$ = 120 °C. Ultimately, the reason is not to confirm the exponential temperature dependence of k but to be able to consider by means of linear regression or also regression of higher order a practically applicable temperature function of the rate constants. Apart from confirmation experiments, the total number of samples to be taken in this four-parameter example is therefore already 30.

Table 4.3 Test plan with four reaction parameters with three settings each

Test #	Parameter A B C D	Sample 10 min	Sample 30 min	Sample 60 min	β^2	σ^2	*SNR*	*SEN*
1	1 1 1 1	$y_1 = \ln\left(\frac{[A]_0}{[A]_1}\right)$						
2	1 2 2 2							
3	1 3 3 3							
4	2 1 2 3							
5	2 2 3 1							
6	2 3 1 2							
7	3 1 3 2							
8	3 2 1 3							
9	3 3 2 1			$y_{27} = \ln\left(\frac{[A]_0}{[A]_{27}}\right)$				

4.2.2 *Irreversible Chemical Reaction of Second Order*

The ideal function of the speed law for one component has the following form:

$$r = k[A]^2 \quad [A]\text{reacts to } \rightarrow [C]$$

The equation for two components (=reactants) is

$$r = k[A][B] \quad a[A] + b[B] \text{ reacts to } \rightarrow c[C] + \cdots$$

The stoichiometry factors are designated by a, b, and c.

The decrease of the reactants concentration with time in case of one component follows from the integration of the differential equations of the reaction kinetics:

$$[A] = \frac{[A]_0}{1 + kt[A]_0}$$

The substitutions with $[A]_0 = 1$ (=100%)

$$y = \frac{1 - [A]}{[A]}$$

$$\beta = k$$

$$M = t$$

lead to the zero-point-proportional equation

$$y = \beta M$$

The application of the dynamic key figures *SNR* and *SEN* evaluates linearity, i.e., the best possible approximation to the second-order speed law. The slope is derived from the preferred choice of parameter settings with low deviation, which is identical to the rate constant.

For two reactants, the time-dependent concentration decreases with

$$\frac{1}{a[B]_0 - b[A]_0} \ln\left(\frac{\frac{[B]}{[B]_0}}{\frac{[A]}{[A]_0}}\right) = kt$$

With the initial concentrations $[A]_0 = [B]_0 = 1$ (=100%)

$$\frac{1}{a-b} \ln\left(\frac{[B]}{[A]}\right) = kt$$

follows the zero-point-proportional equation

$$y = \beta M$$

if the substitutions are applied

$$y = \frac{1}{a-b} \ln\left(\frac{[B]}{[A]}\right)$$
$$\beta = k$$
$$M = t$$

As in the abovementioned cases, the dynamic key figures *SNR* and *SEN* are used to evaluate the greatest linearity with the smallest deviation and the setting of a favorably selectable rate constant.

If $a = b$ is valid for the stoichiometry factors with constant ratio $[A]/[B]$, the concentration–time function for the entire reaction is to be replaced by

$$[A] = \frac{1}{1 + bkt}$$

Synonymous substitutions lead again to the zero-point-proportional equation

$$y = \beta M$$

with the slope

$$\beta = bk$$

4.2.3 Reversible Chemical Reaction of First Order

The chemical reactions discussed so far have only proceeded in one direction from the reactants to the product. This consideration is permissible in the first moment of an incipient reaction, provided that a reaction equilibrium only occurs after a long period of time. For thermodynamic reasons, the equilibrium will always be reached after a sufficiently long time, so that the rate constants become the same in both directions.

$$[A] \underset{k2}{\overset{k1}{\rightleftarrows}} [C]$$

If the concentration $[A]$ approaches its equilibrium concentration $[A]_{\text{equilibrium}}$, the result for the concentration as a function of time is given by

$$[A] - [A]_{\text{equilibrium}} = \left(1 - [A]_{\text{equilibrium}}\right) \mathrm{e}^{-(k_1 + k_2)t}$$

with the initial concentration of $[A]_0 = 1$.

The representation of the zero-point-proportional equation is like the irreversible reaction corrected by the equilibrium concentration with the definitions

$$y = \ln\left(\frac{1 - [A]_{\text{equilibrium}}}{[A] - [A]_{\text{equilibrium}}}\right)$$
$$\beta = k_1 + k_2$$
$$M = t$$
$$y = \beta M$$

4.2.4 Parallel Reaction of First Order

Often a starting substance reacts simultaneously (in parallel) to different products,

$$[A] \xrightarrow{k1} [C]$$
$$[A] \xrightarrow{k2} [D]$$

whereby either only $[C]$ or only $[D]$ is the actual target product. The task of a parameter analysis can be to favor one product path at the expense of the other. The time-dependent reaction process is like the reaction for one product

$$\ln\left(\frac{1}{[A]}\right) = (k_1 + k_2)t \quad [A]_0 = 1$$

resolved to $[A]$

$$[A] = \mathrm{e}^{-(k_1 + k_2)t}$$

with

$$[C] = \frac{k_1}{k_1 + k_2}\left(1 - e^{-(k_1 + k_2)t}\right) \quad [C]_0 = 0 \text{ for } t = 0$$

and

$$[D] = \frac{k_2}{k_1 + k_2}\left(1 - e^{-(k_1 + k_2)t}\right) \quad [D]_0 = 0 \text{ for } t = 0$$

The time-dependent terms are identical. A shift to a larger part, e.g., $[C]$, can therefore only be achieved by a corresponding ratio of the rate constants.

$$\frac{[C]}{[D]} = \frac{k_1}{k_2} \rightarrow \text{Maximum for target product}[C]$$

If for this situation an analysis of the parameter effects is performed, the rate constants $k1$ and $k2$ are calculated from the above relationships of the time-dependent concentrations of the chemical components. The ratio $k1/k2$ is sufficient for a simple evaluation of different test configurations; the logarithmic form, i.e., the *SNR* index, must be used to avoid over- or underestimation for modeling and predicting the progression of the chemical reaction.

$$SNR = 10 \log\left(\frac{k_1^2}{k_2^2}\right)$$

The sensitivity of the reaction, equivalent to the conversion, follows from

$$SEN = 10 \log\left(\frac{k_1^2}{k_0^2}\right)$$

where k_0 stands for the unit of the rate constant with the numerical value of one.

In the first step of data interpretation, the *SNR* index must always be maximized, in the second step, *SEN*, provided that setting values are still available which do not attenuate the numerical value of the *SNR* key figure. The graph shows the time-dependent concentration of the chemical components for two different ratios of the rate constants as an example (Fig. 4.1).

4.2.5 Consecutive Reaction of First Order

The reaction pattern of a single consecutive reaction with negligible reverse reaction is as follows:

$$[A] \xrightarrow{k1} [C] \xrightarrow{k2} [D]$$

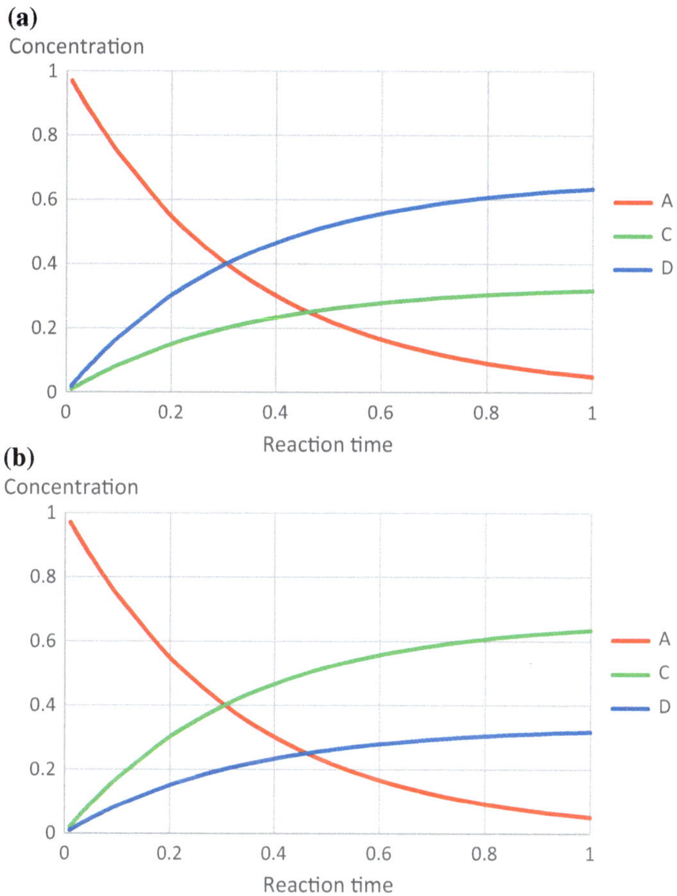

Fig. 4.1 Time-dependent concentration of the products for a parallel reaction of first order with **a** $k_1 = 1/2\ k_2$ and **b** $k_1 = 2\ k_2$

$[A]$ stands for the reactants and $[C]$ for the consecutive product that continues to react to $[D]$. From the progression of the reaction, it can be read immediately that the proportion of the target product $[C]$ depends solely on the ratio of the rate constants k_1/k_2. The constants k_1 and k_2 are calculated from the time-dependent reaction. The concentrations of the components are derived from the kinetic equations.

$$[A] = \mathrm{e}^{-k_1 t} \quad [A]_0 = 1 \text{ for } t = 0$$

$$[C] = \frac{k_1}{k_2 - k_1}\left(\mathrm{e}^{-k_1 t} - \mathrm{e}^{-k_2 t}\right) \quad [C]_0 = 0 \text{ for } t = 0;\ k_1 \neq k_2$$

$$[D] = 1 - \frac{k_2}{k_2 - k_1}\mathrm{e}^{-k_1 t} + \frac{k_1}{k_2 - k_1}\mathrm{e}^{-k_2 t} \quad [D]_0 = 0 \text{ for } t = 0;\ k_1 \neq k_2$$

The rate constant $k1$ follows from the concentration curve of the starting material, the reactants, $k2$ from [C] or rather [D] as control calculation. If it is possible to intervene in the system via chemical parameters, i.e., to change the activation energy, the *SNR* key figure for the highest possible product share [C] can be maximized with the available setting values, while at the same time achieving the highest conversion with the maximum *SEN* value.

$$SNR = 10\log\left(\frac{k_1^2}{k_2^2}\right)$$
$$SEN = 10\log\left(\frac{k_1^2}{k_0^2}\right)$$

The concentration of the assumed target product [C] is displayed time-dependently for two different ratios of the rate constants $k1$ and $k2$ (Fig. 4.2).

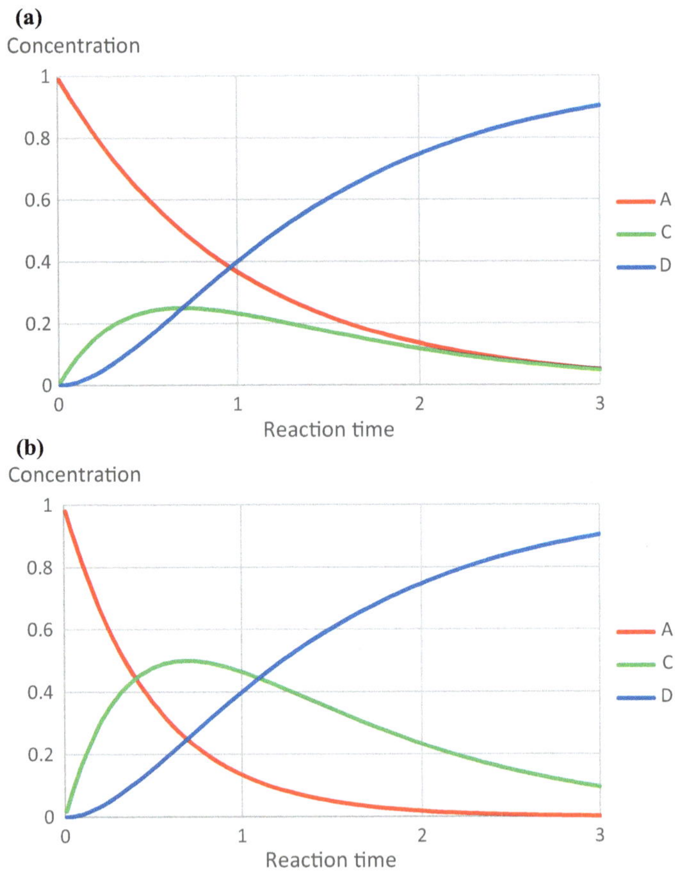

Fig. 4.2 Time-dependent concentration of the components for a first-order consecutive reaction **a** with $k_1 = 1/2\ k_2$ and **b** with $k_1 = 2\ k_2$

The occurrence of the concentration peak depends on the rate constants and the reaction time. For the single consecutive reaction, the reaction time for the maximum concentration of the intermediate product [*C*] follows by derivation and zeroing.

$$t_{\text{peak intermediate product}} = \frac{1}{k_1 - k_2} \ln\left(\frac{k_1}{k_2}\right)$$

Only for the special case $k_1 = 2\ k_2$ the reactant concentration [*A*] and the concentration of the consecutive product [*D*] are equal at the maximum concentration of the intermediate product [*C*].

$$[A] = [D]$$

4.2.6 Double Consecutive Reaction of First Order

Often a desired product results as an intermediate product in a chain of successive reaction stages

$$[A] \xrightarrow{k1} [C] \xrightarrow{k2} [D] \xrightarrow{k3} [E]$$

From the set of kinetic differential equations follows again the time-dependent concentration curves for

$$[A] = \mathrm{e}^{-k_1 t}$$

$$[C] = \frac{k_1}{k_2 - k_1}\left(\mathrm{e}^{-k_1 t} - \mathrm{e}^{-k_2 t}\right)$$

$$[D] = k_1 k_2 \left(\frac{\mathrm{e}^{-k_1 t}}{(k_2 - k_1)(k_3 - k_1)} - \frac{\mathrm{e}^{-k_2 t}}{(k_2 - k_1)(k_3 - k_1)} + \frac{\mathrm{e}^{-k_3 t}}{(k_2 - k_1)(k_3 - k_1)}\right)$$

$$[E] = 1 - \frac{k_2 k_3 \mathrm{e}^{-k_1 t}}{(k_2 - k_1)(k_3 - k_1)} - \frac{k_1 k_3 \mathrm{e}^{-k_2 t}}{(k_2 - k_1)(k_3 - k_1)} + \frac{k_1 k_2 \mathrm{e}^{\ k_3 t}}{(k_2 - k_1)(k_3 - k_1)}$$

$$k_1 \neq k_2 \neq k_3;\ [A]_0 = 1 \text{ for } t = 0;\ [C]_0 = [D]_0 = [E]_0 = 0 \text{ for } t = 0$$

If the concentration [*D*] is to be increased or maximized as an example, this can only be achieved if the ratio of the rate constants k_2/k_3 can be increased by interventions of chemically effective parameters. The key figures for evaluation of the experiments are

$$SNR = 10 \log\left(\frac{k_2^2}{k_3^2}\right)$$

$$SEN = 10 \log\left(\frac{k_2^2}{k_0^2}\right)$$

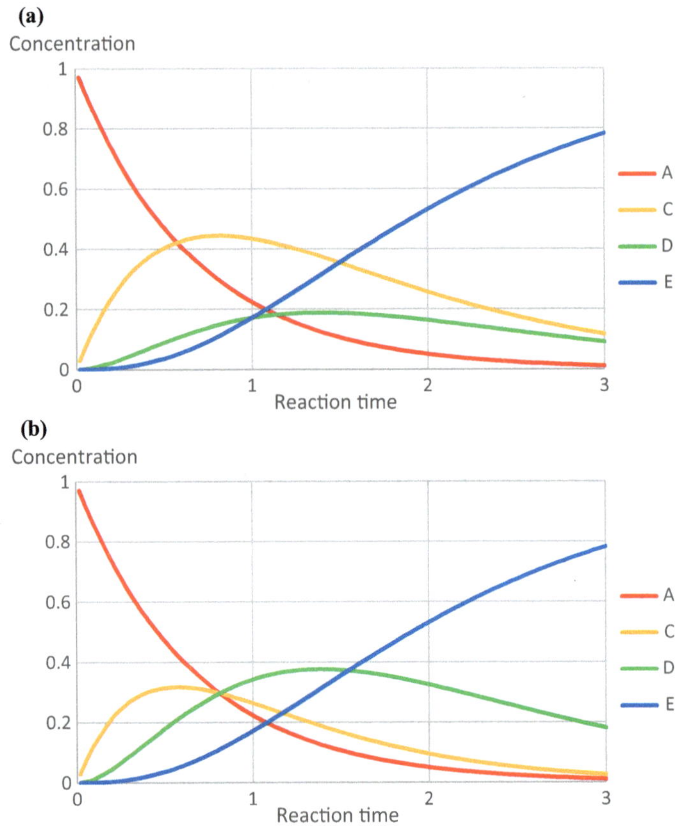

Fig. 4.3 Time-dependent concentrations of the components in a double consecutive reaction of first order **a** with $k_2 = 1/2\ k_3$ and **b** with $k_2 = 2\ k_3$

Access to the rate constants is described using the concentration–time equations given above; k_1 results from $[A] = f(k_1, t)$, k_2, and k_3 with iteration calculations from $[C] = f(k_1, k_2, t)$ and from $[D]$ or $[E] = f(k_1, k_2, k_3, t)$. An example shows the illustration with the target product $[D]$ for two different ratios of the rate constants (Fig. 4.3).

In consecutive reactions of any number of stages of first order, generally formulated, the evaluation

$$SNR = 10 \log\left(\frac{k^2_{\text{intermediate}\ i}}{k^2_{\text{intermediate}\ i+1}}\right)$$

leads to the greatest possible concentration of the ith intermediate product.

Even if the kinetic recording of chemical reactions is reflected in elegant curves, this must not hide the fact that diffusion processes and surface properties lead to more or less massive deviations of the formal-kinetic considerations. For this reason, semi-empirical approaches are often used to describe chemical reactions, taking kinetic models into account. Consequently, experimental fine-tuning within the scope of the confirmation tests is unavoidable, especially in the case of chemical reactions at an industrial scale.

4.3 Optimization Procedures for Time-Dependent Systems

As has already been stated, a satisfactory reaction description is possible with relatively high measuring and calculation effort, although additional fine adjustment is advantageous in connection with confirmation experiments **(1-29*)**. If the validity of the simple exponential approach $r = k[A]^{\alpha}$ to the reaction rate is not given, other concepts must be sought. It is generally observed that at the beginning of the reaction the starting substance (reactant) decreases progressively with the reaction time and the products and by-products increase accordingly. If data acquisition remains in the monotonically increasing or decreasing range, time-dependent behavior can be modeled with an exponential function, provided the exponent is adapted to the measured values in the observed time interval. The term by-product covers all undesirable products (fractions) from parallel as well as possible consecutive reactions and is described by means of regression by a best-fit exponent. From the overall product flow (mass balance) follows at all reaction times:

unused source substance (reactant) + target product + all by-products = 1 (=100%)

Since the sum of the three product flows is always constant, two components, in this case, unconsumed reactant and the by-products, are sufficient for a complete description of time-dependent behavior.

$$\begin{array}{ll} \text{Sourcc substancc(rcactant)} = \mathrm{c}^{-\beta_1 t} & \text{Sourcc substancc(reactant)} - 1 \text{ for } t - 0 \\ \text{Sum of by}-\text{products} - 1 - \mathrm{e}^{-\beta_2 t} & \text{Sum of by}-\text{products} = 0 \text{ for } t = 0 \end{array}$$

The exponents β_1, β_2 are called speed constants to underline the general validity of the considerations. As in the case of reaction kinetics, an increase in the target product will occur if the ratio of the speed constants β_1/β_2 increases accordingly. The already known facts are valid for the corresponding *SNR* and *SEN* key figures.

$$SNR = 10\log\left(\frac{\beta_1^2}{\beta_2^2}\right)$$
$$SEN = 10\log\left(\frac{\beta_1^2}{\beta_0^2}\right)$$

In the *SEN* equation, β_0 stands for the same unit of the numerator with the numerical value one. Just as the characterization of chemical reactions is described with rate constants, similar applications can be found in completely different areas.

The common denominator is particle collisions at the molecular or macroscopic level, which lead to temporal changes. Thus, technical grinding processes in all types of mills are described by the same law.

$$\frac{\text{Particle size through grinding}}{\text{Original particle size}} = e^{-\beta t}$$

The inverse process, grain growth of powder particles with the addition of adhesion promoters, follows the same approach. Physicochemical analysis methods such as chromatography or capillary electrophoresis are of great importance. Here, a temporal breakdown into individual components of a mixture to be analyzed takes place. The decisive point is the same in all cases: To create a robust process with a final product in the form of a relatively narrow particle size distribution or the greatest possible selectivity regarding different molecular weights in analytics. In the latter situation, the retention time is replaced by the reciprocal value of the flow velocities as signal parameters, which can be understood as contact time. Assuming the applicability of an exponential time law, the following figure illustrates this generalized approach in logarithmic scale (Fig. 4.4).

While performing a parameter analysis, the maximum values of the *SNR* and *SEN* key figures are modeled from which the extreme values of the speed constants β_1 = maximum and β_2 = minimum are derived. If confirmatory experiments are valid, an ideal operating point must be set with

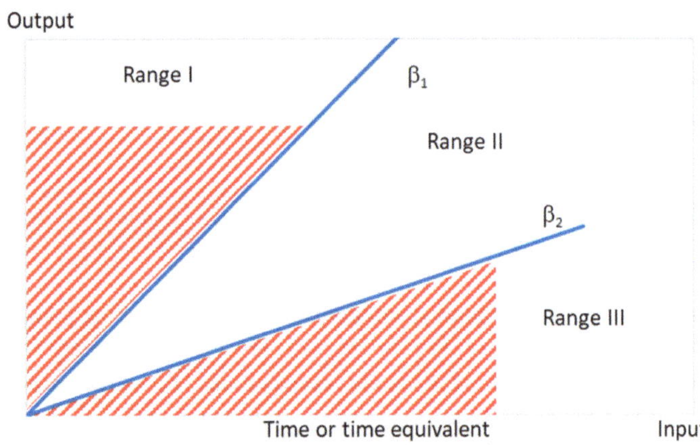

Range I: Fraction of material reacted too far, undesirable properties

Range II: Target product (spread for different molecular weights, narrow particle size distribution etc.)

Range III: Fraction of material unreacted or not reacted sufficiently

Fig. 4.4 Opening of the function window with the speed constants β_1 = maximum and β_2 = minimum

$$\beta_{\text{ideal}} = \frac{\beta_{1\,(\text{maximum})} + \beta_{2\,(\text{minimum})}}{2}$$

The functional range can also be opened for a desired product or certain product properties in single evaluation, where β can be regarded as a static characteristic.

$$LTB = 10\log\left(\frac{\beta_1^2}{\beta_0^2}\right)$$

$$STB = 10\log\left(\frac{\beta_0^2}{\beta_2^2}\right)$$

$$SNR = LTB + STB = 10\log\left(\frac{\beta_1^2}{\beta_2^2}\right)$$

This is the same procedure as in Sect. 2.7, introduced there as "Operating Window". Here, on the other hand, it is applied to exponents, so that the window of the operating function opens further as the signal value increases. For this reason, the term "Dynamic Operating Window" or method of speed optimization can be found in the literature.

4.4 Continuously Running Processes

Continuous processes are characterized by the fact that the concentrations of the substances involved do not change over time. Reactant is added in proportion to how much it is consumed, the target product is removed proportionally, and by-products are discharged before enrichment takes place. For a dynamic parameter analysis, the reciprocal value of the preset flow velocity of the reactant serves as a time-dependent signal parameter. In this way, loop processes or continuous batch processes can be subjected to a parameter analysis.

If this is not possible, a chemical reaction or a similar system can be evaluated by a static approach. For a parameter analysis, two of the three variables are to be measured; the missing one is calculated from the mass balance:

concentration of the unreacted reactant in the output: y_1,
concentration of the unreacted reactant and the target product: y_2, and
concentration of all undesired by-products: $y_3 = 1 - y_2$.

In the first step, the *SNR* evaluation is carried out in favor of the smallest possible fraction of unreacted reactant in the output flow, identical to the highest conversion, whereby the omega transformation is to be used for percentages or fractions of one.

$$SNR_{STB,\text{ reactant}} = 10\log\left(\frac{1-y_1}{y_1}\right)$$

In the second step, the *SNR* index stands for the lowest possible concentration of the sum of all by-products.

$$SNR_{STB,\text{ by-products}} = 10\log\left(\frac{1-y_3}{y_3}\right) = 10\log\left(\frac{1-(1-y_2)}{1-y_2}\right) = 10\log\left(\frac{y_2}{1-y_2}\right)$$

The overall evaluation results from the sum of the *SNR* key figures.

$$SNR_{\text{total}} = SNR_{STB,\text{ reactant}} + SNR_{STB,\text{ by-products}}$$
$$SNR_{\text{total}} = 10\log\left(\frac{1-y_1}{y_1}\cdot\frac{y_2}{1-y_2}\right)$$

SNR_{total} behaves similar to the transformed selectivity but differs in the uptake of the unreacted reactant. This remaining portion is considered for reasons of objective comparability, as it contains potential target product. It ensures that despite different settings of reaction parameters the best possible situation is recognized, although the concentration values of the target product can be the same. Assuming that the speed constants $k1$, $k2$—not accessible here—are present in a ratio of 2:1, a maximum concentration of the target product is obtained if the concentration of unreacted reactant equals the sum of the concentration of all by-products (Sect. 3.2). Since this applies approximately to any speed constant as long as $k_1 > k_2$, the advantage of considering the *SNR* total index lies in comparable predictions for conversion and concentration of the target product.

$$y_1 = y_3$$
$$SNR_{\text{total}} = 10\log\left(\frac{1-y_1}{y_1}\frac{y_2}{1-y_2}\right) = 10\log\left(\frac{1-y_1}{y_1}\frac{1-y_3}{y_3}\right)$$
$$= 10\log\left(\frac{1-y_1}{y_1}\right)^2$$
$$SNR_{STB,\text{ reactant}} = \frac{1}{2}SNR_{\text{total}}$$
$$\text{Concentration of target product} = 1-2y_1$$

The calculated concentration of the target product serves as a guideline. The numerical value is not necessarily suitable as a decisive figure for confirmation experiments, since the assumption of a certain ratio of speed constants is unfounded and, as assumed, information on time-dependent concentration changes is not available.

Exercise

A reaction with the same chemical components is carried out under different reaction conditions. The product concentrations are measured from three tests. Which test gives the best result regarding the highest expected concentration of the target product?

Chemical reaction	Concentration of unreacted reactant (%)	Concentration of target product (%)	Concentration of all by-products (%)
Test #1	30	60	10
Test #2	20	60	20
Test #3	5	60	35

Problem Solution

SNR index of the conversion

$$SNR_{STB,\text{ reactant}} = 10\log\left(\frac{1-y_1}{y_1}\right)$$

SNR_{total} index

$$SNR_{\text{total}} = 10\log\left(\frac{1-y_1}{y_1}\frac{y_2}{1-y_2}\right)$$

Relation for comparison of the tests to calculate the expected conversion

$$SNR_{STB,\text{ educt}} = \frac{1}{2}SNR_{\text{total}}$$

Concentration of the target product

$$\text{Target product} = 1 - 2y_1$$

Chemical reaction	y_1	y_2	$SNR_{STB,\text{ reactant}}$	SNR_{total}	½ SNR_{total}	Conversion (%)	Concentration target product (%)
Test #1	0.30	0.90	3.68	13.22	6.61	17.9	64.2
Test #2	0.20	0.80	6.02	12.04	6.02	20.0	60.0
Test #3	0.05	0.65	12.79	15.48	7.74	14.4	71.2

The settings of test #3 are to be preferred.

4.5 Systems with Complex Signal-to-Noise Ratios

As hitherto described, the function of a system was characterized by an input signal with exactly one resulting output variable (response). In electronic circuits that means with active (diodes, transistors) or passive components (resistors, solenoids, capacitors), however, the response when operating with alternating voltage consists of two coupled variables. This special feature is because not only the amplitude but also the phase shift between input and output or alternating voltage and alternating current are characteristic properties of a circuit function. Apart from signal propagation time, the following phase relationships with the corresponding definitions apply to the ideal passive components listed as examples (Table 4.4).

Depending on the design of an electronic circuit with different components, any phase shift can occur. Circuits of this type are tested with an alternating voltage that varies sinusoidally with time (Fig. 4.5).

Since dealing with trigonometric functions can be cumbersome, the substitution by an exponential function in polar coordinates or a completely equivalent symbolic notation in the form of complex numbers **(1-30*)** is used. The equivalent representations for the amplitude and the phase of a numerical value y are

Table 4.4 Properties of some passive components

Component	Denomination	Phase shift relation	Feature for power consumption
Ohmic resistor	Resistor R [Unit: Ohm, Ω]	AC current and AC voltage in phase Phase shift = 0	Consumed electrical power = Effective power
Solenoid	Inductivity L [Unit: Henry, H]	AC voltage 90° ahead of AC current	Reactive power consumption = 0
Capacitor	Capacity C [Unit: Farad, F]	AC current 90° ahead of AC voltage	Reactive power consumption = 0

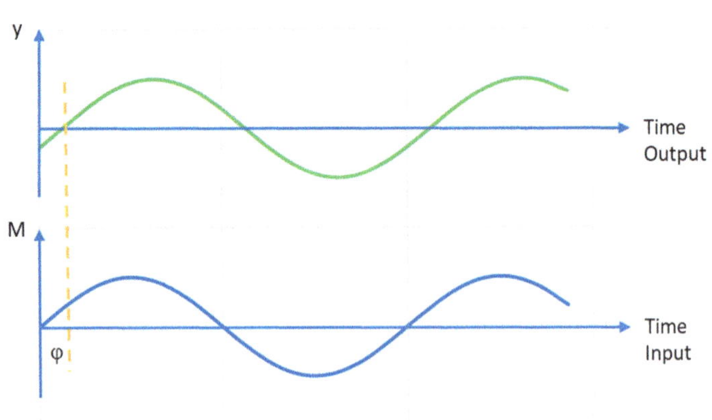

Fig. 4.5 Phase shift between input signal and output (response) of an electronic circuit

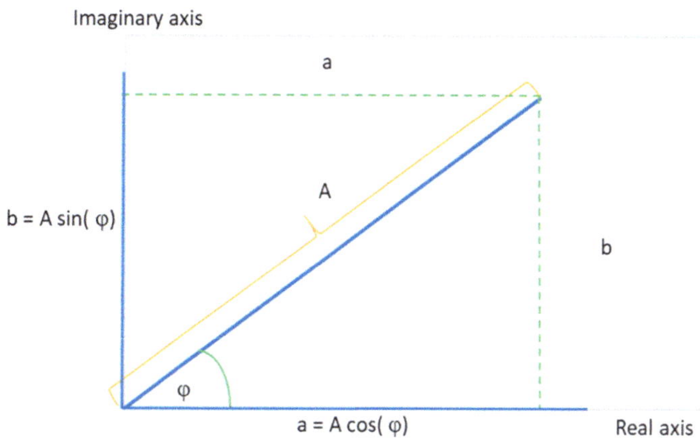

Fig. 4.6 Graphical representation of amplitude and phase of complex numbers

$$y = Ae^{i\varphi} = A(\cos(\varphi) + \mathrm{i}\sin(\varphi)) = a + ib$$

The respective terms and relations are defined in the Gaussian coordinate system (Fig. 4.6).

The conjugate complex number is denominated X^*. When multiplying a complex number by its complex conjugate, the result is the squared amplitude. The square root returns the amplitude of a wave. The phase shift as a trigonometric function of the tangent results from the ratio of the imaginary to the real number.

$$AA^* = (a + ib)(a - ib) = a^2 + b^2 = A^2$$

$$A = \sqrt{a^2 + b^2}$$

$$\tan(\varphi) = \frac{\sin(\varphi)}{\cos(\varphi)} = \frac{b}{a} = \frac{\text{imaginary number}}{\text{real number}}$$

Consequently, there are no imaginary numbers in the end result. If the real and imaginary numerical values of an electronic circuit are known, principally a parameter analysis can be carried out separately. Depending on the requirements of the circuit function, e.g., with the question of minimum phase shift between input and output of all frequencies occurring, the characteristic *STB* of the purely imaginary numbers is to be selected as *SNR* target function. However, if energy consumption is to be minimized, the *SNR* characteristic, also of type *STB*, is to be used for the real numbers to avoid ohmic losses as far as possible. Since in practice a separation of real and imaginary variables is not possible due to parasitic inductivities and capacities, it is advantageous to perform a parameter analysis directly with the complex numbers. Only if both components are considered simultaneously, there is a chance of achieving robust behavior of the electronic function.

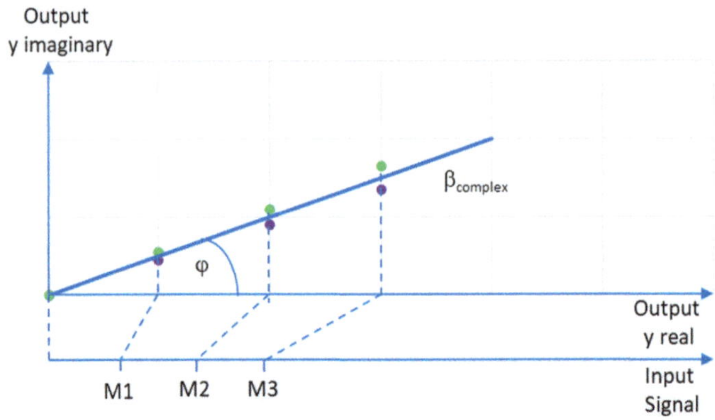

Fig. 4.7 Function of an electronic circuit with real input voltage and complex output current

If a circuit is designed with the intent that the response is proportional to an AC input signal, an AC voltage of variable amplitude is applied to the input for parameter analysis. Different frequencies can be considered as sources of noise, if appropriate, and if needed, different operating temperatures, types of components, and tolerances as well. The phase shift of the response is measured against the momentary phase of the input signal. The latter by definition is without phase shift, i.e., the input value *M* as such is a real number, while the output *y* is complex. The graph illustrates the assumed situation (Fig. 4.7).

The *SNR* key figures for evaluating such a function are gained from the complex slope and complex standard deviation, whereby the result is always a real number because of the multiplication with the conjugated complex number **(1-31*)**.

$$S_2 = \frac{1}{n}\left(M_1 M_1^* + M_2 M_2^* + \cdots + M_n M_n^*\right)$$
$$SNR = 10\,\log\left(\frac{\beta\beta^*}{\sigma\sigma^*} S_2\right)$$
$$SEN = 10\,\log\left(\frac{\beta\beta^*}{\beta_0\beta_0^*}\right)$$

For further explanation, the numerical example of an amplifier circuit is added. The input is an AC voltage of constant frequency, which is gradually increased for three signal values. The total output current is measured with its phase shift with respect to the input voltage. This allows the conversion into an equivalent pair of data in the form of a real and imaginary number. For simplicity, without applying an orthogonal array, two circuits with electronic components of the same values but different parasitic properties are compared (Table 4.5).

$$S_2 = \frac{1}{3}\left(0.1^2 + 0.3^2 + 1.0^2\right) = 0.3667$$

Table 4.5 Complex measurement data of electronic circuits

Electronic circuits	Signal *M*: AC voltage at the input [V] Response *y*: Total AC current [A]		
	*M*1 0.100 + 0i	*M*2 0.300 + 0i	*M*3 1.000 + 0i
Circuit A			
AC current Phase shift	0.2593 34°	0.7489 32°	2.5670 37°
Circuit B			
AC current Phase shift	0.2624 31°	0.7961 32°	2.6599 31°
Circuit A	0.215 + 0.145i	0.635 + 0.397i	2.050 + 1.545i
Circuit B	0.225 + 0.135i	0.675 + 0.422i	2.280 + 1.370i

$$\beta_A = \frac{(0.215+0.145i)0.1+(0.635+0.397i)0.3+(2.050+1.545i)1.0}{0.1^2+0.3^2+1^2}$$

$$\beta_A = 2.0564 + 1.526i$$
$$\beta_A^* = 2.0564 - 1.526i$$
$$\beta_A\beta_A^* = 6.5573$$

$$\sigma_A = \sqrt{\frac{((0.215+0.145i)-(2.0564+1.526i)0.1)^2+\cdots+((2.050+1.545i)-(2.0564+1.526i)1)^2}{3}}$$

$$\sigma_A = 0.01169 - 0.03683i$$
$$\sigma_A^* = 0.01169 + 0.03683i$$
$$\sigma_A\sigma_A^* = 0.001493$$

$$SNR_A = 10\log\left(\frac{6.5573}{0.001493}0.3667\right) = 32.1[\text{dB}]$$

The second electronic circuit *B* is calculated using the same algorithm.

$$\beta_B = 2.2773 + 1.3728i$$
$$\beta_B^* = 2.2773 - 1.3728i$$
$$\beta_B\beta_B^* = 7.0706$$

$$\sigma_B = 0.004798 - 0.005874i$$
$$\sigma_B^* = 0.004798 + 0.005874i$$
$$\sigma_B\sigma_B^* = 0.00005752$$

$$SNR_B = 10\log\left(\frac{7.0706}{0.00005752}0.3667\right) = 46.5[\text{dB}]$$

The electronic circuits differ less in gain than in robustness; circuit B exceeds circuit A by 14.4 [dB] according to an improvement in the ratio β/σ by a factor of 5.

4.6 Systems with More Than One Signal Factor

The technical function of each dynamic system reveals apparently in the response of an input signal. Instead of one single signal, a second signal can be advantageous for adjustment reasons, i.e., tuning and setting of a certain slope. For efficient characterization of system robustness, both signals are combined. The signal causally linked to the function is called control signal M_S, and the second signal for calibration purposes is called process signal M_P (Figs. 4.8 and 4.9).

For an unambiguous system evaluation, the relation of the input signals to the output response must be verified. If both show the same trend, i.e., the response is proportional to both the control and process signal, ideal system behavior is given by the function

$$y = \beta M_S M_P$$

For example, the braking force acting on a brake disc increases with both the hydraulic pressure in the brake cylinder—the control signal—and the surface area of the brake pads. The design of the area size (brake pads) is therefore a simple hardware parameter—the process signal—for adjusting the braking force within a prescribed range.

If, on the other hand, the effects of the signals on the response are inverse, the ideal function is

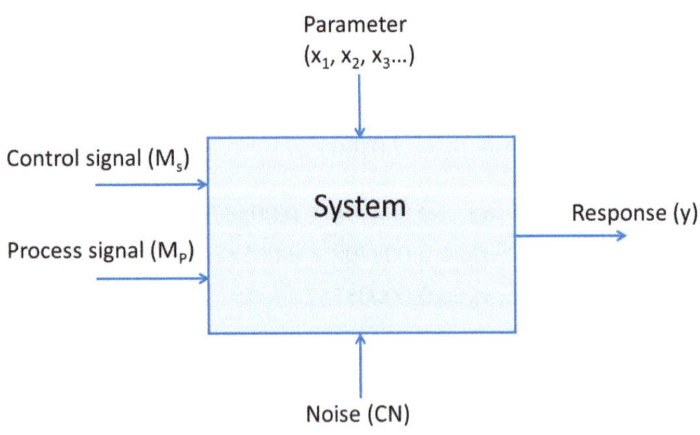

Fig. 4.8 Dynamic system with two input signals

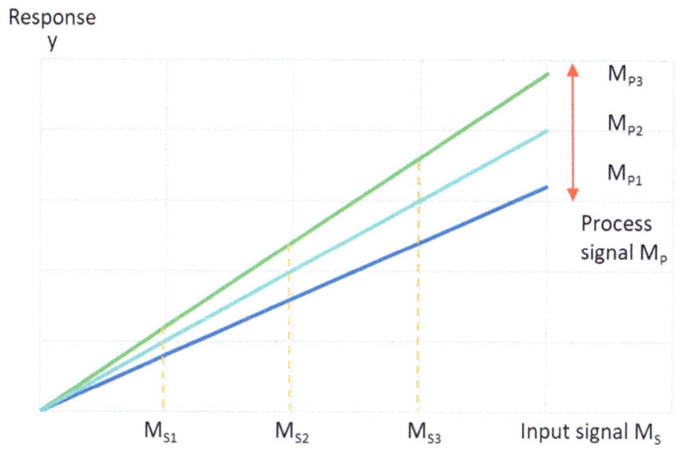

Fig. 4.9 System response with control and process signal

$$y = \beta \frac{M_S}{M_P}$$

For example, in a mechanical grid construction, the deflection in the elastic range depends directly on the load but indirectly proportionally on the cross section. Profiles can be adjusted in their cross section as a process signal so that the individual deflections lie within a certain range and the construction yields under load, but the shape remains approximately unchanged. The combination of two signals combines, from this point of view, an adjustable variable in the design or manufacturing process with the controllability of the dynamic behavior function **(1-32*)**. Applications include as well audio devices in vehicles (voice output navigation device, CD player) to increase the volume with the driving noise depending on the speed, or air conditioners with slightly increased cooling in full sunlight, etc. The attached numerical example explains dealing simultaneously with two signals.

The system under consideration is a continuously variable transmission with a friction wheel drive. The speed of the drive wheel is transmitted to the outlet axis depending on its position (close to or further away from the axis of rotation). The rotational speed of the outlet axis changes with both the speed of the drive motor and its position (Fig. 4.10).

The following relationship applies to the output speed (response):

$$y = \beta M_S M_P$$

y Response = Outlet axis speed.
β Slope = Amplification factor.
M_S Control signal = Speed of the friction wheel.
M_P Process signal = Position of power transmission.

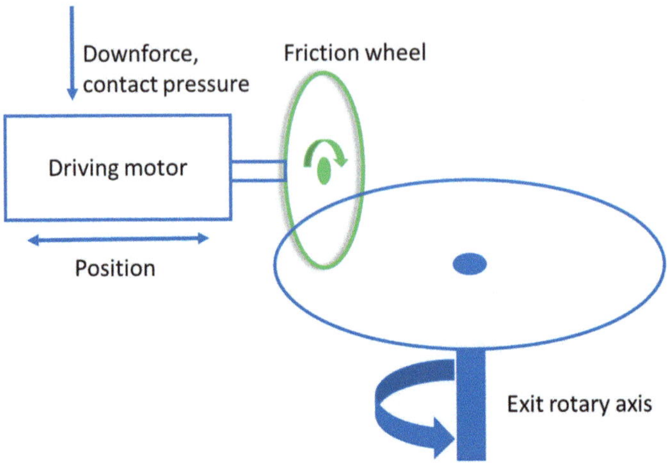

Fig. 4.10 Sketch of a continuously variable transmission with friction wheel drive

Table 4.6 Determination of parameters for testing a continuously variable transmission

Parameter	Levels		
	1	2	3
A: Surface friction wheel	Smooth	Knurled	–
B: Material friction wheel	Elastic	Soft PVC	–
C: Pressure on the drive motor	Low	High	–
CN: Noise sources	New material, ambient temperature	Used material, fridge temperature	–
M_S: Drive speed (rpm)	16	24	32
M_P: Transmission ratio (V)	1: 4	1: 1	4: 1

The structure of the contact surface of the friction wheel, the material used for power transmission, and the contact pressure are tested. Sources of noise are combined to one noise variable (CN) (Table 4.6).

For three control parameters, the application of an orthogonal array L4 with all settings of the signal values and one repetition under noise conditions is sufficient to analyze and improve system robustness (Table 4.7).

Both signal parameters result in nine different setting values, which are re-arranged monotonically increasing in the evaluation table only for reasons of transparency (Table 4.8).

Table 4.7 Performed experiments and measurement data with the orthogonal array L4

Test #	L4			M_S1						M_S2						M_S3					
	A	B	C	M_P1		M_P2		M_P3		M_P1		M_P2		M_P3		M_P1		M_P2		M_P3	
				CN1	CN2	CN1	CN2	CN1	CN2	CN1	CN2	CN1	CN2	CN1	CN2	CN1	CN2	CN1	CN2	CN1	CN2
1	1	1	1	3.9	3.5	14.5	13.0	58	55	5.8	5.4	22	21	79	74	7.5	6.5	27	26	105	101
2	1	2	2	3.9	3.6	14.0	12.0	52	51	5.7	5.2	20	19	71	68	6.5	6.0	22	21	97	91
3	2	1	2	3.9	3.4	13.0	12.0	54	53	5.8	5.1	20	18	83	76	7.0	6.5	26	24	109	105
4	2	2	1	3.5	3.0	12.0	11.0	49	48	5.0	4.0	18	17	69	67	6.0	5.0	22	20	88	79

Table 4.8 Re-arranging of measurement data

Test #	Double signal M																		*SNR*	*SEN*
	M_1 M_SM_P1 4 rpm		M_2 M_SM_P2 6 rpm		M_3 M_SM_P3 8 rpm		M_4 M_SM_P4 16 rpm		M_5 M_SM_P5 24 rpm		M_6 M_SM_P6 32 rpm		M_7 M_SM_P7 64 rpm		M_8 M_SM_P8 96 rpm		M_9 M_SM_P9 128 rpm			
	CN1	CN2	CN1	CN2	CN1	CN2	CN1	CN2	CN1	CN2	CN1	CN2	CN1	CN2	CN1	CN2	CN1	CN2		
1	3.9	3.5	5.8	5.4	7.5	6.5	14.5	13.0	22	21	27	26	58	55	79	74	105	101		
2	3.9	3.6	5.7	5.2	6.5	6.0	14.0	12.0	20	19	22	21	52	51	71	68	97	91		
3	3.9	3.4	5.8	5.1	7.0	6.5	13.0	12.0	20	18	26	24	54	53	83	76	109	105		
4	3.5	3.0	5.0	4.0	6.0	5.0	12.0	11.0	18	17	22	20	49	48	69	67	88	79		

The evaluation of the tests is carried out using the *SNR* and *SEN* indices, which are calculated from the square of the slope, the squared standard deviation, and the dimension factor.

$$S_2 = \frac{1}{18}\left(2M_1^2 + 2M_2^2 + 2M_3^2 + 2M_4^2 + 2M_5^2 + 2M_6^2 + 2M_7^2 + 2M_8^2 + 2M_9^2\right)$$

$$S_2 = \frac{1}{9}\left(4^2 + 6^2 + 8^2 + 16^2 + 24^2 + 32^2 + 64^2 + 96^2 + 128^2\right) = 3518.667$$

$$\beta_1 = \frac{(y_1 + y_2)M_1 + (y_3 + y_4)M_2 + (y_5 + y_6)M_3 + \cdots + (y_{17} + y_{18})M_9}{2\left(M_1^2 + M_2^2 + M_3^2 + \cdots + M_9^2\right)}$$

$$\beta_1 = \frac{(3.9 + 3.5)4 + (5.8 + 5.4)6 + (7.5 + 6.5)8 + \cdots + (105 + 101)128}{2(4^2 + 6^2 + 8^2 + \cdots + 128^2)} = 0.8157$$

$$\sigma_1^2 = \frac{1}{18}\left((y_1 - \beta_1 M_1)^2 + (y_2 - \beta_1 M_1)^2 + \cdots + (y_{17} - \beta_1 M_9)^2 + (y_{18} - \beta_1 M_9)^2\right)$$

$$\sigma_1^2 = \frac{1}{18}\left((3.9 - 0.8157 * 4)^2 + (3.5 - 0.8157 * 4)^2 + \cdots + (101 - 0.8157 * 128)^2\right)$$

$$\sigma_1^2 = 4.7615$$

$$SNR_1 = 10 \log\left(\frac{\beta_1^2}{\sigma_1^2} S_2\right)$$

$$SNR_1 = 10 \log\left(\frac{0.6654}{4.7615} 3518.667\right) = 26.9172[\text{dB}]$$

$$SEN_1 = 10 \log\left(\frac{\beta_1^2}{\beta_0^2}\right)$$

$$SEN_1 = 10 \log\left(\frac{0.6654}{1^2}\right) = -1.7691$$

The further tests are calculated in the same way; the individual results are listed (Table 4.9).

The parameter effects are calculated from the test configurations of the orthogonal array. For easier interpretation, the effects are also displayed graphically (Figs. 4.11 and 4.12; Table 4.10)

Table 4.9 Evaluation of the orthogonal array L4

Test #	Slope β	Standard deviation σ	*SNR*	*SEN*
1	0.8157	2.1821	26.9172 [dB]	−1.7691
2	0.7409	2.2119	25.9633 [dB]	−2.6052
3	0.8308	1.6317	29.6006 [dB]	−1.6102
4	0.6846	2.7976	23.2370 [dB]	−3.2919

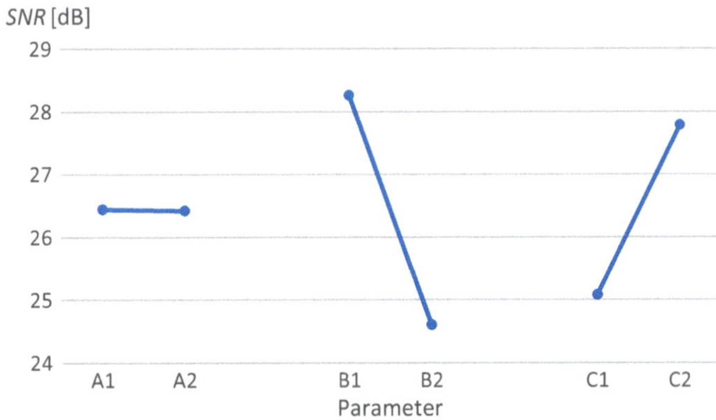

Fig. 4.11 Graphical application of the *SNR* indices of the parameter effects

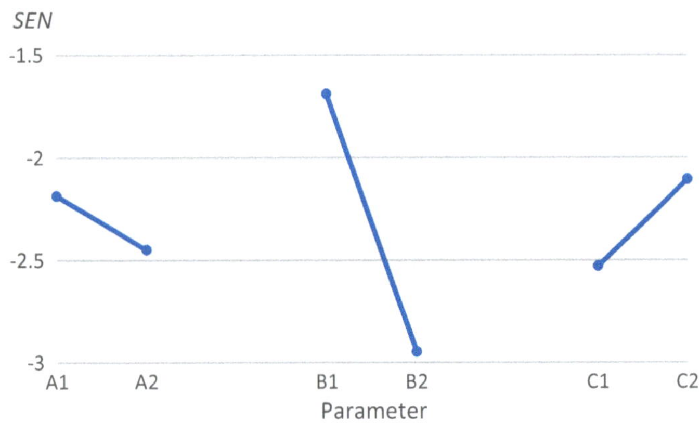

Fig. 4.12 Graphical application of the *SEN* figures of the parameter effects

Table 4.10 Effects of the parameters A, B, C

Parameter setting	Parameter effect	*SNR*	*SEN*
A1	A1 = $(SNR_1 + SNR_2)/2$	26.4403 [dB]	−2.1872
A2	A2 = $(SNR_3 + SNR_4)/2$	26.4188 [dB]	−2.4507
B1	B1 = $(SNR_1 + SNR_3)/2$	28.2589 [dB]	−1.6897
B2	B2 = $(SNR_2 + SNR_4)/2$	24.6001 [dB]	−2.9482
C1	C1 = $(SNR_1 + SNR_4)/2$	25.0771 [dB]	−2.5302
C2	C2 = $(SNR_2 + SNR_3)/2$	27.7820 [dB]	−2.1077
Average	$(SNR_1 + SNR_2 + SNR_3 + SNR_4)/4$	26.4295 [dB]	−2.3189

An advantageous choice to largely compensate for the influence of the sources of noise is

$$SNR_{new} = B1 + C2 - SNR_{average} = (28.2589 + 27.7820 - 26.4295)[dB] = 29.6114[dB]$$
$$SEN_{new} = B1 + C2 - SEN_{average} = -1.6897 - 2.1077 + 2.3189 = -1.4784$$

From this model, the predictions for the dynamic system behavior with the expected average slope and the standard deviation are derived.

$$\bar{\beta}_{new} = 0.8435$$
$$\sigma_{new} = 1.6546$$

The validity range of the prediction can be specified with (Sect. 1.10)

$$\beta_{min./max.} = \bar{\beta}_{prediction} \pm 3\frac{\sigma}{\sqrt{S_2}}$$
$$\beta_{min./max.} = 0.8435 \pm 0.0837$$
$$\beta_{lower\ limit} = 0.76$$
$$\beta_{upper\ limit} = 0.93$$

Since the parameter selection essentially corresponds to test 3, this already counts as confirmation of the prediction. For comparison, the test data is displayed graphically with the improved system behavior (Fig. 4.13).

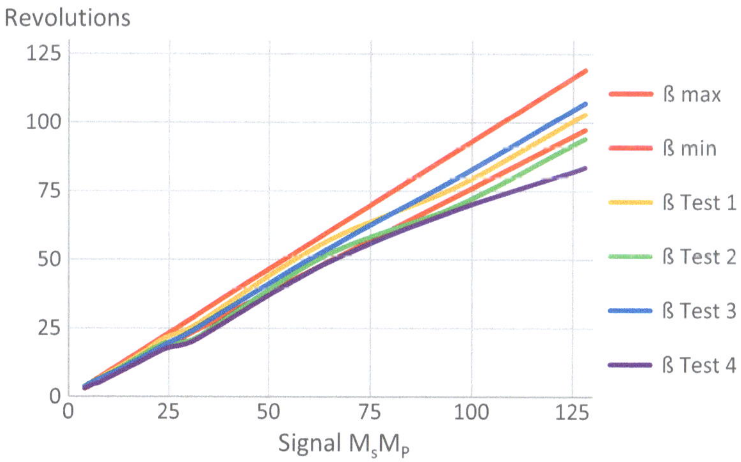

Fig. 4.13 Original system behavior and the result after application of a parameter analysis to improve the properties

As can be seen in the figure, test #3 is practically centered between the slopes for the upper and lower limit value. Thus, the best possible linearity with the smallest deviation is achieved, i.e., the output speed (rpm) reacts reliably to both the control signal and the process signal.

Important items to memorize In time-dependent competing processes (reactions), the result is determined exclusively by the ratio of the reaction constants that describe the respective process.

The importance of the "Dynamic Operating Window" lies in the greatest possible spread of the minimum and maximum slopes (β_{min} and β_{max}), which opens up the functional area of a dynamic system.

The simultaneous consideration of several signal parameters allows a holistic view of a dynamic system; the amount of data increases multiplicatively with the number of signal setting values.

Chapter 5
Nonlinear Systems and Tolerances

Contents

Abstract Not all systems respond in a linear way. Actually, the transfer function for given variables is not only nonlinear but can also cause a stepwise response. This requires additional mathematical calculation resulting in obtaining more information through the corresponding *SNR* indices. The first *SNR* index is applied to focus on the highest response reliability, the second for slope adjustment, while the third and those remaining are used to minimize higher order terms. Experimental data can be lost. If missing data cannot be retrieved and experiments not repeated, data interpretation in the strictly mathematical sense becomes impossible. Nevertheless, an *SNR* index can be calculated and used for iteration according to the applied orthogonal array. Thus, a small percentage in error needs to be sacrificed for data recovering to execute interpretation. Due to the pressure of tight specifications, the issue remains to achieve the highest precision despite existing sources of noise. At that moment, investment must be considered to lower the impact of noise factors as the relation of deviation off-target and monetary loss becomes apparent. This is of the utmost importance as the outcome will be decisive on who stays competitive or goes out of the market.

© Springer Nature Switzerland AG 2019

H. Ruefer, *Living Without Mathematical Statistics*,
https://doi.org/10.1007/978-3-319-99632-5_5

5.1 Standardized Signal-to-Noise Key Figure and Terms of Higher Order

For all systems considered so far, a linear transfer function between input signal and response has been assumed. Nonlinear characteristics are only included to the extent that they can be explicitly mathematically transformed into a linear relationship. If this is not the case, the evaluation of a system can no longer be based on the prerequisite of a direct proportional behavior. In general, transfer characteristics do not have to be linear or continuous to perform special technical functions. For such behavior, the *SNR* indicators for evaluating experiments with the previous algorithms are not applicable.

To be able to detect parameter effects with different contributions of deviation, an experiment is to be carried out with at least one repetition, appropriately under the influence of one noise variable with two setting values. The test result is a pair of values for each input signal *M*, where the state *N*1 represents the larger numerical value (plus tolerance) and *N*2 the smaller numerical value (minus tolerance) (Fig. 5.1).

If no further requirements of physical/chemical nature or other boundary conditions exist, it can be seen that for robust system behavior, the smallest possible distance between *N*1 and *N*2 has to be achieved, irrespective of a linear or nonlinear transfer function. Only then is the influence with respect to this noise source neutralized **(1-33*)**. The system behavior as such is characterized by the central position hereinafter referred to as the standard response with value y_0 to the standard condition *N*0.

$$y_0 = \frac{y_1 + y_2}{2}$$

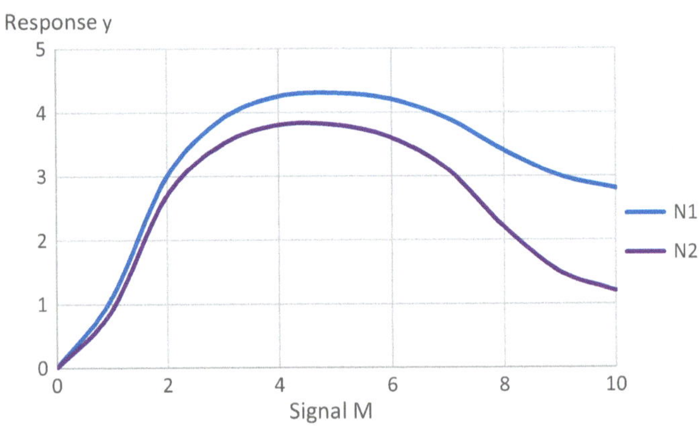

Fig. 5.1 Nonlinear transfer function with one noise source on two levels

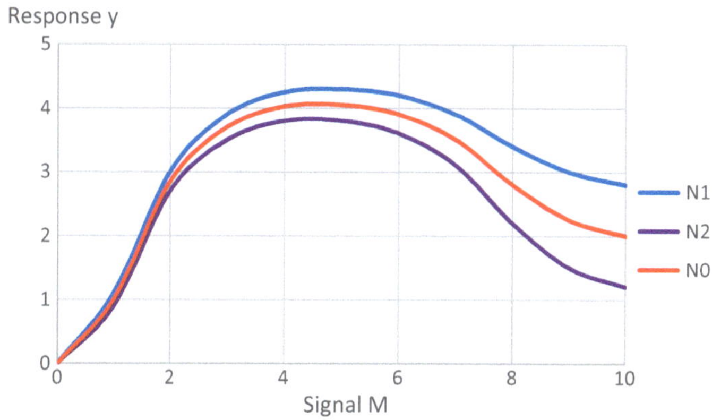

Fig. 5.2 Standardization of a nonlinear transfer function as the average value of two noise levels *N*1 and *N*2

If there are more than two sources of noise, it is advantageous to keep the effort to a minimum by combining them. For this purpose, as previously discussed, the setting values of all sources of noise which increase the average value of the characteristic feature are combined to a common noise level *CN*1, corresponding to those which reduce the average value, to a common noise level *CN*2. Thus, the cumulated noise source with two levels which leads to the greatest spread of the measured values is represented, and the designations *N*1 and *N*2 are retained. The average value as the standard response is justified by the same deviations from *N*0 but with different signs for the same signal value. If a combination of noise sources is not appropriate because the tendency of their effects is not known, y_0 follows as the average value of all noise conditions characterized by the state *N*0. Several noise sources can also be assigned to an orthogonal array and tested. The average value of the test results stands for y_0 with the standard condition *N*0 (Fig. 5.2).

The easiest way to detach from arbitrarily curved transfer characteristics with or without discontinuity is to refer directly to the measured values of conditions *N*1 and *N*2 instead of to the input signal with its abscissa values M_i, i.e., to the standard response y_{0i} under the standard condition *N*0. Simply put, the values of the input signal are substituted by the standard output response. The figures illustrate the conversion of the input signal M_i to the standard response y_{0i} (Fig. 5.3; Table 5.1).

With this standardization rule, the *SNR* key figure of a test is calculated from

$$S_2 = \frac{1}{2n}\left(2y_{01}^2 + 2y_{02}^2 + \cdots + 2y_{0n}^2\right)$$

$$\beta = \frac{(y_{11}y_{01} + y_{12}y_{02} + \cdots + y_{1n}y_{0n}) + (y_{21}y_{01} + y_{22}y_{02} + \cdots + y_{2n}y_{0n})}{2\left(y_{01}^2 + y_{02}^2 + \cdots + y_{0n}^2\right)}$$

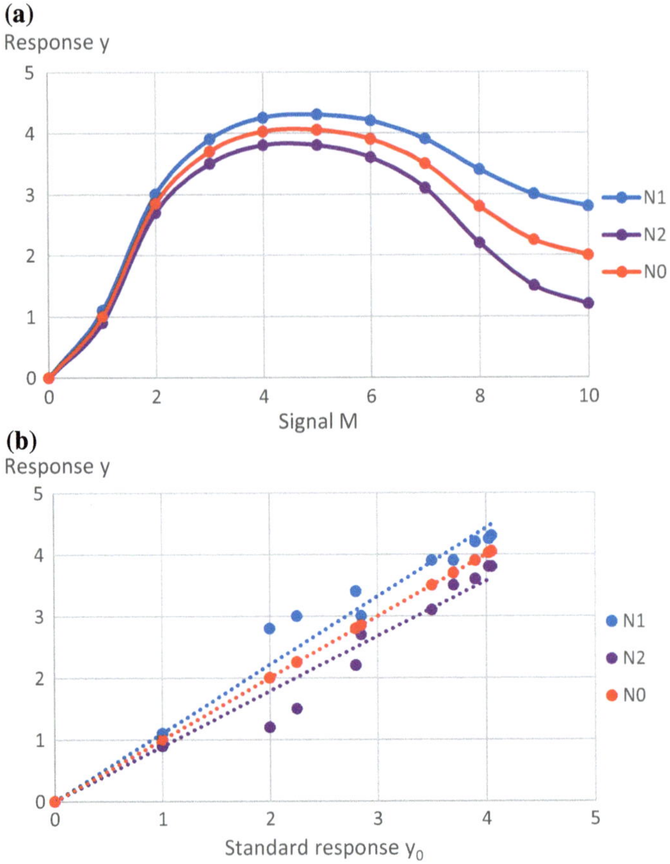

Fig. 5.3 Linearization of any kind of transfer characteristic with substitution of the abscissa (=original input signal) by the ordinates of the standard response. **a** Nonlinear relationship under two noise conditions to determine the standard response. **b** Linearized characteristic under the same noise conditions

Table 5.1 Substitution of the signal values by the standard response for nonlinear systems

Test	Original input signal				
	$M1$	$M2$	$M3$	…	Mn
	New reference to standardized transmission characteristic				
$N0$	y_{01}	y_{02}	y_{03}	…	y_{0n}
$N1$	y_{11}	y_{12}	y_{13}	…	y_{1n}
$N2$	y_{21}	y_{22}	y_{23}	…	y_{2n}

$$\sigma^2 = \frac{\left((y_{11} - \beta y_{01})^2 + \cdots + (y_{1n} - \beta y_{0n})^2\right) + \left((y_{21} - \beta y_{01})^2 + \cdots + (y_{2n} - \beta y_{0n})^2\right)}{2n}$$

$$SNR = 10 \log\left(\frac{\beta^2}{\sigma^2} S_2\right)$$

With reference to the standardized output response, linearization of the transfer characteristic takes place since the abscissa with its input signal is replaced by the identical ordinate values. Consequently, the slope is always one, and the calculations are simplified with

$$S_2 = \frac{1}{n}\left(y_{01}^2 + y_{02}^2 + \cdots + y_{0n}^2\right)$$

$$\beta = 1$$

$$\sigma^2 = \frac{\left((y_{11} - y_{01})^2 + \cdots + (y_{1n} - y_{0n})^2\right) + \left((y_{21} - y_{01})^2 + \cdots + (y_{2n} - y_{0n})^2\right)}{2n}$$

$$SNR = 10 \log\left(\frac{1^2}{\sigma^2} S_2\right)$$

This generally applicable linearization method means that each experimental test needs one repetition under modified noise conditions to calculate the individual standard output response. As a result, the information of different slopes is lost since the slope for all individual standard conditions equals one. For example, if a characteristic relationship is relatively flat with respect to the original input signal, its linearization with the substitution by the ordinate values leads to an increased slope. Conversely, a relatively steep transfer characteristic is linearized in the same way with the similar substitution by the standard responses and equally reset to the slope one.

Both the *SNR* index and the slope or equivalent information are included in an overall rating. However, due to the reference of the ordinate to itself, the numerator of the *S/N* ratio does not contain any information, so that a different approach must be chosen to evaluate the slope. In addition, the previous consideration of the *SEN* key figure no longer applies.

To reveal different behaviors, a defined set of signal values derived from (ideal) physical/chemical laws must be used for all tests carried out. If these are unknown or not applicable, a transfer function can be specified from practical system use instead. The individual standardized ordinate responses, which serve as input signal for the calculation of the *SNR* index, are related to a new ordinate set of an ideal relationship. This represents the new input signal as a target value for all tests performed. With the transformation, the system responses are similarly linearized

Table 5.2 Substitution of the standard response by ideal system behavior (=target values)

Test #	Original input signal				
	*M*1	*M*2	*M*3	...	*Mn*
	New reference of the standard response to ideal transfer function as input signal				
Target or ideal transfer function	m_{01}	m_{02}	m_{03}	...	m_{0n}
Standard response test #1	y_{011}	y_{012}	y_{013}	...	y_{01n}
Standard response test #2	y_{021}	y_{022}	y_{023}	...	y_{02n}
Standard response test #3	y_{031}	y_{032}	y_{033}	...	y_{03n}
and so forth	$y_{0\ldots1}$	$y_{0\ldots2}$	$y_{0\ldots3}$	...	$y_{0\ldots n}$

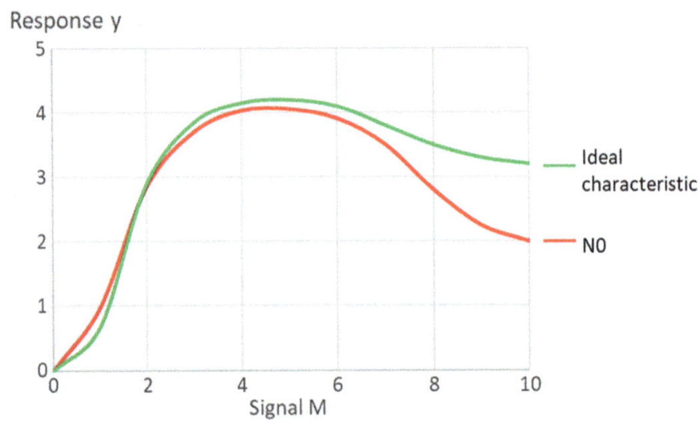

Fig. 5.4 Standard response *N*0 from tests with one or more sources of noise superimposed by an assumed ideal transfer function with respect to the input signal *M*

due to a similar transfer characteristic. All tests to be evaluated are available under the previously determined standard conditions N_{0i}. The ordinate values of the ideal relationship are given the designation

$$y_{\text{ideal},i} = m_{0i} \quad i = 1, 2, 3, \ldots, n$$

used as target values for the abscissa (Table 5.2).

For clarification, the standard response is supplemented in the following figure by the transfer function of ideal system behavior (Fig. 5.4).

After linearization, only this ideal characteristic is available as an exact angle bisector of the slope that equals one; all experimental tests deviate more or less from this (Fig. 5.5).

Ideal system behavior is not always achievable. A maximum efficiency or yield can be limited thermodynamically as well as by existing boundary conditions.

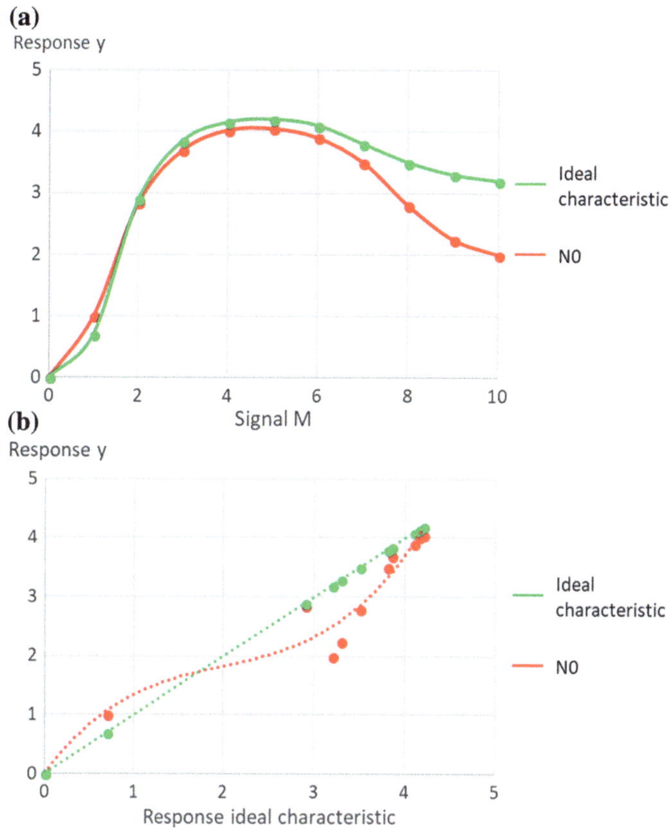

Fig. 5.5 Ideal characteristic and response to standard condition *N*0. **a** As a function of the input signal *M*. **b** Linearized ideal characteristic and standard response *N*0

Table 5.3 Transfer functions for different system behaviors

Defaults	Transfer function	Consequence
Physical/chemical principles	Signal values fixed	Test data are below the ideal transfer function
Practical determinations based on the performance of the system	Signal values not fixed but adapted to system properties	Test data are grouped around real system characteristic
None	Input signal values can be used as average of all standard conditions	Linear slope is omitted as evaluation criterion

Careful consideration must be given which system characteristic is appropriate. The possibilities are summarized in the table (Table 5.3).

Only in the first case, it is advantageous to scale a real transfer function to the ideal slope of value one by adjusting the input signal. If the slope is greater than

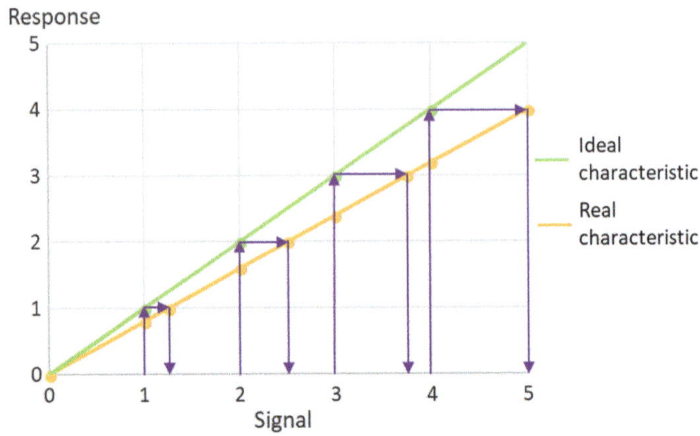

Fig. 5.6 Setting the slope with value one of the real system behavior by scaling

one, the abscissa must be compressed; if it is smaller than one, it must be expanded accordingly. This avoids unnecessary offset in favor of clear data interpretation. Scaling is done with the real slope as a scale factor. If the slope angle is used, the tangent of the angle must be taken; β_0 is the unit factor of the slope with value one. As shown below, the slope can also be adjusted to the target value one by evaluating the linear portion of the slope by selecting suitable parameters with their appropriate setting values (Fig. 5.6).

$$\text{Scaled input signal} \quad m_i = \frac{\beta}{\beta_0} \cdot m_{0i}$$

The original input signal is substituted several times (Table 5.4).

$$M_i \rightarrow y_{0i} \rightarrow m_{0i} \rightarrow m_i$$

To evaluate experimental tests, the *SNR* index must always be determined in the first step, maximized with the parameter effects, and in the second step the slope is

Table 5.4 Ideal response with a scale factor

Test #	Original input signal				
	*M*1	*M*2	*M*3	...	*Mn*
	Ordinate values of a real transfer function as input signal				
Scaled target	m_1	m_2	m_3	...	m_n
Standard response Test #1	y_{011}	y_{012}	y_{013}	...	y_{01n}
Standard response Test #2	y_{021}	y_{022}	y_{023}	...	y_{02n}
Standard response Test #3	y_{031}	y_{032}	y_{033}	...	y_{03n}
and so forth	$y_{0...1}$	$y_{0...2}$	$y_{0...3}$	...	$y_{0...n}$

adjusted as close as possible to the value one without attenuation of the *SNR* index. Here, a new effect becomes visible if the transfer function of the test data deviates from a straight line under standard conditions. This nonlinearity is expressed in the form of a curvature. An additional data analysis leads to the identification of parameter effects that are responsible for nonlinear behavior. If the curvature can be largely eliminated, higher order deviations can also be superimposed as a wave-shaped component. Here, too, higher order analysis can be used to find parameter settings that compensate for oscillating components as far as possible. The order of priority for achieving ideal system behavior is

1. Determine the largest possible *SNR* ratio: *SNR* = Maximum,
2. Adapt the slope of the linear characteristic to the real transfer function: $\beta_1 = 1$,
3. Reduce curvature as far as possible: $\beta_2 = 0$
   Curvature is nonlinear coefficient of second order, and
4. Reduce oscillations as far as possible: $\beta_3 = 0$
   Oscillations are nonlinear coefficients of third or higher order (Fig. 5.7).

Higher order oscillations are described by the nonlinear coefficients of the corresponding order, which should also be brought to zero if necessary. However, based on practical experience, the nonlinear coefficient of curvature assumes such small values that calculation of higher order coefficients is mostly unnecessary. This means it is sufficient to analyze the *SNR* index, the linear slope $\beta 1$, and the coefficient of curvature $\beta 2$ of a nonlinear system. For completeness, the algorithms for calculating the coefficients $\beta 1$, $\beta 2$, and $\beta 3$ are presented. Starting point is the straight-line function of the slope with value one as a power series with the origin as development point. The series consists of the linear term with $\beta 1$, the quadratic term with $\beta 2$, and the cubic term with $\beta 3$ (and similarly higher order terms if needed).

$$y = \beta 1 m + \beta 2\left(m^2 + am\right) + \beta 3\left(m^3 + b_1 m^2 + b_2 m\right) + \beta 4(\cdots)$$

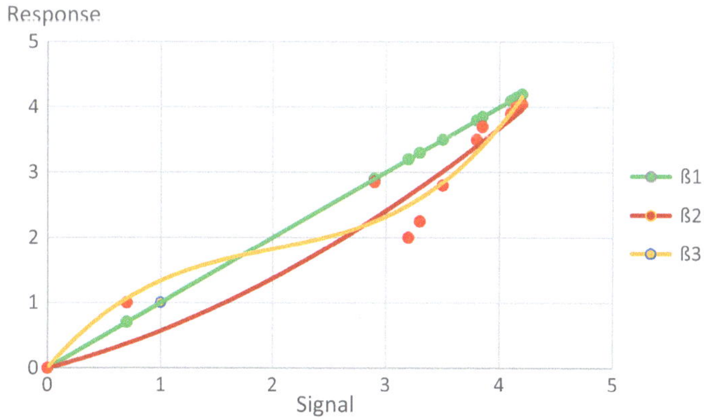

Fig. 5.7 Linear and nonlinear components of a standardized transfer function

To obtain independent coefficients, a, b, etc. are determined from the orthogonality condition, i.e., the pairwise products of the polynomials are set to zero. The polynomial coefficient a can be calculated from the orthogonality of the first and second terms.

$$\text{First and second terms} \quad m(m^2 + am) = 0$$

$$\begin{aligned}\text{Valid for all input signals} \quad & m_1^3 + am_1^2 = 0 \quad i = 1, 2, 3, \ldots, n \\ & m_2^3 + am_2^2 = 0 \\ & m_3^3 + am_3^2 = 0 \\ & \ldots \\ & m_n^3 + am_n^2 = 0\end{aligned}$$

$$\text{Summation} \quad \sum_{i=1}^{n} m_i^3 + a \sum_{i=1}^{n} m_i^2 = 0$$

$$\text{Polynomial coefficient } a \quad a = -\frac{\sum_{i=1}^{n} m_i^3}{\sum_{i=1}^{n} m_i^2}$$

For the sake of simplicity, higher order moments are introduced:

$$\begin{aligned} S_2 &= \frac{1}{n}\left(m_1^2 + m_2^2 + m_3^2 + \cdots + m_n^2\right) \\ S_3 &= \frac{1}{n}\left(m_1^3 + m_2^3 + m_3^3 + \cdots + m_n^3\right) \\ S_4 &= \frac{1}{n}\left(m_1^4 + m_2^4 + m_3^4 + \cdots + m_n^4\right) \\ S_5 &= \frac{1}{n}\left(m_1^5 + m_2^5 + m_3^5 + \cdots + m_n^5\right) \\ &\text{and so forth} \end{aligned}$$

$$\text{Polynomial coefficient } a \quad a = -\frac{S_3}{S_2}$$

The calculations of the polynomial coefficients b_1, b_2 follow from the orthogonality condition of the first and third terms as well as the second and third terms.

$$\text{First and third terms} \quad m\left(m^3 + b_1 m^2 + b_2 m\right) = 0$$

$$\text{Summation} \quad S_4 + b_1 S_3 + b_2 S_2 = 0$$

$$\text{Second and third terms} \quad \left(m^2 - \frac{S_3}{S_2} m\right)\left(m^3 + b_1 m^2 + b_2 m\right) = 0$$

$$\text{Summation} \quad S_5 + b_1 S_4 - \frac{S_3 S_4}{S_2} - b_1 \frac{S_3^2}{S_2} = 0$$

$$\text{Polynomial coefficient } b_1 \quad b_1 = \frac{S_2 S_5 - S_3 S_4}{S_3^2 - S_2 S_4}$$

Introduced in the equation from first and third term,

$$\text{Polynomial coefficient } b_2 \quad b_2 = \frac{S_4^2 - S_3 S_5}{S_3^2 - S_2 S_4}$$

Thus, the power series of the first three components with the calculated polynomial coefficients is completely determined.

$$y = \beta 1 m + \beta 2\left(m^2 - \frac{S_3}{S_2} m\right) + \beta 3\left(m^3 + \frac{S_2 S_5 - S_3 S_4}{S_3^2 - S_2 S_4} m^2 + \frac{S_4^2 - S_3 S_5}{S_3^2 - S_2 S_4} m\right)$$

The quadratic and cubic order coefficients $\beta 1$ and $\beta 2$ are calculated using the same algorithm as the linear slope $\beta 1$.

$$\beta 1 = \frac{y_{01} m_1 + y_{02} m_2 + \cdots + y_{0n} m_n}{m_1^2 + m_2^2 + \cdots + m_n^2}$$

$$\beta 2 = \frac{y_{01}\left(m_1^2 - \frac{S_3}{S_2} m_1\right) + y_{02}\left(m_2^2 - \frac{S_3}{S_2} m_2\right) + \cdots + y_{0n}\left(m_n^2 - \frac{S_3}{S_2} m_n\right)}{\left(m_1^2 - \frac{S_3}{S_2} m_1\right)^2 + \left(m_2^2 - \frac{S_3}{S_2} m_2\right)^2 + \cdots + \left(m_n^2 - \frac{S_3}{S_2} m_n\right)^2}$$

$$\beta 3 = \frac{y_{01}\left(m_1^3 + \frac{S_2 S_5 - S_3 S_4}{S_3^2 - S_2 S_4} m_1^2 + \frac{S_4^2 - S_3 S_5}{S_3^2 - S_2 S_4} m_1\right) + \cdots + y_{0n}\left(m_n^3 + \frac{S_2 S_5 - S_3 S_4}{S_3^2 - S_2 S_4} m_n^2 + \frac{S_4^2 - S_3 S_5}{S_3^2 - S_2 S_4} m_n\right)}{\left(m_1^3 + \frac{S_2 S_5 - S_3 S_4}{S_3^2 - S_2 S_4} m_1^2 + \frac{S_4^2 - S_3 S_5}{S_3^2 - S_2 S_4} m_1\right)^2 + \cdots + \left(m_n^3 + \frac{S_2 S_5 - S_3 S_4}{S_3^2 - S_2 S_4} m_n^2 + \frac{S_4^2 - S_3 S_5}{S_3^2 - S_2 S_4} m_n\right)^2}$$

The denominators of $\beta 1$ and $\beta 2$ can be replaced with the definitions of higher order moments. For $\beta 3$ or higher orders, the pre-calculation with the coefficients appears to be easier.

$$\beta 1 = \frac{y_{01} m_1 + y_{02} m_2 + \cdots + y_{0n} m_n}{n S_2}$$

$$\beta 2 = \frac{y_{01}\left(m_1^2 - \frac{S_3}{S_2} m_1\right) + y_{02}\left(m_2^2 - \frac{S_3}{S_2} m_2\right) + \cdots + y_{0n}\left(m_n^2 - \frac{S_3}{S_2} m_n\right)}{n\left(S_4 - \frac{S_3^2}{S_2}\right)}$$

$$\beta 3=\frac{y_{01}\left(m_1^3+b_1m_1^2+b_2m_1\right)+y_{02}\left(m_2^3+b_1m_2^2+b_2m_2\right)+\cdots+y_{0n}\left(m_n^3+b_1m_n^2+b_2m_n\right)}{\left(m_1^3+b_1m_1^2+b_2m_1\right)^2+\left(m_2^3+b_1m_2^2+b_2m_2\right)^2+\cdots+\left(m_n^3+b_1m_n^2+b_2m_n\right)^2}$$

For the slopes, a clear recognizability of parameter effects is important, so that the equivalent *SNR* representation—the logarithmic form—is not applied. Even if the calculation effort appears to be higher than with linear systems, the procedure of the standardized *SNR* index can always be applied. The decisive advantage of such an approach is the separation into linear and nonlinear components of any transfer function, which can be analyzed, controlled, or compensated independently of each other.

5.2 Analyses of a Nonlinear System

Nonlinear system response can be frequently observed. Typically, the characteristic outcome for batch processes runs through an extreme value as a function of time. The drop-in response to an otherwise rising trend with a steadily increasing signal value is also referred to as a falling or negative system characteristic. Examples can be found in mechanical components (bounce-free switches with snap-through characteristics), welding transformers in electrical engineering, and electronic components such as tunnel diodes, energy-saving lamps, etc.

As a numerical example, a biotechnological fermentation process is treated in more detail without penetrating too deep into the enzyme kinetics. Bacteria (usually Escherichia coli), to which a plasmid, i.e., a gene as an autonomously replicating DNA molecule with promoter, is introduced. The gene is induced by the promoter to express a desired complex organic compound such as an enzyme or protein. The promoter is induced from the outside, i.e., by adding a chemical substance, so that the synthesis of the desired enzyme begins through the given blueprint of the gene, provided the external conditions are suitable. The process as such requires a lead time to multiply the bacteria in a culture medium until the promoter is activated at a certain point in time.

The synthesis process begins when the induction substance is added, and the bacteria are constrained by reducing the food supply. At the same time, biological stress causes a rapid decrease in the number of living cells until a few cells can reproduce by adaptation. From this overall process, consisting of synthesis rate of the desired substance (enzyme) and simultaneous mortality rate of the cells, the aspect of the temporal development of living cells is examined in more detail below. Only the living cells can synthesize the enzyme. The amount of enzyme produced depends on both the bacterial strain and the fermentation conditions, i.e., the food supply and the temperature. In view of a possible increase of the synthesized enzyme quantity/time, the external environment is subjected to a parameter analysis. In doing so, the following parameters are considered:

- the yeast concentration (corresponds to the supply of nitrogen),
- the oxygen partial pressure by variable supply of oxygen,
- the temperature (as constant or temperature ramp, i.e., varies in time), and
- the glucose concentration (corresponds to the energy supply).

Signal parameter is time. The time lag for addition of the induction substance and its concentration represents the noise sources (Table 5.5).

The variables are assigned to the columns of the orthogonal array L9. Samples are taken and characterized every 7 h after inoculation by the inductor substance to record the temporal changes. The tests are evaluated by counting the cell colonies/area as a relative measure (Table 5.6).

Table 5.5 Parameters of the culture medium

Parameter	Levels		
A: Yeast concentration	1	2	3
B: Oxygen partial pressure	1	2	3
C: Temperature	1	2	3
D: Glucose concentration	1	2	3

Table 5.6 Experimental data of living cell colonies measurements with the standard response *N*0

Input signal	0.5 h		7 h		14 h		21 h		28 h		35 h	
Induction condition	*N*1	*N*2	*N*1	*N*2	*N*1	*N*2	*N*1	*N*2	*N*1	*N*2	*N*1	*N*2
Test #1	314	647	176	519	0.3	66.7	0.2	0.6	1.9	0.6	9.3	4
Test #1 *N*0	480.5		347.5		33.5		0.4		1.25		6.65	
Test #2	62	533	0.4	9.9	4.3	0.34	32	1.6	24.5	0.51	144	0.03
Test #2 *N*0	297.5		5.15		2.32		16.8		12.505		72.015	
Test #3	166	481	10.1	138	2.4	52	22.1	28	189	16.8	401	4.7
Test #3 *N*0	323.5		74.05		27.2		25.05		102.9		202.85	
Test #4	102	297	0.02	9.6	3.7	2.2	58.2	1	132	1.22	148	0.23
Test #4 *N*0	199.5		4.81		2.95		29.6		66.61		74.115	
Test #5	167	312	13.2	22.2	0.7	4.1	1.7	1.5	27.7	23.6	116	148
Test #5 *N*0	239.5		17.7		2.4		1.6		25.65		132	
Test #6	58	579	0.13	26	0.33	0.8	2.2	12.1	5.4	72	89	137
Test #6 *N*0	318.5		13.065		0.565		7.15		38.7		113	
Test #7	216	437	8.6	92	1.9	45	11.2	21.3	32.8	8.5	76.5	7.8
Test #7 *N*0	326.5		50.3		23.45		16.25		20.65		42.15	
Test #8	216	742	0.3	46	0.4	3.6	0.8	17.9	1.1	145	1.5	329
Test #8 *N*0	479		23.15		2		9.35		73.05		165.25	
Test #9	342	664	0.5	478	0.4	66	22	8.3	89	11.8	139	14.8
Test #9 *N*0	503		239.25		33.2		15.15		50.4		76.9	

Table 5.7 Calculation of *SNR* key figures from the signal values substituted by the standard response *N*0 for each test

Test #	Dimension factor S_2	Squared standard deviation σ^2	*SNR* [dB]
1	58800.78	9707.37	7.82
2	15693.82	10173.92	1.88
3	27206.59	12699.37	3.31
4	8439.7	3347.09	4.02
5	12627.3	923.27	11.36
6	19321.85	11622.84	2.21
7	18691.56	2627.77	8.52
8	43785.37	16959.61	4.12
9	53339.18	14898.62	5.54

The signal values are to be substituted by the standard response *N*0 for each test. The *SNR* index is calculated using the dimension factor and the squared standard deviation (Table 5.7).

$$S_2 = \frac{1}{n}\left(y_{01}^2 + y_{02}^2 + \cdots + y_{0n}^2\right)$$

$$\sigma^2 = \frac{\left((y_{11} - y_{01})^2 + \cdots + (y_{1n} - y_{0n})^2\right) + \left((y_{21} - y_{01})^2 + \cdots + (y_{2n} - y_{0n})^2\right)}{2n}$$

$$SNR = 10\log\left(\frac{1^2}{\sigma^2} S_2\right)$$

The parameter effects are determined from the setting levels of the orthogonal array L9. The *SNR* key figures of the parameter effects are displayed.

Since the force exerted on the cells cannot be reduced, the largest number of living cells from all tests carried out is assumed to be the ideal transfer function at all points in time. The coefficients for linear and nonlinear behavior are calculated from the standard response *N*0 to the new signal values *m* of the ideal function (Table 5.8).

$$\beta 1 = \frac{y_{01}m_1 + y_{02}m_2 + \cdots + y_{0n}m_n}{nS_2}$$

$$\beta 2 = \frac{y_{01}\left(m_1^2 - \frac{S_3}{S_2}m_1\right) + y_{02}\left(m_2^2 - \frac{S_3}{S_2}m_2\right) + \cdots + y_{0n}\left(m_n^2 - \frac{S_3}{S_2}m_n\right)}{n\left(S_4 - \frac{S_3^2}{S_2}\right)}$$

The parameter effects are shown in the illustrations.

Preferred settings are based on the highest possible *SNR* index (Fig. 5.8). In addition, the evaluation of the linear coefficient $\beta 1$ must be considered

Table 5.8 Linear and quadratic coefficients related to the ideal system function

Signal ideal transfer function	m_1 742	m_2 519	m_3 66.7	m_4 58.2	m_5 189	m_6 401	Linear slope $\beta 1$	Quadratic slope $\beta 2$
Test #								
1	480.5	347.5	33.5	0.4	1.25	6.65	0.529	0.001138
2	297.5	5.15	2.32	16.8	12.51	72.02	0.250	0.000816
3	323.5	74.05	27.2	25.05	102.9	202.9	0.373	0.000109
4	199.5	4.81	2.95	29.6	66.61	74.12	0.190	0.000278
5	239.5	17.7	2.4	1.6	25.65	132	0.239	0.000383
6	318.5	13.07	0.565	7.15	38.7	113	0.289	0.000697
7	326.5	50.3	23.45	16.25	20.65	42.15	0.285	0.000888
8	479	23.15	2	9.35	73.05	165.3	0.438	0.001014
9	503	239.3	33.2	15.15	50.4	76.9	0.528	0.001031

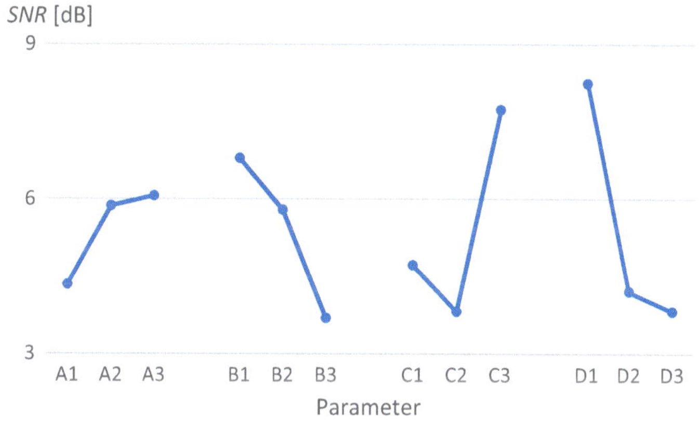

Fig. 5.8 *SNR* indices of the parameter effects in the fermentation process

simultaneously, i.e., it must be adjusted as close as possible to the numerical value one, provided there are no opposing effects (Fig. 5.9). The quadratic coefficient $\beta 2$ is already so small that an adjustment close to zero is no longer important (Fig. 5.10). The temporal development of the living cells in the fermentation process appears advantageous with the following settings:

$$A3 + B1 + C3 + D1$$

The predictions were confirmed and resulted in a 40% increase in the synthesis rate of the enzyme. In addition, the environmental conditions are stable and favorable so that other bacterial strains can be tested for their synthesis efficiency without further changes or adaptations.

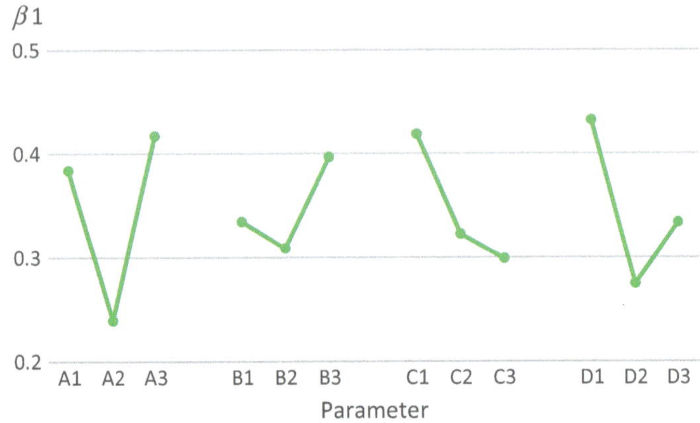

Fig. 5.9 Parameter effects of the linear coefficient $\beta 1$

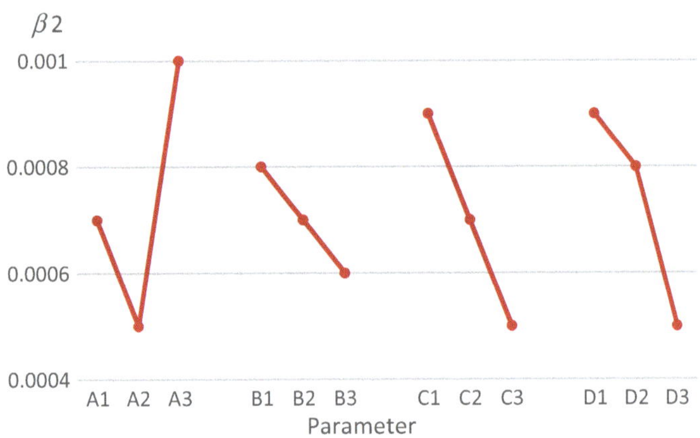

Fig. 5.10 Parameter effects of the quadratic coefficient $\beta 2$

5.3 Incomplete Data—Sequential Iteration

Imponderability is part of everyday life in experimental technique. Analytical samples may be contaminated, and plant components may have been damaged by corrosion or measured values may not be recorded if they are outside the measuring range. If a test system has already been dismantled or a failed test can no longer be repeated for time and cost reasons, one or the other data point is missing within the framework of an orthogonal array to be processed. Strictly speaking, it is not possible to evaluate the overall parameter effects, since each test contains part of the overall information. Conversely, however, the tests performed also contain partial information about the missing test data, albeit with less reliability. Instead of doing without an evaluation, it is much better to derive the missing data from the existing data. Thus, nothing more stands in the way of an evaluation, whereby in the worst case smaller deviations in predictions from model calculations must be accepted. Missing measured values are determined by an iteration procedure. Starting from the average value of all available data, the missing data point can be approximated with a sufficient number of iteration steps until the parameter effects do not change within a set accuracy. The procedure is explained below using a numerical example that can be transferred in the same way to other characteristics or several data points. To test the accuracy of the iteration procedure, a complete experimental design is taken from which one data point is removed. After each iteration step, the immediate comparison shows how close the approximate value is to the true value.

The logarithmic scale, i.e., the *SNR* index, has proven useful to begin with. To model the missing data point, only the parameters with relatively large effect amplitudes need to be considered; this is usually about half of all parameters or a smaller number. The starting point—the zero-iteration step—is to fill the data gap with the average value from all other data. The greatest effects used to model the configuration (=test #) of the missing measurement value is taken from the numerical or graphical evaluation. This represents a first improved value that is used for the first iteration step. The principle, Evaluation–Modeling–Deployment, is repeated until convergence or sufficient accuracy occurs.

If the trend of the iterated numerical value is not uniform (monotonic), the number of parameters used for modeling can be reduced. If the original data loss occurred because the measured values exceeded or fell below the expected range, the starting point for the zero-iteration step is not the average value. Instead, the largest numerical value plus 3 [dB] must be selected if the value is exceeded, and the smallest value minus 3 [dB] if it is not reached. All further steps are carried out as described.

For reason of comparison, the complete data set is given. In the adjacent column, one data point is removed and replaced by the average value in the next column. These data are used for an initial evaluation, of which the effect variables are plotted (Table 5.9).

Only the strongest parameter effects are appropriate for the iteration steps. In the numerical example shown, the parameters *A*, *C*, and *D* are used (Fig. 5.11).

The missing data point is test #10 with the configuration specified by the orthogonal array.

$$\text{Test}\,\#10: A2 + B1 + C1 + D3 + E3 + F2 + G2 + H1$$

The setting values of the parameters with the highest effect amplitudes must be selected according to the test configuration.

$$SNR = A2 + C1 + D3 - 2\emptyset$$

If the iterated new value exceeds or falls short of the previous value by less than 5%, this result can be used for the final evaluation. Usually, the iterated value to be used converges already after a few consecutive steps, as becomes clear in the application of eight iteration steps (Fig. 5.12).

Table 5.9 Iterative calculation of one missing data point

Test #	*SNR*	Test #10 missing	0. Iteration	1. Iteration	2. Iteration	3. Iteration
1	6.77	6.77	6.77	6.77	6.77	6.77
2	12.24	12.24	12.24	12.24	12.24	12.24
3	12.58	12.58	12.58	12.58	12.58	12.58
4	5.33	5.33	5.33	5.33	5.33	5.33
5	10.61	10.61	10.61	10.61	10.61	10.61
6	15.34	15.34	15.34	15.34	15.34	15.34
7	10.50	10.50	10.50	10.50	10.50	10.50
8	15.91	15.91	15.91	15.91	15.91	15.91
9	8.62	8.62	8.62	8.62	8.62	8.62
10	**14.51**	–	**12.37**	**14.19**	**14.79**	**14.99**
11	13.19	13.19	13.19	13.19	13.19	13.19
12	14.22	14.22	14.22	14.22	14.22	14.22
13	12.83	12.83	12.83	12.83	12.83	12.83
14	15.02	15.02	15.02	15.02	15.02	15.02
15	14.52	14.52	14.52	14.52	14.52	14.52
16	15.68	15.68	15.68	15.68	15.68	15.68
17	10.21	10.21	10.21	10.21	10.21	10.21
18	16.72	16.72	16.72	16.72	16.72	16.72
		Mean **12.37**	*A*2, *C*1, *D*3 **14.19**	*A*2, *C*1, *D*3 **14.79**	*A*2, *C*1, *D*3 **14.99**	*A*2, *C*1, *D*3 **15.06**

Fig. 5.11 Graphical representation of the parameter effects using the average value for the missing data point

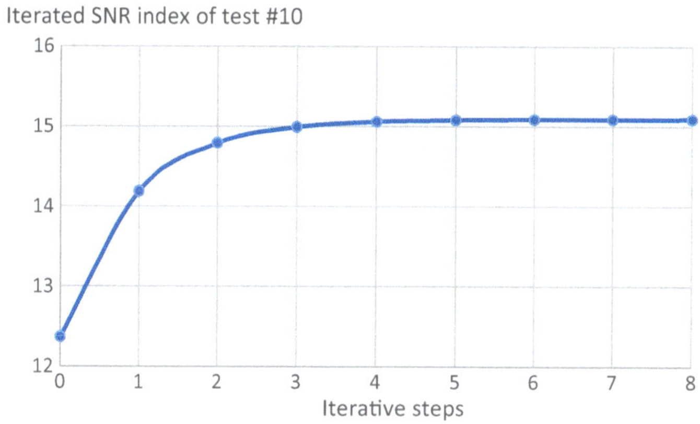

Fig. 5.12 Saturation characteristics of iterated values

Table 5.10 Comparison of the results of the iterated data point with the original data

2. Iteration	$A2 + C3 + D3 - 2\,\emptyset = 17.67$ [dB]
Original data	$A2 + C3 + D3 - 2\,\emptyset = 17.63$ [dB]

Practical experience shows that 1–2 iteration steps are sufficient after the first evaluation with the average data value used, especially since the convergent final value is usually above or below the true value. The same parameters are used for a model prediction, but with the largest *SNR* indices. The table shows a comparison of the results from the second iteration and the original data (Table 5.10).

On the logarithmic scale, the relative deviation of the result of the complete data set for this numerical example is 0.2%. Transformed back to the linear scale this corresponds to a relative error of 0.5%.

5.4 Loss Function

In order to fulfill an intended function, compliance with certain critical parameters is indispensable. For example, if one considers an organic synthesis reaction, the desired product occurs at a set temperature in the required quantity and purity. If the temperature of the thermostat used deviates to lower or higher values, undesired side reactions increase until the originally intended reaction ceases completely if the deviations are sufficiently large. This also applies to mechanical systems where tolerances are exceeded or undershot and to electrical devices where deviations in voltage or frequency become too great.

Above all, however, this is valid in the medical field when administering too high or too low a dose for the anesthesia of an upcoming surgery or other drug treatments. In the latter case, critical limit values are described by the term lethal dose (LD). If functional failure is detected in 50% of all cases, the short form LD50 is used. The identical definition can be applied to any other system, where specifications with a lower and an upper limit value correspond to the equivalent of the lethal dose LD50 in the case of bilateral characteristics. Although a certain characteristic meets the specification, a deviation from the ideal value can already lead to a malfunction. This means that a characteristic value is not good per se if it lies within the specified range, but it must be considered that an increasing distance to the ideal value comes with a continuously increasing loss. No deficit occurs at the target value itself, i.e., the loss is zero. At the specified limit values, however, a process malfunction or product failure can be expected. The characterization of the current value of parameters which are decisive for the fulfillment of a certain function is carried out with the loss function L. If the ideal value is set to the zero point, the loss function can be developed as a power series around the origin as a starting point **(1-34*)**.

$$L(y) = k(y - m)^2$$

L Loss function.
k Loss coefficient.
y Characteristic response.
m Ideal value.

The loss coefficient k results from the distance Δ from the ideal value m. When the limit values are reached, total loss occurs sooner or later, i.e., costs A arise through renewed operation of a process while maintaining critical values or replacement of the product as such.

$$A = k((m + \Delta) - m)^2$$

$$\text{with } y = m + \Delta \quad \text{or} \quad y = m - \Delta$$

$$A = k\Delta^2$$

$$k = \frac{A}{\Delta^2}$$

$$L(y) = \frac{A}{\Delta^2}(y - m)^2$$

The loss function is illustrated in the figure.
The loss is a function of three variables (Fig. 5.13):

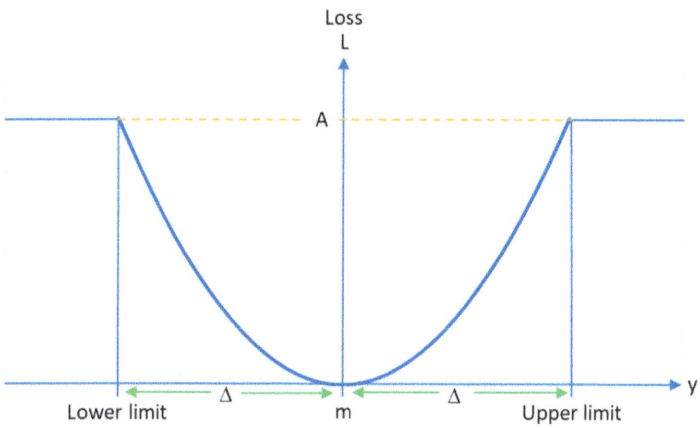

Fig. 5.13 Loss function L as a function of the characteristic y with upper and lower specification limits

$$A = \text{Process or product costs}$$
$$(y-m)^2 = \text{Deviation of the characteristic response from the target value}$$
$$\Delta = \text{Maintaining a system function within upper and lower specification limit.}$$

For comparative evaluations, the loss coefficient can be set to one. Setting and maintaining a critical response depends on the variables involved. With the requirement of the smallest possible loss, the characteristic response is centered on the target value. To avoid dealing with numerical values close to zero, instead it is advisable to find a maximum value with the reciprocal loss by superimposing favorable parameter effects. Corresponding effects can interact both additive and multiplicative, such as percentages. To avoid misjudgments, a logarithmic scale for parameter effects is introduced which, mathematically speaking, adds the exponents of the effects in the case of multiplicative behavior. With an additional scaling factor of 10, the loss analysis stands for nothing else than the *SNR* index.

$$SNR = 10 \log\left(\frac{L_0}{L}\right)$$

L_0 Currency unit (e.g., €, $, ¥).

The characteristic considered with the loss function can be one-sided, double-sided, symmetrical, or asymmetrical. For a double-sided value, there is a nominal value, and for a one-sided value there is a value as small as possible, *STB*, or as large as possible, *LTB*, which is entered as a reciprocal value. Instead of the evaluation of individual states, i.e., process or product properties, the loss function can also be calculated from the average value and squared standard deviation of an ensemble **(1-35*)**.

$$L = k\left(\sigma^2 + (\bar{y} - m)^2\right)$$

Thus, all manifestations of a loss description can be traced back to the *S/N* or *SNR* key figures **(1-36*)**. The decisive advantages of using the loss function are

- any technical feature, no matter how complicated, is transformed into a monetary unit;
- the minimum loss, independent of any specification leads to focusing critical characteristic responses to the ideal value, synonymous with a maximum *SNR* index; and
- technical and economic aspects are in a simple relationship.

5.5 Tolerances in Static Systems

The term "robust system" refers to a condition with the greatest possible compensation of noise factors by means of advantageous parameter setting values. Using nonlinear parameter effects, dispersion margins are greatly reduced, resulting in stable and reliable system behavior. However, if there are more demanding technical requirements to further reduce residual deviations, this can no longer be achieved at zero cost using higher quality materials, components with lower tolerances, greater measuring accuracy, etc. if the existing system parameters are exhausted. A higher precision requires the use of investment. If money is no question, any small tolerances of system characteristic features can be realized for special applications. In most cases, however, not only is the budget limited but also competitive pressure compels to return to reality. In addition to the intrinsic parameter possibilities, the question arises as to the relationship between the additional expenses required for improving precision and reliability. The facts are shown schematically (Fig. 5.14).

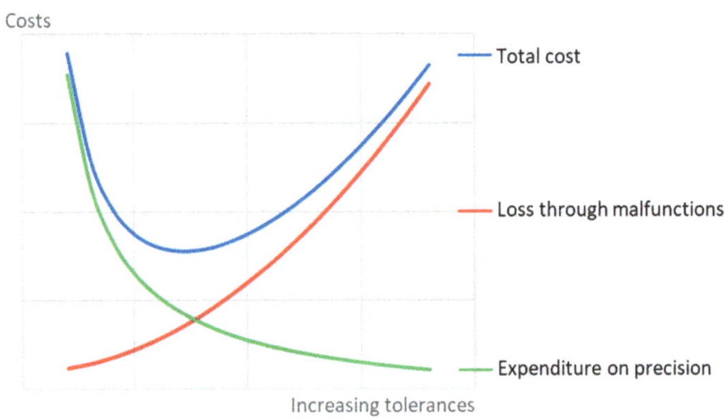

Fig. 5.14 Economical–technical optimization: Cost situation due to the superposition of improving precision and increasing system malfunctions caused by larger tolerances

The real function curve is composed of smaller steps, which, however, does not change an always existing minimum with overlapping of both loss variables. If a parameter analysis with three (or more) setting values shows an extremely large *SNR* index, then it appears promising to examine possible tolerances for these setting values more closely. In order not to specify each parameter beforehand, often at considerable expense to reduce dispersion, a tolerance design provides the necessary information on the most effective parameters. To know the upcoming investment, an economic and technical optimum can be modeled up to which a system improvement appears to be meaningful for realization. Further reduction of tolerances beyond the minimum of the total costs will be omitted for competitiveness. If necessary, a tolerance analysis must be carried out in ever more complex steps until the best possible trade-off between technical perfection and effort has been found.

For a more precise output response with a lower deviation, only the linear portion of the effects of a tolerance analysis proves reliable, not higher order components. Consequently, the linear and the square component of the parameter effects have to be calculated, so that three setting values must be chosen. As a prerequisite, a parameter analysis is carried out with maximization of the *SNR* key figures that serve as the best possible starting point for the tolerance analysis. For this state, the advantageous parameter settings are set to level 2, and the symmetrical distances Δ are assigned to levels 1 and 3 **(1-37*)**.

$$\Delta = \pm\sqrt{\frac{3}{2}}\sigma$$

The dispersion value refers to the individual manufacturer specifications, e.g., for electronic components (±5%), device tolerances (±10%), or similar, whereby it is common practice that the covered total range corresponds to 6 times of the standard deviation. This idea is based on dealing with Gauss-like bell curves in industrial manufacturing processes to comply with certain product specifications. If 3 times of the standard deviation is permitted on both sides of the average, the production yield is close to 100%.

Here is a numerical example: A resistor of 1 kΩ as a passive electronic component has a tolerance of ±5%. The true value is therefore in the range of 950–1050 Ω. If this component is subjected to a tolerance analysis, the most exact resistor values for new settings to be tested are to be selected for test purposes only with

$$\text{Tolerance} \quad 100\,\Omega = 6\sigma \rightarrow \sigma = \frac{100}{6}\Omega$$

$$\begin{aligned}&\text{Level 2} \quad 1000\,\Omega\\ &\text{Level 1} \quad 1000\,\Omega - \sqrt{\tfrac{3}{2}}\cdot\tfrac{100}{6}\,\Omega = 980\,\Omega\\ &\text{Level 3} \quad 1000\,\Omega + \sqrt{\tfrac{3}{2}}\cdot\tfrac{100}{6}\,\Omega = 1020\,\Omega\end{aligned}$$

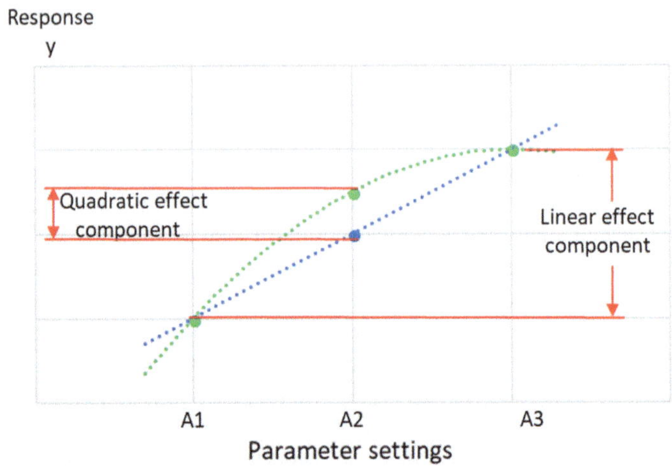

Fig. 5.15 Separating the parameter effect into linear and quadratic components

The three setting values are equidistant and monotonically increasing (or decreasing). The linear effect of the tolerance analysis is given by the response difference of the first level against the third level, the quadratic component results from the difference of level 2 to the linearly interpolated output variable in the middle between levels 1 and 3 (Fig. 5.15).

To be independent of the sign, the squared form is used.

$$\begin{array}{ll} \text{Linear component} & A_{\text{linear}} = (A1 - A3)^2 \\ \text{Quadratic component} & A_{\text{quadratic}} = (A2 - 1/2\,(A1 + A3))^2 \end{array}$$

By carrying out a tolerance analysis, not only the greatest effects become visible but with the decomposition also those with the greatest linear component. Only the latter behave reliably and are promising for a narrowing of tolerances.

The analytical experimental part must be supplemented by a cost analysis. The narrowing of tolerances can be differently efficient and expensive, so that only model calculations show with which combinations under existing boundary conditions an actual technical–economic optimum can be achieved. A more detailed numerical example is given to illustrate the successive steps. As usual, given boundary conditions such as specifications, production costs, quantities, investments, and amortization period can be adapted to individual conditions at any time.

The starting point is a metallization process. Large epoxy plates are immersed in a chemical solution. A uniform metal layer thickness of 200 μ ± 10 μ must be maintained. The production costs per panel amount to 50$, whereby 250,000 pieces per year are produced on one line. It is to be expected that a new specification of 200 μ ± 5 μ will have to be met in the foreseeable future.

Table 5.11 Parameter and setting values for the metallization process

Parameter	Levels		
	1	2	3
A: (Error column)	–	–	
B: Etching time (s)	10	20	30
C: Activation time (min)	1	3	7
D: Speed of rotation (rpm)	3	6	9
E: Salt concentration (g/l)	20	30	40
F: Concentration ratio of 2 components	1.2/0.8	1/1	0.8/1.2
G: Temperature (°C)	80	70	60
H: Flow rate (l/h)	0.5	1.0	1.5

The questions to be answered are as follows:

To what extent can the process be improved by parameter analysis?
Are investment funds necessary due to a new specification?
If investment is unavoidable, what is the most cost-effective solution for a cost analysis considering an amortization period of 1 year?

For the existing process, a parameter analysis is carried out first. For this purpose, seven parameters with three setting values each are available (Table 5.11).

With seven variables, an orthogonal array of size L18 is required. Since column A with two setting values is not assigned a technical value, it is available for error interpretation. The table shows the experimentally measured values and the characteristic key figures derived from them. Noise factors determined by the system geometry are combined with the respective deviation trend to level 1 with smaller (CN1) and level 2 with larger values (CN2) (Table 5.12).

The *SNR* and *SEN* key figures of the orthogonal array L18 are evaluated numerically with graphical presentation of the results (Figs. 5.16 and 5.17; Tables 5.13 and 5.14).

The effect on column A stands for an error analysis. Similarly, effect amplitudes fall into the error category; all considerably larger amplitudes are to be interpreted as effects. From experience, about half of the parameters analyzed should be used for a model calculation, provided that they clearly stand out from the noise background. For achieving the smallest relative deviations, the configuration appears to be advantageous

$$C3 + D2 + E2 + F3$$

The same setting is to be used as the basis for the average value of the layer thickness, whereby *C* is to be classified more as error size, whereas *G* can be used for adjustment in the range *G*1 to *G*2. The resulting average layer thicknesses are calculated in advance with

Table 5.12 Experimental data with *SNR* and *SEN* key figures for analysis of parameter effects

Test #	CN1	CN2	y	σ^2	$\frac{y^2}{\sigma^2}$	*SNR* [dB]	*SEN*
1	112	139	125.5	182.25	86.42	19.37	41.97
2	104	109	106.5	6.25	1814.76	32.59	40.55
3	149	156	152.5	12.25	1898.47	32.78	43.67
4	78	88	83	25	275.56	24.40	38.38
5	501	528	514.5	182.25	1452.46	31.62	54.23
6	129	144	136.5	56.25	331.24	25.20	42.70
7	119	132	125.5	42.25	372.79	25.71	41.97
8	91	102	96.5	30.25	307.84	24.88	39.69
9	241	284	262.5	462.25	149.07	21.73	48.38
10	98	107	102.5	20.25	518.83	27.15	40.21
11	99	114	106.5	56.25	201.64	23.05	40.55
12	393	412	402.5	90.25	1795.08	32.54	52.10
13	108	124	116	64	210.25	23.23	41.29
14	150	173	161.5	132.25	197.22	22.95	44.16
15	108	114	111	9	1369.00	31.36	40.91
16	732	757	744.5	156.25	3547.39	35.50	57.44
17	90	101	95.5	30.25	301.50	24.79	39.60
18	93	102	97.5	20.25	469.44	26.72	39.78

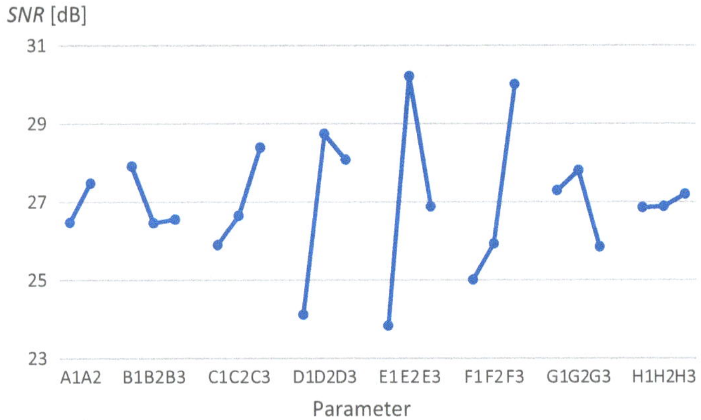

Fig. 5.16 Graphical illustration of *SNR* indices of parameter effects

$$D2 + E2 + F3 + G2$$

The results of this proposal are to be compared with the standard operational settings, all at level 2

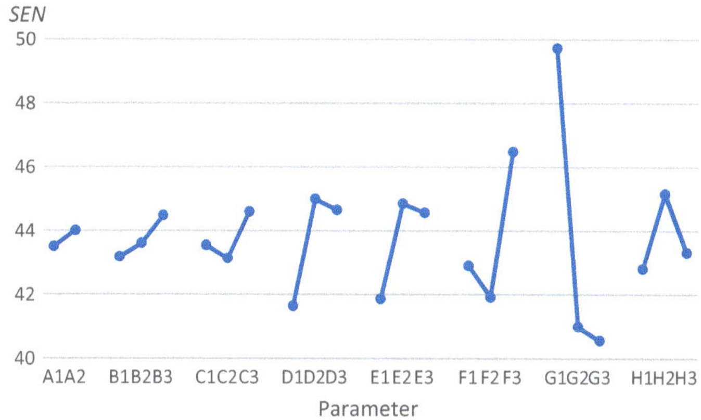

Fig. 5.17 Sensitivity of parameter effects presented as *SEN* indices

Table 5.13 Evaluation of the *SNR* key figures of the orthogonal array L18

Parameter	*A* [dB] error	*B* [dB]	*C* [dB]	*D* [dB]	*E* [dB]	*F* [dB]	*G* [dB]	*H* [dB]
Level 1	26.48	27.91	25.89	24.12	23.83	25.00	27.29	26.85
Level 2	27.48	26.46	26.65	28.74	30.21	25.92	27.80	26.88
Level 3		26.56	28.39	28.08	26.88	30.00	25.84	27.20

Table 5.14 Evaluation of the *SEN* indices of the orthogonal array L18

Parameter	*A* error	*B*	*C*	*D*	*E*	*F*	*G*	*H*
Level 1	43.50	43.18	43.54	41.63	41.86	42.89	49.71	42.80
Level 2	44.00	43.61	43.13	44.99	44.85	41.91	40.99	45.15
Level 3		44.48	44.59	44.65	44.56	46.46	40.56	43.31

Table 5.15 Predictions of the model calculations

	SNR [dB]	*SEN*	$\bar{y}$	σ^2	σ	3σ
Existing process *C2* + *D2* + *E2* + *F2*	30.59	G1 − G2	Adjustment	34.94 μ²	5.91 μ	17.73 μ
Improved process *D2* + *E2* + *F3* + *G2*	36.41	46.02	200 μ	9.13 μ²	3.02 μ	9.06 μ

$$C2 + D2 + E2 + F2$$

The table shows the calculation of the proposed configurations from the back-transformation of the logarithmic scale (Table 5.15).

Table 5.16 Tolerance specifications for different parameters

Parameter	Tolerance range
Temperature controller: 20–100 °C	±2 °C
Weighing/dosing accuracy	±5%
Sensors	±1%

The existing process can be adjusted to the target value with the settings $G1 - G2$, but the deviation exceeds the specified limit values of ±10 μ, which can be seen from the 3σ figure. The improved settings after the parameter analysis lead to a stable process that meets the previous requirements. With the planned change of the specification with permissible deviations of ±5 μ, however, the same problem arises again. In this situation, a way to reduce intrinsic process tolerances must be sought. Instead of narrowing the specifications of all parameters through cost expenditure, a tolerance analysis provides the decisive indications for the most cost-effective success. The specified tolerances of the components (=parameters) in use so far are (Table 5.16).

For the setting values to be selected, the offset for the best possible setting must be determined at level 2. The specified range usually covers 6σ, i.e.,

$$\sigma = \text{specified range}/6$$

This determines the setting values for the tolerance analysis on three levels (Table 5.17).

For all parameters, the new settings must be realized only once to carry out the experiments, even if they are cumbersome (Tables 5.18 and 5.19).

For a clear detection of the effect amplitudes, the original measurement data is evaluated. The parameter effects are plotted (Fig. 5.18).

The following table contains the component decomposition of the effects and their process-related relative effect contribution as the ratio of the squared standard deviation to the sum of the total squared deviations (Table 5.20).

From the graph, F and G can be read as the strongest parameter effects, whereby their quadratic proportions are relatively small. Only F and G are suitable for narrowing tolerances, all other effects appear in the range of the error component and are combined with column A (Table 5.21).

To achieve tighter tolerances, more valuable materials, measuring methods or similar, hereinafter referred to as grades, must be used. Before an investment decision is made, however, the available options must be evaluated in model calculations. Only in connection with the costs a decision can be made, combined with the greatest technical benefit. For parameters F and G, higher grades with corresponding prices are listed in the table. The currency unit is to be read as $.

Table 5.17 Settings of three levels for tolerance analysis

Level 1	Level 2	Level 3
$m - \sqrt{\frac{3}{2}} \cdot \sigma$	m	$m + \sqrt{\frac{3}{2}} \cdot \sigma$

Table 5.18 Tolerance design

Parameter	Levels		
	1	2	3
A: (Error column)	–	–	
B: Etching time (s)	9.8	10	10.2
C: Activation time (min, s)	6′50″	7	7′10″
D: Speed of rotation (rpm)	5.98	6	6.02
E: Salt concentration (g/l)	29.4	30	30.6
F: Concentration ratio of 2 components	0.78/1.22	0.8/1.2	0.82/1.18
G: Temperature (°C)	69.2	70	70.8
H: Flow rate (l/h)	0.49	0.50	0.51

Table 5.19 Test data after carrying out the experiments

Test #	1	2	3	4	5	6	7	8	9
Average layer thickness $\bar{y}[\mu]$	112	200	303	271	147	161	246	249	134
Test #	10	11	12	13	14	15	16	17	18
Average layer thickness $\bar{y}[\mu]$	212	284	123	245	122	264	144	142	241

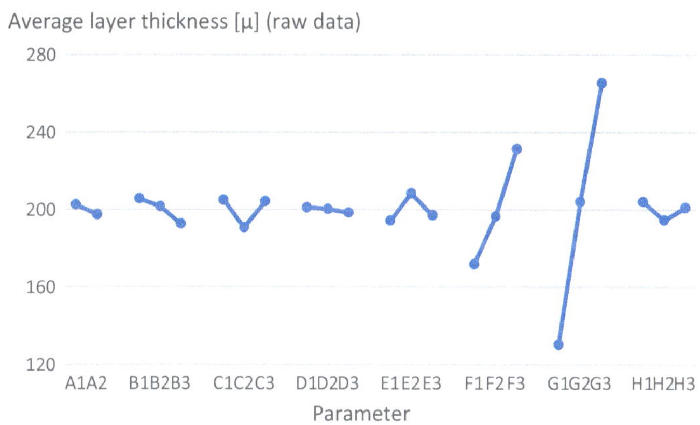

Fig. 5.18 Parameter effects of tolerance analysis

Assuming an annual production of 250,000 units on one machine, the costs can be allocated to the production unit (Table 5.22).

The grades allow 12 possible combinations. With the linear effects of the parameters, the range of deviation of the process is reduced according to their effective portion.

Table 5.20 Data evaluation with component decomposition of parameter effects

Parameter	Effects	Level 2 linear interpolated	Decomposition of the effects	Linear and quadratic components	Relative effective portion
*A*1	202.56		*A*, linear	26.12	0.001161
*A*2	197.44				
*B*1	205.67	199.17	*B*, linear	169.00	0.007510
*B*2	201.67		*B*, quadratic	6.25	0.000278
*B*3	192.67				
*C*1	205.00	204.67	*C*, linear	0.44	0.000020
*C*2	190.67		*C*, quadratic	196	0.008710
*C*3	204.33				
*D*1	201.17	199.83	*D*, linear	7.11	0.000316
*D*2	200.33		*D*, quadratic	0.25	0.000011
*D*3	198.50				
*E*1	194.33	195.75	*E*, linear	8.03	0.000357
*E*2	208.50		*E*, quadratic	162.56	0.007224
*E*3	197.17				
*F*1	172.00	201.67	*F*, linear	3520.44	0.156441
*F*2	196.67		*F*, quadratic	25.00	0.001111
*F*3	231.33				
*G*1	130.33	197.92	*G*, linear	18270.03	0.811882
*G*2	204.17		*G*, quadratic	39.06	0.001736
*G*3	265.50				
*H*1	204.17	202.67	*H*, linear	9.00	0.000400
*H*2	194.67		*H*, quadratic	64.00	0.002844
*H*3	201.17				
			Sum	22503.30	1

Table 5.21 Selected parameters for tolerance analysis

Parameter	Effective portion
F, linear component	0.16
G, linear component	0.81
Total error	0.03

$$\sigma^2_{\text{new}} = \sigma^2_{\text{existing}}\left(\left(\frac{\Delta_{F,\text{new}}}{\Delta_{F,\text{existing}}}\right)^2 \cdot F,\ \text{linear} + \left(\frac{\Delta_{G,\text{new}}}{\Delta_{G,\text{existing}}}\right)^2 \cdot G,\ \text{linear} + \text{total error}\right)$$

The calculation can be extended with more parameters in the same way. The loss function, which considers the distance to specification limits, as an additional cost factor, must be added to the real costs (Sect. 4.4) (Table 5.23).

Model calculations must be carried out for all grades (Table 5.24).

Table 5.22 Investment costs for the acquisition of higher grades with their corresponding technical features

	Parameter F: weight balance			Parameter G: thermostat		
Grade	Precision	Investment costs	Based on the number of production units	Precision (°C)	Investment costs	Based on the number of production units
a: existing	±5%	0	0	±2	0	0
b: improved	±1%	50,000	0.2	±1	75,000	0.3
c: superior	±0.1%	150,000	0.6	±0.5	250,000	1
d: premium	–	–	–	±0.2	625,000	2.5

Table 5.23 Proportional composition of a total cost analysis

Total cost	Monetary loss depending on the squared deviation + investment
Total cost/unit	Production costs/(Δ specification)^2 $\cdot \sigma^2$ + Investment cost/unit
Total cost/unit	50\$/100 $\mu^2 \cdot \sigma^2$ + Investment (\$)/250,000

The total cost as a function of the technical perfection is easier to recognize in the graphical illustration (Fig. 5.19).

The graph and table show the greatest effect with the least effort. Even with the improved grade for both parameters, the new specification of 200 μ ± 5 μ is met at the lowest investment costs (Table 5.25).

The starting point for a tolerance analysis is always the use of the most cost-effective equipment (material, measuring method, etc.). The analysis shows where improvements, i.e., investments, need to be made. This makes the process more expensive and consequently the manufacturing costs, which turns in the product price. On the other hand, carrying out a process with high precision in all parameters could undermine competitiveness.

5.6 Tolerances in Dynamic Systems

In most cases, tolerance analyses are carried out in static systems but are not limited to these. In a dynamic system, the sensitivity with respect to tolerances can be different for each signal value, i.e., a separate analysis must be performed for each signal value. This is not only cumbersome, but the result is also not obvious for determining the best compromise for all signal values. Therefore, it is better to keep the dynamic design with the same parameter setting values as for the static analysis.

Table 5.24 Cost–benefit ratios derived from tolerance analysis

Parameter *F* Weight balance Precision	Parameter *G* Thermostat Precision	Squared standard deviation $\sigma^2[\mu^2]$	Cost *F* per unit	Cost *G* per unit	Sum invest per unit	Cost within specification limits Loss function	Total cost per unit	$\sigma[\mu]$	$3\sigma[\mu]$
Original process		34.94	0	0	0	17.47	17.47	5.91	17.73
0.05	2	9.13	0	0	0	4.56	4.56	3.02	9.06
0.05	1	3.57	0	0.3	0.3	1.78	2.08	1.89	5.67
0.05	0.5	2.18	0	1	1	1.09	2.09	1.48	4.43
0.05	0.2	1.79	0	2.5	2.5	0.90	3.40	1.34	4.01
0.01	2	7.760	0.2	0	0.2	3.88	4.08	2.79	8.36
0.01	1	2.20	0.2	0.3	0.5	1.10	1.60	1.48	4.45
0.01	0.5	0.81	0.2	1	1.2	0.40	1.60	0.90	2.70
0.01	0.2	0.42	0.2	2.5	2.7	0.21	2.91	0.65	1.95
0.001	2	7.70	0.6	0	0.6	3.85	4.45	2.77	8.32
0.001	1	2.14	0.6	0.3	0.9	1.07	1.97	1.46	4.39
0.001	0.5	0.75	0.6	1	1.6	0.38	1.98	0.87	2.60
0.001	0.2	0.36	0.6	2.5	3.1	0.18	3.28	0.60	1.81

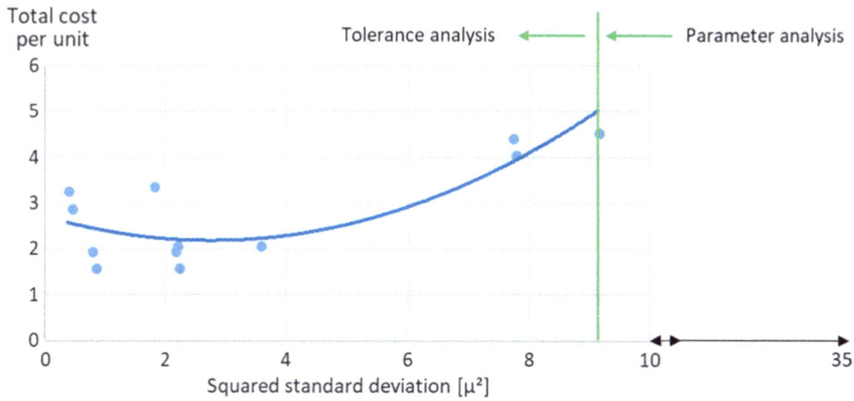

Fig. 5.19 Graphical representation of the cost–benefit relation, i.e., total costs as a function of the squared standard deviation, for determining the minimum cost

Table 5.25 Result of the tolerance analysis

Parameter *F* weight balance Precision	Parameter *G* thermostat Precision	Squared standard deviation $\sigma^2 [\mu^2]$	Cost *F* per unit	Cost *G* per unit	Sum Invest per unit	Cost within specification limits loss function	Total cost per unit	$\sigma[\mu]$	$3\sigma[\mu]$
0.01	1	2.20	0.2	0.3	0.5	1.10	1.60	1.48	4.45

The best possible setting found after the previous parameter analysis is set as level 2, the setting values 1 and 3 at intervals of $\pm 1.22\sigma$. The same orthogonal array is to be used as a test design for all signal values. For each test, the slope β is determined. This must be compared with the best possible numerical value from the result of the parameter analysis, ideally the numerical value one. Only the difference between the individual slope and the previously obtained slope is evaluated. The numerical values are grouped around zero, either in the positive or negative range. Although a dynamic system is available, the evaluation is carried out with the static *S/N* ratio of the squared distance in relation to the average value zero. Without repeating the measurement, it is sufficient to observe the quadratic distance from zero.

$$\begin{aligned}&\text{Result}\quad \Delta = \beta_{i,\text{tolerance design}} - \beta_{\text{parameter analysis}}, \qquad i: \text{Test}\,\#\\&\text{If}\quad \beta_{\text{parameter analysis}} = 1\\&\text{follows}\quad \Delta = \beta_{i,\text{tolerance design}} - 1\end{aligned}$$

$$\text{Evaluation}\quad \frac{S}{N} = \frac{\Delta^2}{\beta_0^2}$$

The parameter effects are calculated from the orthogonal array, whereby only the linear component is decisive for an effective narrowing of the critical process/

product characteristics. To reduce the tolerances over the entire signal range, the parameters with the highest amplitudes must be used.

5.7 Tolerances in Nonlinear Systems

Even if a tolerance analysis is more demanding, the application possibilities are versatile. For example, the case study analyzing a nonlinear system (Sect. 4.2) to achieve the highest *SNR* index shows opposite direction with some parameter effects with respect to the linear coefficient $\beta 1$. Since the parameter configuration has top priority to achieve a maximum *SNR* index, sensitivity and linearity are not independently adjustable. A way out of this dilemma is to carry out a tolerance analysis. The settings for the largest *SNR* index are retained and assigned to the medium parameter level. The tolerance analysis reveals the effects that provide the greatest contribution to the adjustment of coefficients $\beta 1 \rightarrow 1$ and $\beta 2 \rightarrow 0$, thus maintaining the *SNR* indices as close as possible to their ideal values.

The implementation requires higher grades for selected parameters, with which decisive improvements in sensitivity and linearity can be achieved. Overall, dealing with tolerances is relatively complex, so that, whenever possible, simulation calculations should preferably be carried out.

5.8 Tolerances for Coupled Parameters

Since the spreading of the parameter settings in a tolerance design is relatively small and equidistant, the effects are usually linear or slightly curved. However, if pairs of variables are strongly coupled, the effect analysis shows extremely nonlinear behavior, recognizable at a minimum or maximum value at the position of the nominal value or also at the size of the error amplitude, if one or more free columns are available and are interpreted. It is not possible to calculate costs and benefits based on linear tolerance reduction. Using an additional orthogonal array, decoupling and a clear interpretation of appropriate tolerance specifications can be achieved. In the usual first orthogonal array, the variables assume the role of the noise factors, a second orthogonal array changes the tolerance specifications and evaluates the resulting dispersion with the *SNR* index.

$$SNR = 10 \log\left(\frac{y_0^2}{\sigma^2}\right)$$

If the *SNR* index remains practically constant for different tolerance specifications, this variable cannot be used to tighten specifications. The effort is only worthwhile if a strong dependency of the *SNR* index occurs **(1-3)**. For an easy understanding, the procedure is outlined using the smallest orthogonal array with

Table 5.26 Definition of the setting values for a tolerance design with three parameters

Parameter	Levels		
	1	2	3
A	$-\sqrt{\frac{3}{2}}\sigma_A$	Nominal value N_A	$+\sqrt{\frac{3}{2}}\sigma_A$
B	$-\sqrt{\frac{3}{2}}\sigma_B$	Nominal value N_B	$+\sqrt{\frac{3}{2}}\sigma_B$
C	$-\sqrt{\frac{3}{2}}\sigma_C$	Nominal value N_C	$+\sqrt{\frac{3}{2}}\sigma_C$

Table 5.27 Tolerance design

Test #	A	B	C	D	y_i
1	$-\sqrt{\frac{3}{2}}\sigma_A$	$-\sqrt{\frac{3}{2}}\sigma_B$	$-\sqrt{\frac{3}{2}}\sigma_C$	–	
2	$-\sqrt{\frac{3}{2}}\sigma_A$	N_B	N_C	–	
3	$-\sqrt{\frac{3}{2}}\sigma_A$	$+\sqrt{\frac{3}{2}}\sigma_B$	$+\sqrt{\frac{3}{2}}\sigma_C$	–	
4	N_A	$-\sqrt{\frac{3}{2}}\sigma_B$	N_C	–	
5	N_A	N_B	$+\sqrt{\frac{3}{2}}\sigma_C$	–	
6	N_A	$+\sqrt{\frac{3}{2}}\sigma_B$	$-\sqrt{\frac{3}{2}}\sigma_C$	–	
7	$+\sqrt{\frac{3}{2}}\sigma_A$	$-\sqrt{\frac{3}{2}}\sigma_B$	$+\sqrt{\frac{3}{2}}\sigma_C$	–	
8	$+\sqrt{\frac{3}{2}}\sigma_A$	N_B	$-\sqrt{\frac{3}{2}}\sigma_C$	–	
9	$+\sqrt{\frac{3}{2}}\sigma_A$	$+\sqrt{\frac{3}{2}}\sigma_B$	N_C	–	

three parameters. For more parameters, correspondingly larger orthogonal arrays must be applied. For the best possible performance of an intended function, the nominal values are determined by means of a parameter analysis. The nominal values plus/minus the tolerance specifications result in the setting values for the tolerance design (Table 5.26).

An orthogonal array L9 is used as a test design with three levels (Table 5.27).

With no more than three variables, the fourth column serves for an error calculation, which occurs with an amplitude like the parameter effects if interactions cannot be neglected. The parameters are decoupled by reference to the tolerance itself, which is specified as a variable in a second orthogonal array. If the setting is limited to two values, the dispersion value can be halved, for example, to reduce specification limits (Table 5.28).

If there are no more than three variables with two setting values each, the orthogonal array L4 is sufficient in size. With three setting values each, nine noise levels of the original standard tolerance design are to be superimposed (Table 5.29).

Table 5.28 Parameter-specific definition of tolerances

Parameter	Levels	
	1	2
A	σ_A	$\sigma_A/2$
B	σ_B	$\sigma_B/2$
C	σ_C	$\sigma_C/2$

As an example, the evaluation shown in the figure is assumed (Fig. 5.20).

Parameter *A* appears relatively insensitive to a halving of the tolerance. Accordingly, its qualitative upgrading does not pay off. Parameter *B* reacts most strongly of all, i.e., an investment in limiting specification limits will be noticeable in the product function. Parameter *C* can only be used if the function improvement achieved is insufficient.

Instead of halving the tolerances, any specifications are permissible, so that the reverse question can also be answered as to how far specifications can be opened without jeopardizing the function of the product.

5.9 Safety Factor

The loss function should always be used when technical features or specifications are related to costs. Process or production costs should not only be seen in terms of improving functionality but also in terms of avoiding loss in the event of functional failure. The loss is caused by a deviation of the ideal value of a mostly technical quantity.

$$L = \frac{A_0}{\Delta_0^2}(y - m)^2$$

A_0 Total product cost.
Δ_0 Limit value for guaranteeing a function.
y Actual characteristic value.
m Target (ideal value).

Deviations of a certain characteristic feature in the production process are at the most within the upper and lower limits specified by the user. A manufacturer's tolerance requires a safety distance from the limit values to be always on the safe side (Fig. 5.21).

The tolerances can be one-sided, double-sided, symmetrical, or asymmetrical. The safety factor Φ is defined as the ratio of tolerances from the user's point of view, i.e., a function to be fulfilled, and the capabilities of the manufacturing process.

Table 5.29 Overall test design with variable tolerances

				C	$-\sqrt{\frac{3}{2}}c_C$	N_C	$+\sqrt{\frac{3}{2}}\sigma_C$	N_C	$+\sqrt{\frac{3}{2}}\sigma_C$	$-\sqrt{\frac{3}{2}}\sigma_C$	$+\sqrt{\frac{3}{2}}\sigma_C$	$-\sqrt{\frac{3}{2}}\sigma_C$	N_C	
				B	$-\sqrt{\frac{3}{2}}c_B$	N_B	$+\sqrt{\frac{3}{2}}\sigma_B$	$-\sqrt{\frac{3}{2}}\sigma_B$	N_B	$+\sqrt{\frac{3}{2}}\sigma_B$	$-\sqrt{\frac{3}{2}}\sigma_B$	N_B	$+\sqrt{\frac{3}{2}}\sigma_B$	
				A	$-\sqrt{\frac{3}{2}}c_A$	$-\sqrt{\frac{3}{2}}\sigma_A$	$-\sqrt{\frac{3}{2}}\sigma_A$	N_A	N_A	N_A	$+\sqrt{\frac{3}{2}}\sigma_A$	$+\sqrt{\frac{3}{2}}\sigma_A$	$+\sqrt{\frac{3}{2}}\sigma_A$	
				Test	1	2	3	4	5	6	7	8	9	
Test	A	B	C	Simulation calculations										*SNR*
1	σ_A	σ_B	σ_C		y_{11}	y_{12}	y_{13}	y_{14}	y_{15}	y_{16}	y_{17}	y_{18}	y_{19}	
2	σ_A	$\sigma_B/2$	$\sigma_C/2$		y_{21}	y_{22}	y_{23}	y_{24}	y_{25}	y_{26}	y_{27}	y_{28}	y_{29}	
3	$\sigma_A/2$	σ_B	$\sigma_C/2$		y_{31}	y_{32}	y_{33}	y_{34}	y_{35}	y_{36}	y_{37}	y_{38}	y_{39}	
4	$\sigma_A/2$	$\sigma_B/2$	σ_C		y_{41}	y_{42}	y_{43}	y_{44}	y_{45}	y_{46}	y_{47}	y_{48}	y_{49}	

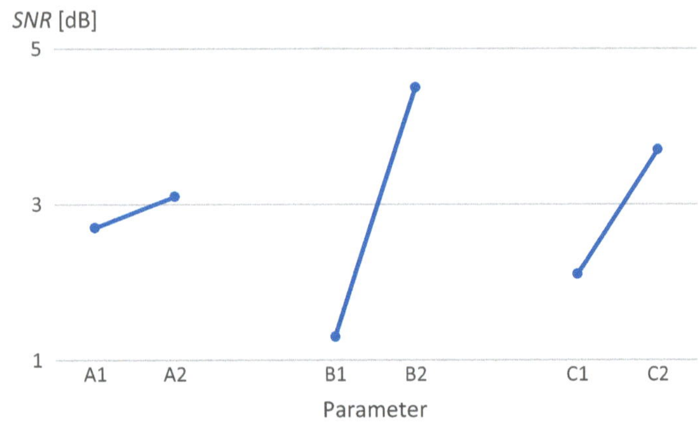

Fig. 5.20 Evaluation of a decoupled tolerance design

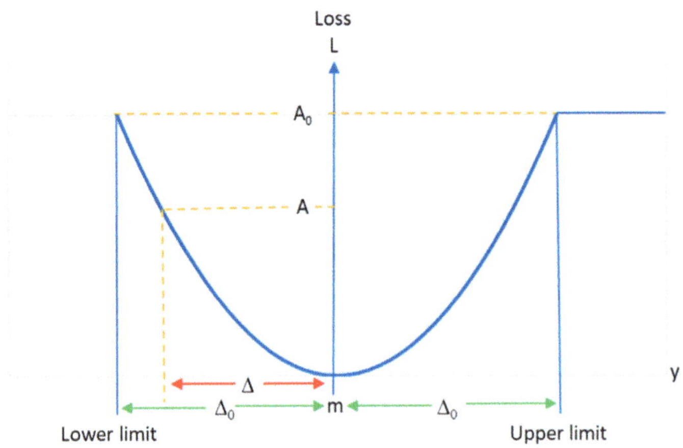

Fig. 5.21 Loss function with manufacturer and customer tolerances

$$\Phi = \frac{\Delta_0}{\Delta} \quad \text{(reciprocal definition for the situation of an } LTB \text{ key figure)}$$

The loss A of a manufactured unit at a distance Δ of the target value corresponds to the costs incurred for rework on the manufacturer side.

$$A = \frac{A_0}{\Delta_0^2}\Delta^2$$

$$\frac{\Delta_0}{\Delta} = \sqrt{\frac{A_0 = \text{Loss if functionality range is exceeded}}{A = \text{Loss if manufacturer tolerance is exceeded}}}$$

Experience shows that a safety factor of 1.05–1.1 is considered sufficient for short-lived products. For critical products with a high hazard potential, however, the safety factor must be 3–4 (load on braided steel cable for cable cars, high-pressure equipment, etc.). As a numerical example, the one-sided specification of an engine piston suitable for the bores of an engine block is given.

$$\text{Diameter} = 100\,\text{mm} \quad \Delta_+ = 0 \quad \Delta_- = 0.2\,\text{mm}$$

$$\text{Loss engine block} = 1000. - € = A_0$$

$$\text{cost of rework} = 100. - € = A$$

$$\sqrt{\frac{A_0 = 1000. - €(\text{Loss engine block})}{A = 100. - €(\text{cost of rework})}} = \frac{\Delta_0}{\Delta} = 3.16 = \Phi$$

Internal manufacturer specification for the lower tolerance limit:

$$\Delta_- = \frac{0.2\ \text{mm}}{3.16} = 60\,\mu$$

Based on the cost analysis, an internal minus tolerance of 60 μ must be specified instead of the minus tolerance of 200 μ for the cylinder supplier.

Important items to memorize Any nonlinear (and also noncontinuous) characteristic of a dynamic system can be linearized.

In addition to the dispersion and the coefficient of the linear slope component, nonlinear coefficients of a higher order can be derived.

The prerequisite for a quantitative cost–benefit analysis is the performance of a tolerance analysis.

Chapter 6
Special Applications

Contents

Abstract A few mostly numerical examples show how to deal with challenging situations. If a system is affected by several different factors, such as components, subsystems, combinations, or options, an orthogonal array is applied to use minimal data as a generally rule. With the ongoing development aiming at higher system complexity, both, soft- and hardware fall within this category. It is not practicable to test all combinations. Assigning on-off situations as factors to an orthogonal array of adequate size is considerably easier and faster with unambiguous results. If a decision needs to be made it suffices to choose the result with the higher *SNR* index. For investigation and analysis, a system can be stressed to its limits if the model of prediction is confirmed. There is a chance to unravel basic technical, physical, or chemical relations to substitute empirical constants.

6.1 Identification of Defective Components

All types of machines, from clockwork to helicopters, may experience unpredictable malfunctions or failures on one unit or another, although most of the equipment is functioning properly. The result is often a tedious trial and error search, especially when the error is rare. Instead, systematically designed combinations of good and questionable components make it possible to clearly determine which parts are responsible for the deficits. The prerequisite is that at least one device is in perfect working order, as well as a device with recurrent defects such as unwanted

© Springer Nature Switzerland AG 2019

H. Ruefer, *Living Without Mathematical Statistics*,
https://doi.org/10.1007/978-3-319-99632-5_6

Table 6.1 Assignment of components to be tested of 2 machines

Component	Device with faultless system function disassembled in individual components Level 1	Device with faulty system function disassembled in individual components Level 2
A	1	2
B	1	2
C	1	2
D	1	2
E	1	2
F	1	2
G	1	2
etc.		

vibrations, which can lead to resonant oscillations followed by destruction, high noise levels, overheating or the like. The relevant parts of both devices are disassembled into their individual components. Setting level 1 defines the components of the intact device, setting level 2 stands for that of the unreliable device (Table 6.1).

The advantageous application of orthogonal arrays comes into play here. Thus, an orthogonal array L12 is sufficient for 11 or less components. For more individual parts an L16 or a correspondingly larger one can be used. Each test to be performed represents a machine, in this case composed of intact and questionable individual parts according to the default values of the orthogonal array. The function of each machine must be checked for one or more deficient characteristics, typically the failure frequency. The evaluation of the average value sets the malfunctions in relation to the components (=parameters), which in turn can be disassembled and examined more precisely for tolerances, material problems, inhomogeneity or other. Table 6.2 shows a numerical example.

Test 1 can be omitted, as these settings are identical to the intact device function. The graphic representation of the difference of the effect amplitudes facilitates the identification of defective components, in this case parts *B* and *G* (Fig. 6.1).

Components *B* and *G* must be replaced and checked critically for compliance with existing specifications before installation.

6.2 Mixtures/Solutions from Several Components

In the production of mixtures or solutions, 2 boundary conditions must be met.

1. The sum of all components (*A*, *B*, *C*, *D*, …) equals 100% (=1)

$$A + B + C + D + \ldots = 1$$

Table 6.2 Testing of 12 machines composed of 11 individual parts

Test #	*A*	*B*	*C*	*D*	*E*	*F*	*G*	*H*	*I*	*J*	*K*	Malfunctions per time
1	1	1	1	1	1	1	1	1	1	1	1	0
2	1	1	1	1	1	2	2	2	2	2	2	2
3	1	1	2	2	2	1	1	1	2	2	2	0
4	1	2	1	2	2	1	2	2	1	1	2	4
5	1	2	2	1	2	2	1	2	1	2	1	3
6	1	2	2	2	1	2	2	1	2	1	1	5
7	2	1	2	2	1	1	2	2	1	2	1	1
8	2	1	2	1	2	2	2	1	1	1	2	3
9	2	1	1	2	2	2	1	2	2	1	1	0
10	2	2	2	1	1	1	1	2	2	1	2	2
11	2	2	1	2	1	2	1	1	1	2	2	3
12	2	2	1	1	2	1	2	1	2	2	1	5

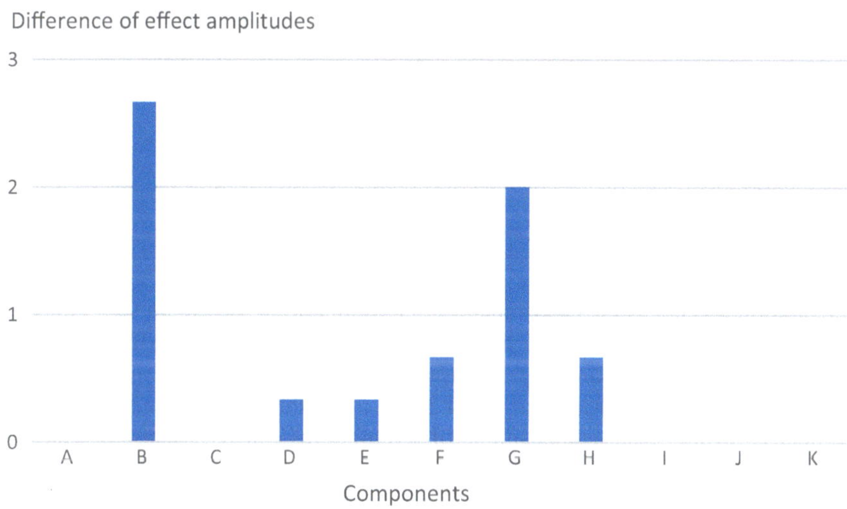

Fig. 6.1 Graphical illustration of the effect difference from setting levels 1 and 2 to identify intact and faulty components

2. The concentration of the individual components is between 0 and 100% (=1)

$$0 < A < 1$$
$$0 < B < 1$$
$$0 < C < 1$$
$$0 < D < 1 \quad \text{and so forth}$$

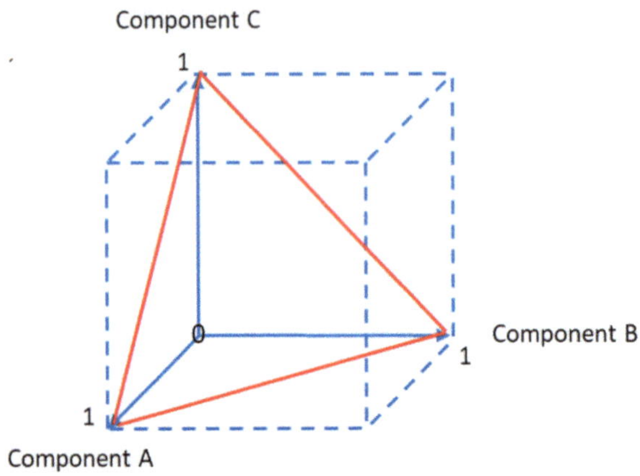

Fig. 6.2 Mixing diagram for 3 components

Table 6.3 Determination of the concentration ranges for 4 mixture components

Parameter	Level 1 Low concentration [%]	Level 2 Medium concentration [%]	Level 3 High concentration [%]
Component *A*	30	40	50
Component *B*	20	40	60
Component *C*	1	2	5
Component *D*	5	10	15

Common practice is the determination of a minimum or maximum value for one or more components to test only meaningful experimental mixtures for given reasons such as to avoid mixing gaps, explosive concentrations, etc. Compliance with all conditions is most clearly arranged if no more than 3 components must be considered. In this case, compositions of the mixtures can be visualized within a triangle, spanned as side diagonals of a 3-dimensional cube with the concentration values 0–1 for the 3 orthogonal axes (Fig. 6.2).

Although intuitively obvious, this approach is limited to a maximum of 3 components. Since 4 or far more components must be considered, e.g. the recipe of baking powder or the mixture of cement (aggregates), the use of an orthogonal array of suitable size offers a generally valid access without limitation of the number of components. Variables are the components that are assigned to the columns. As an example, an application for 4 components with 3 levels defined as settings for the concentration ranges low, medium, and high is listed (Table 6.3).

The orthogonal array L9 is sufficient for 4 parameters, the concentrations are shown in the parameter table. The totals of each row are all different from 100%. By multiplication with the reciprocal factor, the recipes for each test are gained (Tables 6.4 and 6.5).

Table 6.4 Experimental design for mixtures with correction factor

Test #	*A*	*B*	*C*	*D*	Sum	Reciprocal factor
1	30	20	1	5	56	1.7857
2	30	40	2	10	82	1.2195
3	30	60	5	15	110	0.9091
4	40	20	2	15	77	1.2987
5	40	40	5	5	90	1.1111
6	40	60	1	10	111	0.9009
7	50	20	5	10	85	1.1765
8	50	40	1	15	106	0.9434
9	50	60	2	5	117	0.8547

Table 6.5 Test formulations

Test #	*A*	*B*	*C*	*D*	Characteristic response	*SNR*
1	53.57	35.71	1.79	8.93	Such as: Rising of a baking mixture Sense of taste Strength of cement and so forth	
2	36.59	48.78	2.44	12.2		
3	27.27	54.55	4.55	13.64		
4	51.95	25.97	2.6	19.48		
5	44.44	44.44	5.56	5.56		
6	36.04	54.05	0.9	9.01		
7	58.83	23.53	5.88	11.77		
8	47.17	37.74	0.94	14.15		
9	42.74	51.28	1.71	4.27		

After the tests have been carried out and the measured values have been recorded, they are evaluated using the corresponding *SNR* index. The actual concentrations of the components are determined with the corresponding reciprocal factor from the best conditions.

6.3 Testing of Hardware and Software

Modern devices are characterized by a multitude of adjustment possibilities to meet the most diverse user requirements. In addition to cameras, photocopiers and TV sets, even washing machines have a variety of programs with additional options for temperature, spin rpm, switch-on times, etc. Even more when dealing with software, every user is free to create his requirements in the form of individual function links. Depending on the software, there is an almost incalculable number of link options, all of which should work without restrictions, without a system crash, or a connected printer which only makes illegible documents, or transmission errors in videos or images even sometimes followed by mysterious error messages, to name but a few examples. In the case of fewer variants, i.e. simple hardware, all of them can be tested for their functionality. With 10 or more related options, typically for software, the number of links and thus the test effort increases exponentially. In these situations, orthogonal arrays are advantageous to at least detect links that cause malfunctions with the smallest possible number of tests **(1-4)**. If the critical combinations are known, they can be specifically improved for both hardware and software. The basic procedure is explained using a numerical example.

The user's choices are assigned to the columns of an orthogonal array. If continuous values or a larger number range exist for a presetting, it is sufficient to limit itself to 3 setting values of the beginning, middle, and end of the scale. In the case of discrete values, 3 of them can also be taken as an example. Options in the sense of on/off, on the other hand, always have only 2 setting values. These specifications determine the size of an orthogonal array to be used, which, if appropriate, can also be adjusted with the level number. All pairwise parameter links are examined. Effects between 3 and more parameters cannot be excluded, but experience shows that these play a subordinate role. After the parameter table has been defined and assigned to the columns of a suitable orthogonal array, the tests are evaluated with the specifications:

$$\text{Function fulfilled as expected} \rightarrow 0$$

$$\text{Problem occurred, malfunction} \rightarrow 1$$

For this type of application of orthogonal arrays, pure or partial orthogonality is meaningless. The only decisive factor is the evaluation of each existing combination of pairwise parameters with all setting values. In the example, 8 choices, 1 time to 2 levels and 7 times to 3 levels, are assigned to an L18 (Tables 6.6 and 6.7).

Due to the size of the orthogonal array L18, each combination of two parameters with 3 setting values appears twice. Each variant must be found line by line and evaluated normally, i.e. the results must be added and divided by 2. On the first column is parameter A with 2 settings. Pairwise combinations with all other parameters occur 3 times, i.e. the sum of the results must be divided by 3. If a combination always causes a malfunction, the effect has the numerical value one. From the total number of 231 variants, the links with the numerical value one can be recognized very quickly in graphical or tabular form (Fig. 6.3).

If only a few malfunctions occur during the test, critical combinations can be determined more easily by comparing the configurations of the rows of the orthogonal array directly in pairs. Problems occurred in tests #3 and #10. Only those settings that appear simultaneously in both test rows can be responsible for the malfunction (Table 6.8).

All pairwise combinations of $B1$, $D3$, and $E3$ lead to the same result. As a minimum requirement, the resulting possible links must be examined in detail regarding their hardware or software function before the product can be released.

Table 6.6 Orthogonal array L18 for testing hardware or software settings

Test #	*A*	*B*	*C*	*D*	*E*	*F*	*G*	*H*	Result
1	1	1	1	1	1	1	1	1	0
2	1	1	2	2	2	2	2	2	1
3	1	1	3	3	3	3	3	3	0
4	1	2	1	1	2	2	3	3	0
5	1	2	2	2	3	3	1	1	0
6	1	2	3	3	1	1	2	2	0
7	1	3	1	2	1	3	2	3	0
8	1	3	2	3	2	1	3	1	0
9	1	3	3	1	3	2	1	2	0
10	2	1	1	3	3	2	2	1	1
11	2	1	2	1	1	3	3	2	0
12	2	1	3	2	2	1	1	3	0
13	2	2	1	2	3	1	3	2	0
14	2	2	2	3	1	2	1	3	0
15	2	2	3	1	2	3	2	1	0
16	2	3	1	3	2	3	1	2	0
17	2	3	2	1	3	1	2	3	0
18	2	3	3	2	1	2	3	1	0

Table 6.7 Number of paired parameter links in the orthogonal array L18

Paired links	Quantity
*A*1*B*1, *A*1*B*2, *A*1*B*3, *A*2*B*1, *A*2*B*2, *A*2*B*3 and corresponding *C*, *D*, *E*, *F*, *G*, *H*	42
*B*1*C*1, *B*1*C*2, *B*1*C*3, *B*2*C*1, *B*2*C*2, *B*2*C*3, *B*3*C*1, *B*3*C*2, *B*3*C*3 and *D*, *E*, *F*, *G*, *H*	54
*C*1*D*1, *C*1*D*2, *C*1*D*3, *C*2*D*1, *C*2*D*2, *C*2*D*3, *C*3*D*1, *C*3*D*2, *C*3*D*3 and *E*, *F*, *G*, *H*	45
*D*1*E*1, *D*1*E*2, *D*1*E*3, *D*2*E*1, *D*2*E*2, *D*2*E*3, *D*3*E*1, *D*3*E*2, *D*3*E*3 and *F*, *G*, *H*	36
*E*1*F*1, *E*1*F*2, *E*1*F*3, *E*2*F*1, *E*2*F*2, *E*2*F*3, *E*3*F*1, *E*3*F*2, *E*3*F*3 and *G*, *H*	27
*F*1*G*1, *F*1*G*2, *F*1*G*3, *F*2*G*1, *F*2*G*2, *F*2*G*3, *F*3*G*1, *F*3*G*2, *F*3*G*3 and *H*	18
*G*1*H*1, *G*1*H*2, *G*1*H*3, *G*2*H*1, *G*2*H*2, *G*2*H*3, *G*3*H*1, *G*3*H*2, *G*3*H*3	9
Sum	231

6.4 Dealing with Qualitative Characteristics

If no measuring method exists or no measuring instrument is available for the test evaluation, a result can still be subjectively evaluated based on experience. The description, e.g. optical gloss, a taste or tactile sample, should be assigned to as clearly distinguishable categories as possible. In the case of 10 samples, the number of assignments is typically found in characteristic classes as outlined in Table 6.9.

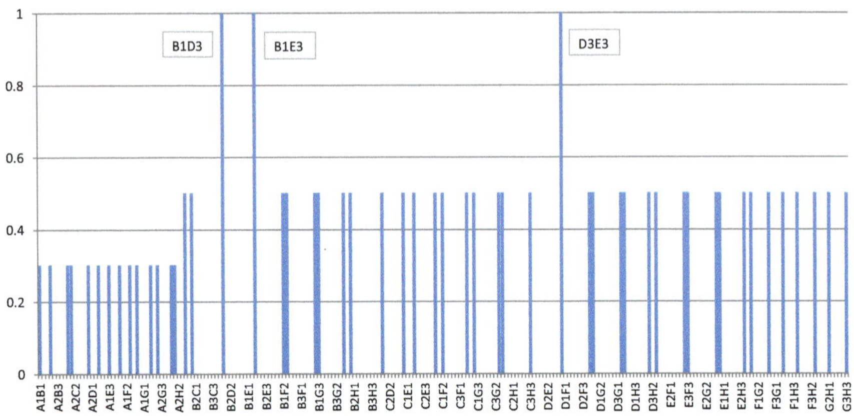

Fig. 6.3 Evaluation of all pairwise links in an orthogonal array L18

Table 6.8 Test configurations with malfunction

Test #	*A*	*B*	*C*	*D*	*E*	*F*	*G*	*H*	Result
3	1	1	3	3	3	3	3	3	1
10	2	1	1	3	3	2	2	1	1
Coincidence	–	*B*1	–	*D*3	*E*3	–	–	–	

Table 6.9 Products are divided into categories or classes according to their appearance

Quality class	Excellent	Good	Normal	Bad
Number of samples	3	1	4	2

Table 6.10 Product classes in cumulative form

Cumulated quality class	Excellent	Excellent good	Excellent good normal	Total
Number of samples	3	4	8	10

The number can decrease or increase repeatedly from class to class, depending on the assessment. Monotonous behavior can be achieved if the categories are cumulated, i.e. each subsequent category is added to the previous one (Table 6.10).

The categories or classes represent basic values on a continuous evaluation scale, which is also referred to as quasi-continuous. The more categories an overall appearance allows, the more continuous is the scale and the more reliable is the data interpretation. For comparisons, test repetitions or parameter analysis of an orthogonal array, the contents of the same cumulative classes are added and normalized with the total number of samples as a fraction or percentage (Table 6.11).

Table 6.11 Standardization of the cumulated product classes

Cumulated quality class	Excellent	Excellent good	Excellent good normal	Total
Fraction	0.3	0.4	0.8	1

The evaluation of the data is already given in this form, but not the modeling for predicting the results of a certain parameter configuration. Dealing with fractions or percentages requires the transformation to the logarithmic scale before effects may be added and subtracted. The *SNR* index to be applied is,

$$SNR = 10\log\left(\frac{p}{1-p}\right)$$

The goal is to reduce the bad fraction or rejects by selecting suitable parameters and to accumulate the largest possible sample fraction in the most valuable class. With the setting values of the largest parameter effects, all cumulated classes are modeled, transformed back and separated into the individual components of each category by successive subtraction. To confirm the model, the predictions for all categories or classes must be proven.

Exercise

Molded parts are visually assessed. Since not all appear flawless, 3 parameters, the filling position, the temperature, and the filler are each examined at 2 setting values. The orthogonal array L4 is used, whereby 20 samples are provided for each test. Can the proportion of good and acceptable parts be increased by suitable parameter settings?

Test #	*A*	*B*	*C*	Good	Acceptable	Rejects
1	1	1	1	12	5	3
2	1	2	2	14	4	2
3	2	1	2	13	3	4
4	2	2	1	15	4	1

Problem Solution

Conversion of measured values into cumulative classes

Test #	*A*	*B*	*C*	Good	Good acceptable	Total
1	1	1	1	12	17	20
2	1	2	2	14	18	20
3	2	1	2	13	16	20
4	2	2	1	15	19	20

Evaluation of the orthogonal L4 in fractions or percentages.

Parameter	Good	Good acceptable	Total
$A1$	(12 + 14)/40 = 0.65	(17 + 18)/40 = 0.875	1
$A2$	(13 + 15)/40 = 0.70	(16 + 19)/40 = 0.875	1
$B1$	0.625	0.825	1
$B2$	0.725	0.925	1
$C1$	0.675	0.90	1
$C2$	0.675	0.85	1

Result:

Settings for the largest fraction in the most valuable class: $A2$ + $B2$
Settings for largest fraction in cumulative class: $B2$ + $C1$
Advantageous setting for reducing rejects and increasing the number of units to higher classes:

$$A2 + B2 + C1$$

This parameter configuration has already been carried out as test #4, i.e. the confirmation does exist. For the sake of completeness, the modeling is carried out with the *SNR* key figures using the Omega transformation

Parameter	Good [dB]	Good acceptable [dB]
$A1$	2.69	8.45
$A2$	3.68	8.45
$B1$	2.22	6.73
$B2$	4.21	10.91
$C1$	2.84	9.54
$C2$	2.84	7.53
Average	3.08	8.60

The strongest effects originate from parameters A and B.

Class type good: $SNR = (3.68 + 4.21 - 3.08)[\text{dB}] = 4.81[\text{dB}]$
Back-transformed: 0.75
Class type good + acceptable: $SNR = (8.45 + 10.91 - 8.60)[\text{dB}] = 10.76[\text{dB}]$
Back-transformed: 0.92
The representation in each category takes place as a fraction, which is transferred to the number of pieces.

Category	Good	Acceptable	Rejects
Fraction	0.75	0.17	0.08
Pieces	15	3–4	1–2

If parameter C is added with the setting $C1$, the proportion in the class good remains at just under 0.75 but rises in the cumulative class to 0.937. Thus, the fraction in the acceptable class is 0.187 and results in 4 pieces rounded up, as confirmed with test #4.

6.5 Quantitative Modelling

To achieve certain goals such as critical product properties, yields, throughput, substance purity, detection limits or other characteristics, multiparameter experiments are performed using an orthogonal array. The data is interpreted, and an advantageous parameter configuration is chosen depending on the result. The confirmation experiment shows how well the model proposal describes reality or even deviates from it. Sometimes it is desirable to penetrate deeper into the interrelationships, i.e. to understand which laws underlie the response behavior. This can be achieved under the following conditions, if

- measured values with relatively small deviation exist
- possible interactions between all parameter pairs are negligibly small compared to the individual parameter effects
- the quadratic correlation coefficient for selecting the transfer function practically equals one
- parameter setting values and response are continuous data
- a purely orthogonal array is used, and the parameters have at least 3 setting values.

With more than 3 settings, higher-order polynomials could be adjusted, but simple correlations such as hyperbolic or power functions are preferred to explain the basic procedure. Based on an experimental parameter analysis, the conditions to be considered are explained. In the two-parameter system to be investigated, two different masses are connected with a thread running over a wheel. The side with the heavier mass moves downwards, whereby it is investigated to what extent the movement can be slowed down without standing still (Fig. 6.4).

The starting height above the ground is 1 m, where mass B (=parameter B) is the heavier mass; the time from the moment the mass B is released until it hits the ground is measured. For investigation of the parameter effects, the following specifications are made (Table 6.12):

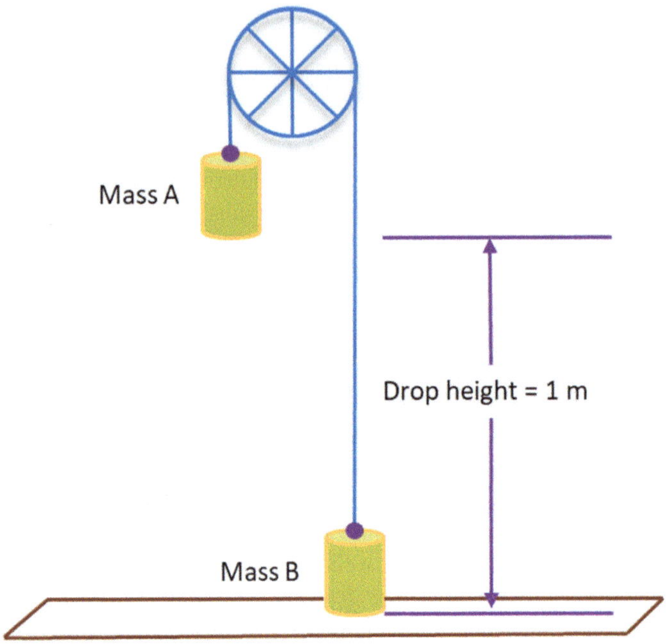

Fig. 6.4 Sketch of the system to be analyzed consisting of 2 parameters

Table 6.12 Setting values of the parameter

Parameter	Level 1	Level 2	Level 3
A: mass *A* [g]	180	200	220
B: mass *B* [g]	225	250	275

$$m_A = 200 \text{ g} \pm 20 \text{ g}$$
$$m_B = 250 \text{ g} \pm 25 \text{ g}$$

To detect possible interactions, only a purely orthogonal array can be used, whereby 2 additional parameter columns without technical assignment per parameter pair combination must be provided with 3 setting values. With 2 variables, the orthogonal array L9 with 4 columns is just large enough. The table shows the test data with the settings and the measurement results (Table 6.13).

In the orthogonal array L9, 4 columns with 3 setting values each are available. Of these, 2 technical quantities the masses *A* and *B* are assigned. If there is also an effect on the blank columns, this is due to an interaction between the existing variables. The graphical evaluation of the *SNR* (*LTB*) index is displayed in Fig. 6.5.

*A*1, *A*2, *A*3 and *B*1, *B*2, *B*3 stand for the parameter effects. From the slope behavior the fall time increases with larger mass *A* and smaller mass *B*, always

Table 6.13 Orthogonal array L9 with setting values and experimental results

Test #	Mass *A* [g]	Mass *B* [g]	Fall time [s]		*S/N*	*SNR (LTB)* [dB]
			Lowest reading	Highest reading		
1	180	225	1.35	1.36	1.836	2.639
2	180	250	1.12	1.13	1.266	1.023
3	180	275	0.98	0.99	0.97	−0.132
4	200	225	1.86	1.87	3.478	5.413
5	200	250	1.36	1.37	1.863	2.702
6	200	275	1.14	1.15	1.311	1.176
7	220	225	4.27	4.3	18.361	12.639
8	220	250	1.79	1.8	3.222	5.081
9	220	275	1.36	1.37	1.863	2.702

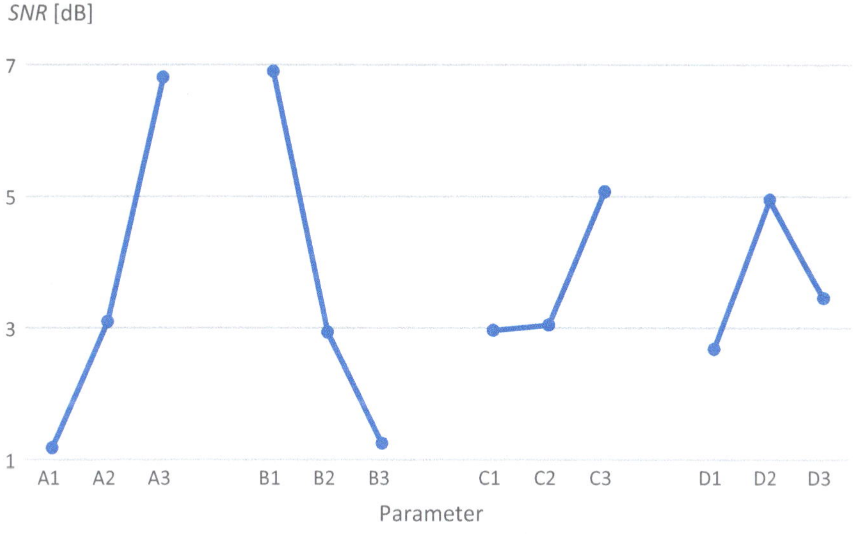

Fig. 6.5 Test evaluation with 2 variables *A* and *B* with 3 setting values each

under the condition that the mass *A* is smaller than the mass *B*. However, effects on the blank columns *C* and *D*, which indicate a hidden relationship between *A* and *B*, are also noticeable. Since the proportion of interaction is by no means negligible, a way must first be found to eliminate or at least decisively weaken it. A promising approach is to recognize the driving force in a physical (or chemical, biological) system. In the example chosen, the movement can only start when one side is heavier than the other. For this reason, the system can be described from a kinetic view point. Instead of the isolated masses on each side, one variable can be determined just as well from the mass difference, the other variable from the total

Table 6.14 New definition of the setting values for decoupling the variables

Parameter	Level 1	Level 2	Level 3
A: mass difference: Δm[g]	5	10	15
B: total mass: sum [g]	$200 + \Delta m$	$400 + \Delta m$	$600 + \Delta m$

Table 6.15 New approach of the experimental design with measurement results

Test #	*A*	*B*	Fall time [s] y	y^2/y_0^2	*SNR* [dB]
1	5	205	2.91	8.468	9.278
2	5	405	4.09	16.728	12.234
3	5	605	5.02	25.2	14.014
4	10	210	2.08	4.326	6.361
5	10	410	2.90	8.41	9.248
6	10	610	3.54	12.532	10.98
7	15	215	1.71	2.924	4.66
8	15	415	2.38	5.664	7.532
9	15	615	2.90	8.41	9.248

mass, decoupled by the mass difference. To remain in a similar range of numbers, the smallest mass difference is defined as 5 g, the total mass increasing from 200 g (Table 6.14).

Considering the respective setting value of the mass difference, parameter *B* is decoupled from parameter *A*. If—as shown in the first approach—the relative data deviation is consistently negligible ($\sigma/y < 0.01$), a repetition of the experiment can be omitted. For the equivalent *SNR* index, the square of the measured value in the logarithmic scaling is taken.

$$SNR = 10 \log\left(\frac{y^2}{y_0^2}\right); \quad y_0 = 1\left[s^2\right]$$

Table 6.15 shows the new approach with the test data.

The parameter effects as *SNR* indicators and the fall times result from the analysis of the orthogonal array L9 (Table 6.16, Fig. 6.6).

The fall time increases with decreasing mass difference and increasing total mass. The interaction columns show practically no effect, i.e. the kinetic approach describes the true nature of system behavior with now independent variables *A* and *B*.

By meeting the requirements

- negligible deviation of data,
- independent, i.e. interaction-free parameters,

the best match for a transfer function must be found. The response y is transformed back from the logarithmic representation and possible relations from the variables

Table 6.16 *SNR* key figures and response

Parameter levels	Level settings	*SNR* [dB]	Response *y* Fall time [s]
*A*1	5	11.842	3.909
*A*2	10	8.863	2.774
*A*3	15	7.146	2.277
*B*1	210	6.766	2.179
*B*2	410	9.671	3.045
*B*3	610	11.414	3.721
*C*1	–	9.263	–
*C*2	–	9.281	–
*C*3	–	9.307	–
*D*1	–	9.258	–
*D*2	–	9.291	–
*D*3	–	9.302	–

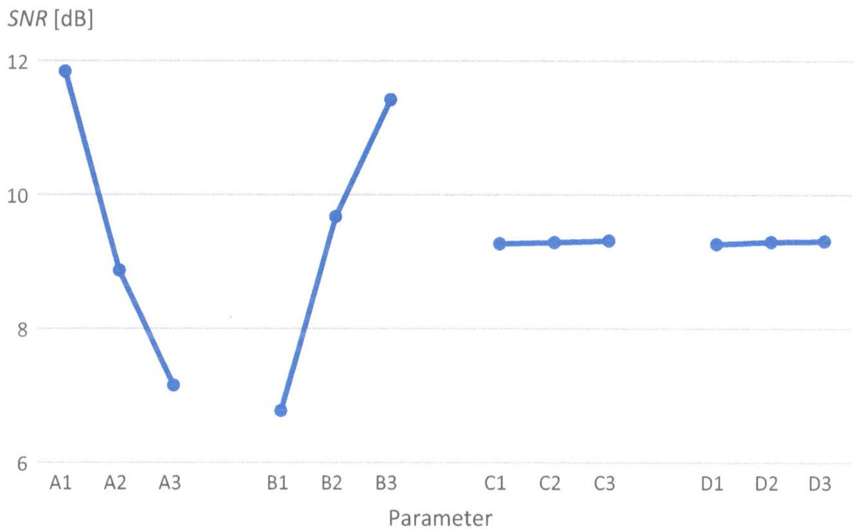

Fig. 6.6 Illustration of *SNR* key figures of the parameters *A* and *B* with interactions *C* and *D*

and the response are scrutinized via the correlation of their data. Since only 3 setting values exist, any polynomials seem to fit. However, the decisive criterion is that no deviations occur regarding the chosen functional approach, equivalent to the squared correlation coefficient in the range from,

$$0.999 < r^2 < 1$$

If a linear relationship is assumed for parameter A or an exponential approach is chosen, or a polynomial of second degree is adjusted, the correlation coefficient is in the range of,

$$r^2 = 0.95\ldots0.98$$

Only the power function

$$y \sim x_1^a \quad x_1: \text{ numerical value of the variable } A$$

with

$$a = -0.49$$

Results in a correlation coefficient $r^2 = 0.9999$

The same procedure for parameter B applies here only in the case of another power function

$$y \sim x_2^b \quad x_2: \text{ numerical value of the variable } B$$

with

$$b = 0.50$$

Results in a correlation coefficient $r^2 = 0.9999$

The variable A stands for the mass difference, the variable B for the total mass and the response y for the fall time, hereinafter referred to as t. The dependency of the fall time follows from both power functions, whereby due to unavoidable, although small measurement deviation it is assumed that the exponents may be rounded.

$$t \sim \Delta m^{-0.5}$$

$$t \sim m^{0.5}$$

Together

$$t = c\sqrt{\frac{m}{\Delta m}}$$

c = constant of proportionality

Averaged over all 9 tests, the result is $c = 0.453$ [s]

$$t = 0.453\sqrt{\frac{m}{\Delta m}}\ [\mathrm{s}]$$

Further tests must be carried out to verify the quantitative model.

An additional challenge is to reduce a found regularity to more fundamental values, from which the numerical factor of proportionality constants can be explained or even substituted. If this approach is also successful, there is no need for confirmation experiments. In the system under consideration, the side with the larger mass starts to move after it is released until it hits the ground with increasing speed. The masses are subject to accelerated movement in the gravitational field at the earth's surface. For a uniformly accelerated movement, the path-time law applies:

$$s = \frac{1}{2} b t^2$$

s path;
b acceleration;
t time.

If the squared time enters the above equation from the relation found, the following results:

$$s = \frac{1}{2} b c^2 \frac{m}{\Delta m}$$

resolved for acceleration,

$$b = \frac{2s}{c^2} \frac{\Delta m}{m}$$

For a better understanding of the coefficient, the experimental design for each test is extended by the numerical value $2s/c^2$ (Table 6.17).

Since the numerical value for $2s/c^2$ comes very close to the acceleration due to gravity at the surface of the earth, the conclusion is that the accelerations calculated from the tests are nothing else but the acceleration due to gravity, modified by the mass ratio $\Delta m/m$

Table 6.17 Calculation of the coefficient $2s/c^2$ from the proportionality constant c

Test #	A	B	Fall time y	c [s]	$2s/c^2$ [m/s^2]
1	5	205	2.91	0.454	9.70
2	5	405	4.09	0.454	9.70
3	5	605	5.02	0.456	9.62
4	10	210	2.08	0.454	9.70
5	10	410	2.90	0.453	9.75
6	10	610	3.54	0.453	9.75
7	15	215	1.71	0.452	9.79
8	15	415	2.38	0.452	9.79
9	15	615	2.90	0.453	9.75
			Average	0.453	9.75

$$g = \frac{2s}{c^2}$$

g = gravitational acceleration: 9.75...9.85 m/s^2

Fall times for any mass ratios and distances result from the law:

$$t = \sqrt{\frac{2s}{g}\frac{m}{\Delta m}}$$

By defining the parameters with causal force or energy relation, interaction-free behavior can be analyzed. Conversely, reliable and clear results from the interpretation of data can always be guaranteed if the natural laws are considered in advance or if at least plausible assumptions are incorporated.

6.6 Measurement and Time Series

The notation time series is used when a reference to the same variable is recorded over several periods. These are time periods, otherwise they are called measurement series, usually considered for any period, e.g. material behavior depending on an increasing temperature, force, voltage or similar. In case of recording repeated time periods, these can be compared with each other. If something changes from period to period, a trend can be modeled linearly or non-linearly. It is crucial to characterize appropriately a behavior over a period for the purpose of comparability. Frequently encountered time series refer to a product function, such as the service life of light bulbs, electronic components, mechanical compression, tensile or bending tests of various materials, etc. For this purpose, the test objects are subjected to test cycles until at least half of them fail **(1-38*)**.

6.6.1 Durability Test of Two Materials

For example, 2 different plastics $A1$, $A2$ from 2 suppliers (or the same plastics from 2 different production sites) are exposed to mechanical stress cycles. In principle, 2 samples each are sufficient as test objects, if one combines all negative production tolerances, the other all positive ones. Since this information or selection is usually not accessible, the use of several samples, e.g. 5 pieces or more each from different batches is used instead. If a number of 40 cycles is specified as repeatable mechanical load, corresponding to the LD50 limit, more than 40 load cycles must be applied. A sample failure depending on the load cycle is registered. If the function is maintained the record is one; zero stands for the malfunction, i.e. material fracture of the sample (Table 6.18).

Table 6.18 Samples in the life cycle test

Cycles	1	…	35	36	37	38	39	40	41	42	43	44	45	46	Life cycle
*A*1: 1	1	1	1	1	1	1	0	0	0	0	0	0	0	0	38
*A*1: 2	1	1	1	1	1	1	1	1	1	0	0	0	0	0	41
*A*1: 3	1	1	1	1	1	1	1	1	1	1	1	1	0	0	44
*A*1: 4	1	1	1	1	1	1	1	1	1	1	1	1	1	0	45
*A*1: 5	1	1	1	1	1	1	1	1	1	1	1	1	1	1	46
Survival rate *A*1	1	1	1	1	1	1	0.8	0.8	0.8	0.6	0.6	0.6	0.4	0.2	
*A*2: 1	1	1	1	1	1	0	0	0	0	0	0	0	0	0	37
*A*2: 2	1	1	1	1	1	1	1	0	0	0	0	0	0	0	39
*A*2: 3	1	1	1	1	1	1	1	1	1	0	0	0	0	0	41
*A*2: 4	1	1	1	1	1	1	1	1	1	1	1	0	0	0	43
*A*2: 5	1	1	1	1	1	1	1	1	1	1	1	1	1	0	45
Survival rate *A*2	1	1	1	1	1	0.8	0.8	0.6	0.6	0.4	0.4	0.2	0.2	0	

The horizontal reading of the data characterizes the individual life span, the vertical the survival rate. Both views are equivalent. The evaluation period begins before the first observed functional failure. No information is available before. Each load cycle could be considered as a cumulative sample class with qualitative characteristics (Sect. 5.4). Since such test series usually are not applied to shift samples to higher grades, a comparison of the average survival rate or average life cycles above the evaluation interval or, if given, at the specification limit is sufficient. Since the numerical range covers zero to one, the characteristic parameter is to be scaled as for percentages, i.e. the omega transformation is to be applied. The result appears as a difference of both test series, whereby a difference greater than 3 [dB] in the logarithmic scale stands for a clear effect (Table 6.19).

A difference greater than 3 [dB] in the applicable *SNR* indicators points to intrinsic material differences with respect to their long-term elasticity behavior.

Table 6.19 Comparative evaluation of the life cycle test

Period of evaluation 37–46 cycles	
Average survival rate	$SNR = 10\log\left(\frac{p}{1-p}\right)$
*A*1: 0.68	*A*1: 3.3 [dB]
*A*2: 0.50	*A*2: 0 [dB]
Difference	

6.6.2 Evaluation of a Medicative Therapy

Apart from material testing, the same topic is of enormous importance in the pharmacological and medical fields. An important example is the therapeutic window already mentioned (Sect 2.7). Depending on a constantly increasing dosage, the *SNR* index to be used fulfills the condition of an objective evaluation of a loss of function. Instead of a dosage, a matter may also be the assessment of a therapy over a time period **(1-5, 1-39*)**. The following clinical case study explains the procedure with the difficulties that can occur in long-term studies **(1-6)**. Under standard medical practice (double-blind testing), 44 patients with chronic hepatitis are divided into 2 groups of similar age structure and composition. The test group contains the active ingredient (prednisolone), the placebo group is used for control. The further life time is recorded in months from the beginning of the treatment (Table 6.20).

For a simplified evaluation without taking the actual patient age into account, the average survival time, normalized with the maximum recording interval, appears appropriate, although in many cases the true patient age still to be reached is generally unknown. The results of the 15-year study are as follows (Table 6.21).

A difference of more than 4 [dB] indicates a long-term effect of the therapy. For a final consideration, further time-resolved physiological parameters regarding their effect are to be included in an overall consideration.

Table 6.20 Patients' lifetime from the start of the therapy in months

Test Group											
Patient	1	2	3	4	5	6	7	8	9	10	11
Months	2	6	12	54	56[a]	68	89	96	96	125[b]	128[b]
Patient	12	13	14	15	16	17	18	19	20	21	22
Months	131[b]	140[b]	141[b]	143	145[b]	146	148[b]	162[b]	168	173[b]	181[b]
Control Group											
Patient	1	2	3	4	5	6	7	8	9	10	11
Months	2	3	4	7	10	22	28	29	32	37	40
Patient	12	13	14	15	16	17	18	19	20	21	22
Months	41	54	61	63	71	127[b]	140[b]	146[b]	158[b]	167[b]	182[b]

[a]Patient does not provide any further data
[b]Patient is living at the end of the study

Table 6.21 Results of the long-term study on drug testing

	Average lifetime in the documented period [months]	Average lifetime normalized over recording time	$SNR = 10\log\left(\frac{p}{1-p}\right)$ [dB]
Test group	109.5	0.60	1.80
Control group	64.7	0.36	−2.58
Difference			4.38

Exercise

The possibility of shortening the length of stay in the intensive care unit by means of a therapeutic approach is being investigated in a clinical case study.

The theoretically necessary time to spend in the intensive care unit is determined from the critical physiological parameter values. The test and control groups consist of 19 patients each.

Length of stay in ICU (Intensive Care Unit) in hours

Test group	Control group
16	44
48	24
74.5	44
32	44
32	72
44	44
32	44
44	44
32	44
44	44
24	32
32	96
44	24
32	32
24	32
44	32
32	48
24	32
32	44

Can this data be used to justify a therapeutic effect?

Problem Solution

An effect can be verified when the ICU time of the test group is adequately shorter than that of the control group. The time average serves as decisive response. Standardized with the longest stay of 4 days, the *SNR* index allows an objective comparison.

Evaluation of the ICU time of the clinical case study

	Average length of stay in ICU [h]	Average length of stay in ICU standardized with the maximum value	$SNR_{p(\mathrm{STB})} = 10\log\left(\frac{p-1}{p}\right)$ [dB]
Test group	36.1	0.38	2.19
Control group	43.2	0.45	0.88
Difference			1.31

The difference between the patient groups appears too small to be considered as effect of the therapy.

Important items to memorize Questions that appear to be inaccessible in a first attempt for a parameter analysis are to be considered again with the following aspects:

- increasing abstraction,
- use of a different nomenclature, i.e. not insisting to a special technical description,
- modification of the characteristics,
- use of partial orthogonal arrays to generate relatively fewer but extreme variants

Sometimes unusual ways emerge, which make a problem transparent.

Part II
Pattern Recognition and Diagnosis

Chapter 7
Dealing with Correlated Data

Contents

Abstract There is no interaction between parameters regarding systems designed for a technical purpose, or the interaction is well known and can be decoupled. In contrast, chains of chemical reactions which are controlled by enzymes with feedback mechanisms exist in the field of biology with an emphasis on humans. To analyze such a system by assigning variables to an orthogonal array can be misleading due to the numerous interactions which confound the parameter effect. Thus, the first step necessary is not to analyze parameters to obtaining the system response but to recognize the appearance of objects or patients or in a more generic way, a pattern. If slightly different patterns can be distinguished, the identification of parameters responsible for a specific pattern needs to be the second step which must be taken. The first step called pattern recognition, invented by P.C. Mahalanobis, considers all paired parameter correlations. Regarding the second step, orthogonal arrays and *SNR* indices are applicable for further analysis. By means of the *SNR* key figures, the individual parameter effect contribution to a specific pattern is thus investigated known as root-cause-analysis.

7.1 Correlated Variables

In the first part of the introduction to Genichi Taguchi's experimental methodology, a model proposal proves to be correct if there are no or only minor interactions between pairs of variables. If the amplitude of an interaction is not negligible,

© Springer Nature Switzerland AG 2019

H. Ruefer, *Living Without Mathematical Statistics*,
https://doi.org/10.1007/978-3-319-99632-5_7

measures such as a more suitable energy-related characterization variable or decoupling of the parameter setting values must be taken. However, if most of the parameters used are related to each other, a different approach must be taken.

As early as in the 1930s, the Indian mathematician C. P. Mahalanobis developed an algorithm named after him to mathematically amplify seemingly minor differences in characteristics of objects resulting from a multitude of interrelated variables **(2–1)**. In this way, the objects can be classified and, depending on the result, assigned to a predefined group or separated if necessary. The parameters responsible for the critical deviations of the objects can be analyzed as explained later. Mahalanobis' approach consists of introducing a dimension-free metric that provides a result of exactly one characteristic value for every object (also product, element or individual person) in a selected group. The metric is defined in such a way that their average has the numerical value one. With this definition, all group members form the so-called unit space. The further a product or individual is away from it with its numerical value, i.e., the greater the distance becomes, the more its properties differ from those of this unit group **(2–1*)**.

The introduction of a threshold value allows a clear distinction to be made between belonging or not belonging to the previously defined group. The advantage of such an approach is that the principle itself characterizes an almost unlimited number of parameters with exactly one result, called Mahalanobis' distance (Mahalanobis Distance = MD). The square of the Mahalanobis distance has the notation D^2. The metric of the D^2 scale is always positive greater than zero. The individual steps for calculating the squared Mahalanobis distance are introduced numerically using for the sake of comprehension an example with two parameters, which afterward can be extended to multiparameter systems. Here, the weight and height of a typical group of persons, also referred to as "normal" group, with characteristics of persons outside this standard group are considered. The parameters are not independent of each other, since the weight increases with the height, although dispersion occurs. To display numerical values of the parameters in a comparable and dimension-free manner, all data are standardized to the average value zero and the standard deviation with the numerical value one.

$$y_{i,j} = \frac{x_{i,j} - \bar{x}_j}{\sigma_j}$$

$i = 1, 2, \ldots n$, number of items (objects, individuals);
$j = 1, 2, \ldots k$, number of parameters.

In Table 7.1, the original data is complemented by the standardized values.

If, assuming a Gaussian normal distribution, each parameter is considered individually, all individuals (1–12) are included in the unit group within a deviation radius of $\pm 3\sigma$. The relative distance of the standardized values to the center describes the Euclidean distance known from geometry. Pairs of values of the individuals within the circle area do not differ from the unit group when the Euclidean distance is used (Fig. 7.1).

Table 7.1 Weight and height of the unit group (individuals 1–10) and 2 other individuals with standardized data

Individuals unit group i = 1–10	Weight [kg] ($x_{i,1}$)	Height [m] ($x_{i,2}$)	Standardized weight ($y_{i,1}$)	Standardized height ($y_{i,2}$)
1	55	1.61	−1.1286	−1.7619
2	60	1.75	−0.5699	−0.08390
3	50	1.69	−1.6873	−0.8030
4	65	1.72	−0.01117	−0.4435
5	68	1.77	0.3240	0.1558
6	62	1.66	−0.3464	−1.1626
7	75	1.81	1.1062	0.6352
8	76	1.86	1.2180	1.2345
9	61	1.82	−0.4581	0.7551
10	79	1.88	1.5532	1.4742
Average	65.10	1.757	0	0
Standard deviation	8.95	0.08343	1	1
Individuals not belonging to the unit group i = 11–12				
11	67	1.62	0.2123	−1.6420
12	77	1.73	1.3297	−0.3236

$$\sqrt{y_{i,1}^2 + y_{i,2}^2} = 3$$

If the parameters are correlated, i.e., there is a relation between weight and height, the Euclidean distance is incapable of recognizing individuals that fall out of a normal range. The consideration of the interaction, on the other hand, leads to a compression of the circle, which depends on the strength of the existing relationship. Geometrically, the circle of the Euclidean distance is transformed into an ellipse (Fig. 7.2).

The strength of a linear relationship is characterized by the correlation coefficient r, which is written for standardized variables:

$$r_{y1,y2} = \frac{\overline{y_1 \cdot y_2}}{\sigma_{y1} \cdot \sigma_{y2}} = \overline{y_1 \cdot y_2}$$

In this example, y_1 stands for the data series of standardized weights, y_2 for standardized heights. Due to the definition, the correlation coefficient is scale independent. Complete independence regarding a linear relation of the variables under consideration exists for,

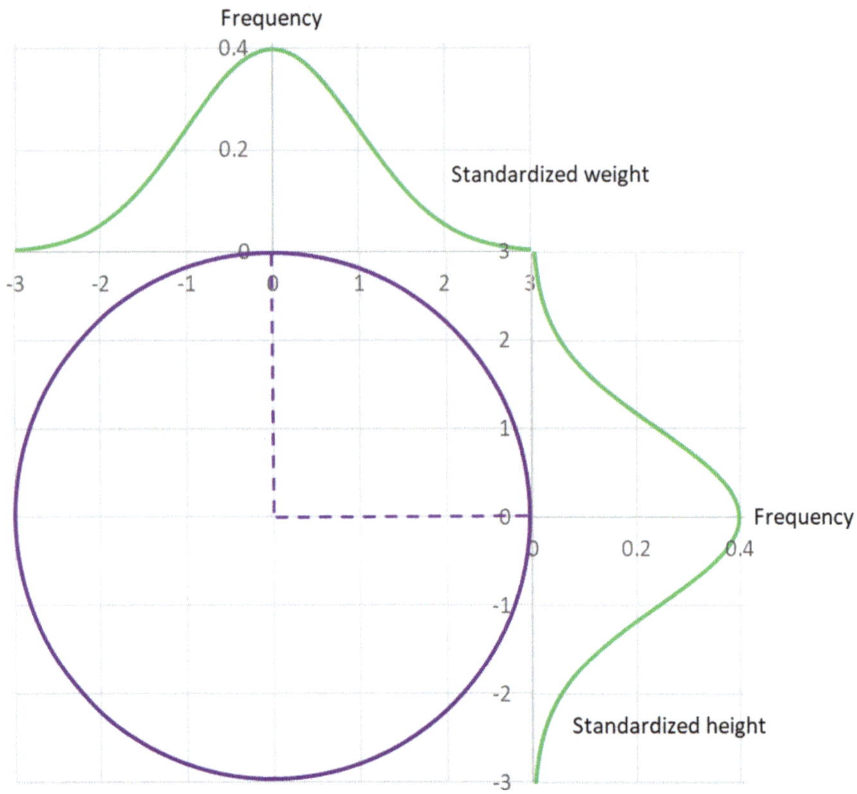

Fig. 7.1 Position of the individuals (value pairs) within a deviation circle with radius 3σ

$$r_{12} = 0$$

Parameter 1 and 2 are also called orthogonal to each other, the enclosing ellipse degenerates into a circle. If the parameters are identical except for a proportionality constant, then applies,

$$r_{12} = \pm 1$$

Parameters 1 and 2 are collinear, the enveloping ellipse degenerates into a straight line with a slope of $\pm 45°$ **(2–2*)**. With three parameters an ellipsoid is created, with four or more the corresponding higher dimensional structure, which encloses the points of the unit group, in general is called Mahalanobis space.

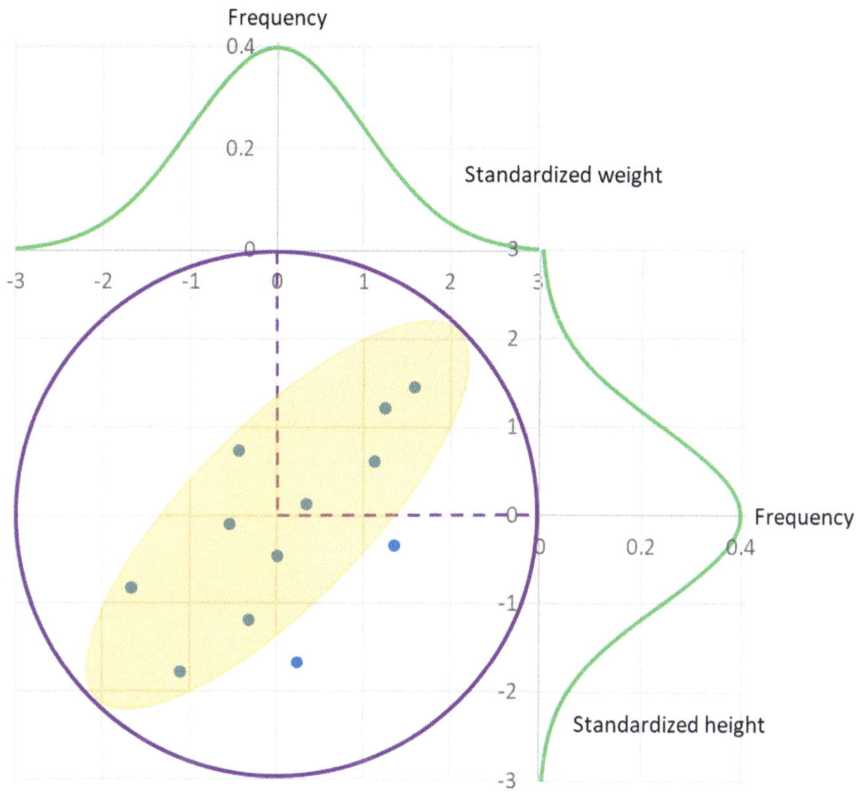

Fig. 7.2 Individuals of the unit group are located within an elliptic deviation area, others not belonging to the unit group are outside

7.2 The Mahalanobis Algorithm

Using the above example, the Mahalanobis algorithm is treated in single steps. Since the strength of the interaction is of crucial importance, the correlation coefficient plays a central role. With the data of the two parameters resulting in a value of,

$$r_{12} = 0.8$$

For multiparameter systems, it is advantageous to transfer the respective correlation coefficient from paired parameters into a matrix. Since the correlation of a parameter with itself is one, the result is always a symmetrical square matrix with ones for all diagonal elements. R stands for the correlation matrix.

$$R = \begin{pmatrix} r_{11} & r_{12} \\ r_{21} & r_{22} \end{pmatrix}$$
$$R = \begin{pmatrix} 1 & 0.8 \\ 0.8 & 1 \end{pmatrix}$$

In the next step, the inverse of the correlation matrix must be calculated, abbreviated with the symbol R^{-1}. Since the product of both, correlation matrix and its inverse, results in the unit matrix E, this allows to derive the elements a_{ij} **(2–3*)**.

$$R \cdot R^{-1} = E$$

$$\begin{pmatrix} r_{11} & r_{12} \\ r_{21} & r_{22} \end{pmatrix} \cdot \begin{pmatrix} a_{11} & a_{12} \\ a_{21} & a_{22} \end{pmatrix} = \begin{pmatrix} 1 & 0 \\ 0 & 1 \end{pmatrix}$$

$$\begin{pmatrix} 1 & 0.8 \\ 0.8 & 1 \end{pmatrix} \cdot \begin{pmatrix} 2.7773 & -2.2217 \\ -2.2217 & 2.7773 \end{pmatrix} = \begin{pmatrix} 1 & 0 \\ 0 & 1 \end{pmatrix}$$

$$R^{-1} = \begin{pmatrix} 2.7773 & -2.2217 \\ -2.2217 & 2.7773 \end{pmatrix}$$

The definition of the squared Mahalanobis distance used here is calculated as the product sum of the elements of the inverse correlation matrix and the standardized parameter values divided by the number of parameters.

$$D^2 = \frac{1}{k} \sum_{i=1}^{k} \sum_{j=1}^{k} a_{ij} y_i y_j$$

k number of parameters;
y_i, y_j standardized parameters;
a_{ij} matrix elements of R^{-1}.

For two parameters, the squared Mahalanobis distance D^2 results in,

$$D^2 = \frac{1}{2}(a_{11}y_1y_1 + a_{12}y_1y_2 + a_{21}y_2y_1 + a_{22}y_2y_2)$$

$$D^2 = \frac{1}{2}\left(a_{11}\left(y_1^2 + y_2^2\right) + 2a_{12}y_1y_2\right)$$

$$\begin{aligned} D_1^2 &= \frac{1}{2}\left(2.7773\left(1.1286^2 + 1.7619^2\right) - 2 \cdot 2.2217 \cdot (-1.1286) \cdot (-1.7619)\right) \\ &= 1.6617 \end{aligned}$$

Table 7.2 Squared Mahalanobis distance D^2 for all individuals (1–12)

Individuals of unit group i = 1–10	Standardized weight ($y_{i,1}$)	Standardized height ($y_{i,2}$)	Squared Mahalanobis distance D^2
1	−1.1286	−1.7619	1.6617
2	−0.5699	−0.08390	0.3545
3	−1.6873	−0.8030	1.8385
4	−0.01117	−0.4435	0.2623
5	0.3240	0.1558	0.06735
6	−0.3464	−1.1626	1.1489
7	1.1062	0.6352	0.6985
8	1.2180	1.2345	0.8357
9	−0.4581	0.7551	1.8518
10	1.5532	1.4742	1.2808
Average	0	0	1
Standard deviation	1	1	
Individuals not belonging to the unit group i = 11–12			
11	0.2123	−1.6420	4.5813
12	1.3297	−0.3236	3.5567

Table 7.3 Distinguishability of elements of unknown allocation by the threshold value D_T^2

Threshold $D_T^2 = 3$	Allocation
$D^2 < 3$	Elements belong to the unit group
$D^2 \geq 3$	Elements are outside from the unit group

The D^2 values of all other individuals are calculated accordingly and are added to Table 7.2.

Regarding a multidimensional extension of the parameter space, both the representation and the calculation in vector and matrix form are preferred **(2-4*)**.

$$D^2 = \frac{1}{k} Y R^{-1} Y^T$$

k number of parameters;
Y $(y_1, y_2, \ldots y_k)$ individual vector;
Y^T transposed vector;
R^{-1} inverse correlation matrix.

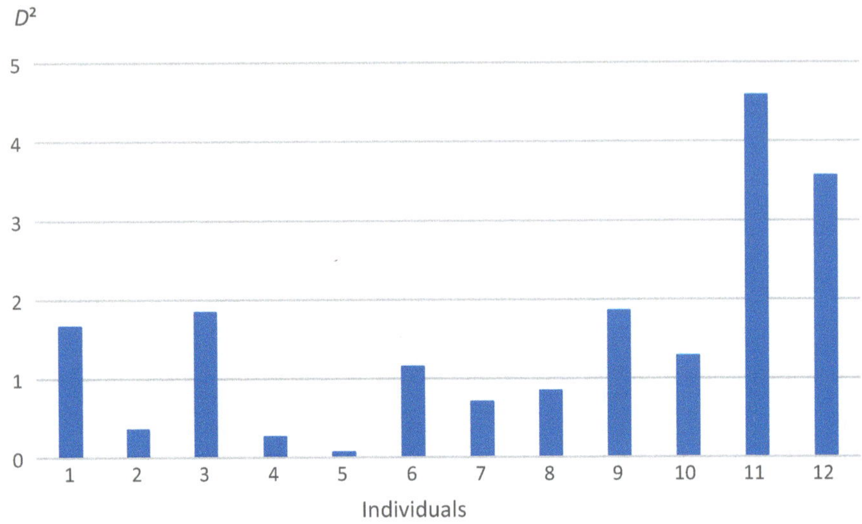

Fig. 7.3 Presentation of the squared Mahalanobis distances for all individuals (1–12)

Fig. 7.4 Task of a diagnostic system

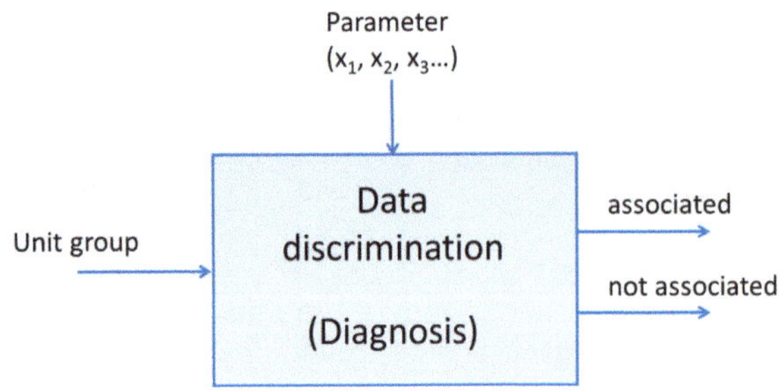

If a pattern deviates slightly from the unit group, D^2 grows strongly. The larger the numerical value, the larger are the differences in relation to the Mahalanobis space. A threshold value as a criterion can be defined for checking whether an individual belongs to the unit group or not. One way of determining the threshold is to make use of the deviation of individuals in the unit group. Even if the D^2 values of the unit group do not necessarily follow a Gaussian distribution, it can be assumed that all individuals are no further than 3σ from the center of the unit space. Since the squared standard deviation is equal to the average of the squared Mahalanobis distances, i.e., identical to one, the triple value stands for the threshold that determines the allocation of the individuals **(2–5*)** (Table 7.3).

Thus, the individuals #11 and #12 clearly stand out from the unit group (Fig. 7.3).

Table 7.4 Unit group of 34 patients and 5 individuals not belonging to the unit group characterized by clinical data of 11 parameters

Patient	*A*	*B*	*C*	*D*	*E*	*F*	*G*	*H*	*I*	*J*	*K*
1	8.8	776.6	100.99	57.756	39.716	0.4348	1557.8	118.44	12.09	21.63	184.82
2	9.5	589.67	60.182	27.356	62.457	1.3652	1204.8	148.76	15.36	34.47	254.42
3	7.2	743.3	152.87	45.211	75.799	1.1905	908.05	152.51	15.07	36.53	229.89
4	7.9	1887.4	88.76	60.173	70.325	2.1645	1361.8	167.83	22.36	33.33	205.43
5	8.3	1518.4	131.6	51.466	90.741	5.1471	1674.3	101.63	21.76	30.62	262.89
6	7.1	322.31	64.876	7.2	73.896	2.4	1132.2	221.49	19.68	42.17	177.69
7	9	1082.8	105.3	9.4203	63.86	0.6623	2251.7	187.37	16.14	32.98	218.54
8	6.7	861.75	94.009	13.825	84.653	2.4752	1802	178.71	28.22	39.17	188.94
9	6.9	647.06	101.1	36.397	64.314	1.2048	1831.4	343.53	27.45	25.49	231.97
10	7.5	706.67	120	64	53.68	1.7778	1178.6	265.48	31.60	38.49	198.41
11	8.4	827.46	54.577	33.099	78.486	0.3984	1015	257.75	27.49	25.50	215.38
12	9.1	1185.6	202.65	114.77	53.405	0.7576	905.3	229.27	26.88	27.96	218.92
13	8.4	678.57	66.429	21.509	35.094	2.6415	1346.4	241.43	25.69	29.64	146.43
14	8.6	359.26	54.815	24.815	70.8	0.4	537.04	305.6	15.20	19.20	266.98
15	8.8	725.69	70.139	29.091	91.473	1.0909	1458.2	357.75	17.45	28.00	250.91
16	9.1	606.62	48.475	20.956	56.41	1.1029	1194.9	340.44	15.02	22.79	244.07
17	7.8	875.52	63.445	19.748	59.402	0.8403	2327.7	435.9	25.64	37.82	211.04
18	9.7	345.91	37.107	9.7484	47.687	1.0601	974.84	193.29	7.47	20.44	236.75
19	8.9	308.77	38.511	11.93	65.942	1.0526	757.89	255.09	10.87	28.26	197.41
20	8.5	488.81	65.085	22.034	67.6	0.8	1417.9	282.84	14.00	30.80	145.96
21	6.9	896.15	125.38	59.048	81.498	4.4053	1630.8	430.95	18.06	59.91	234.46
22	8	439.33	63.786	23.045	41.423	15.481	818.93	229.63	10.83	32.51	222.22
23	6.7	517.51	87.16	22.179	69.397	1.4778	1194.6	157.64	11.82	42.36	287.94
24	7.7	1097.3	61.868	35.019	110.17	13.983	2521.7	265.61	24.58	59.32	229.57

(continued)

Table 7.4 (continued)

Patient	*A*	*B*	*C*	*D*	*E*	*F*	*G*	*H*	*I*	*J*	*K*
25	7.5	589.35	85.171	18.631	50.439	6.1404	889.73	84.649	17.54	32.89	159.70
26	8.4	765.8	46.097	28.996	57.48	1.1858	996.05	147.64	12.60	35.18	193.31
27	8.5	530.47	88.189	42.294	54.61	0.3937	1031.5	92.913	9.93	25.53	208.66
28	9.3	1140.4	127.66	23.759	47.35	0.7092	609.93	142.55	12.72	25.89	180.85
29	7.7	636.04	107.69	74.205	71.66	1.7094	1692.3	158.97	17.81	34.82	260.68
30	8.7	1353.5	93.949	44.586	79.167	1.5152	1600	94.697	10.61	28.03	257.96
31	7.9	893.38	37.132	11.448	51.563	0.7353	861.95	106.73	13.97	27.21	202.02
32	8.4	1595.3	55.642	31.518	60.843	0.7937	1464.3	78.968	10.32	26.59	195.60
33	7.7	746.09	41.079	21.875	47.951	3.125	780.08	174.22	21.31	38.11	278.22
34	7.7	767.93	60.944	41.35	55.052	1.6878	907.17	179.4	12.20	30.80	214.59
Average	8.1559	809.02	82.432	34.072	64.245	2.4209	1289.3	209.7	17.64	32.48	218.02
Sigma	0.8001	366.19	36.355	22.035	16.084	3.3507	476.51	94.029	6.36	8.94	35.02
35	7.7	332.06	68.702	29.008	41.026	30.709	1622.1	348.47	21.79	30.53	255.73
36	7.3	419.8	73.038	32.917	71.163	17.917	744.03	129.35	17.67	136.18	220.74
37	11.6	663.23	84.88	19.259	35.688	1.0309	505.15	70.103	13.75	27.15	247.42
38	9.3	852.52	128.84	164.04	84.615	1.0791	2063.7	187.64	11.72	31.84	250.68
39	7.8	980.54	317.12	151.39	78.486	9.3385	2117.9	130.08	33.47	51.00	227.14

Table 7.5 Standardization of patient data

Patient	*A*	*B*	*C*	*D*	*E*	*F*	*G*	*H*	*I*	*J*	*K*
1	0.8051	−0.0886	0.5105	1.0748	−1.5251	−0.5927	0.5634	−0.9705	−0.8729	−1.2134	−0.948
2	1.68	−0.599	−0.612	−0.3048	−0.1112	−0.3151	−0.1774	−0.648	−0.3587	0.2222	1.0393
3	−1.1947	−0.1795	1.9376	0.5055	0.7184	−0.3672	−0.8001	−0.6082	−0.4043	0.4523	0.3388
4	−0.3198	2.9449	0.1741	1.1845	0.378	−0.0765	0.1521	−0.4453	0.7419	0.095	−0.3596
5	0.1801	1.9371	1.3524	0.7894	1.6474	0.8136	0.8079	−1.1493	0.6478	−0.2085	1.2812
6	−1.3197	−1.3291	−0.4829	−1.2195	0.6	−0.0062	−0.3297	0.1254	0.3206	1.0828	−1.1517
7	1.0551	0.7476	0.629	−1.1188	−0.024	−0.5248	2.0196	−0.2375	−0.2357	0.0557	0.015
8	−1.8197	0.144	0.3185	−0.9189	1.2689	0.0162	1.0759	−0.3295	1.6633	0.7476	−0.8303
9	−1.5697	−0.4423	0.5136	0.1055	0.0043	−0.3629	1.1376	1.4233	1.5427	−0.7819	0.3985
10	−0.8198	−0.2795	1.0334	1.3582	−0.6569	−0.1919	−0.2324	0.5932	2.1954	0.6717	−0.5598
11	0.3051	0.0504	−0.7662	−0.0442	0.8854	−0.6036	−0.5756	0.511	1.5489	−0.7811	−0.0752
12	1.18	1.0284	3.3069	3.6624	−0.674	−0.4964	−0.8059	0.2082	1.4532	−0.5061	0.0257
13	0.3051	−0.3562	−0.4402	−0.5701	−1.8125	0.0659	0.1199	0.3375	1.2661	−0.3176	−2.0442
14	0.5551	−1.2282	−0.7596	−0.4201	0.4075	−0.6031	−1.5787	1.0199	−0.3836	−1.4852	1.3981
15	0.8051	−0.2276	−0.3331	−0.2261	1.6929	−0.3969	0.3544	1.5746	−0.0291	−0.5013	0.9392
16	1.18	−0.5527	−0.934	−0.5953	−0.4871	−0.3933	−0.1982	1.3905	−0.4121	−1.0834	0.7438
17	−0.4448	0.1816	−0.5222	−0.6501	−0.3012	−0.4717	2.1792	2.4057	1.2581	0.596	−0.1993
18	1.93	−1.2647	−1.2467	−1.1039	−1.0295	−0.4061	−0.6599	−0.1745	−1.5985	−1.3465	0.5348
19	0.9301	−1.3661	−1.2081	−1.0049	0.1055	−0.4083	−1.1152	0.4827	−1.0645	−0.4722	−0.5884
20	0.4301	−0.8744	−0.4772	−0.5463	0.2086	−0.4837	0.2699	0.7778	−0.5723	−0.1883	−2.0575
21	−1.5697	0.2379	1.1815	1.1335	1.0727	0.5922	0.7166	2.3531	0.0664	3.0666	0.4696
22	−0.1948	−1.0096	−0.5129	−0.5004	−1.419	3.8977	−0.9871	0.212	−1.0702	0.0029	0.12
23	−1.8197	−0.7961	0.13	−0.5398	0.3203	−0.2814	−0.1989	−0.5537	−0.9146	1.1047	1.9965
24	−0.5698	0.7872	−0.5656	0.043	2.8553	3.4506	2.5864	0.5947	1.0907	3.0006	0.3299

(continued)

Table 7.5 (continued)

Patient	*A*	*B*	*C*	*D*	*E*	*F*	*G*	*H*	*I*	*J*	*K*
25	−0.8198	−0.5999	0.0754	−0.7008	−0.8584	1.1101	−0.8386	−1.3299	−0.015	0.0459	−1.6654
26	0.3051	−0.118	−0.9995	−0.2304	−0.4206	−0.3686	−0.6154	−0.66	−0.7926	0.3012	−0.7056
27	0.4301	−0.7607	0.1584	0.3731	−0.5991	−0.605	−0.5411	−1.242	−1.2124	−0.7773	−0.2672
28	1.43	0.905	1.2441	−0.4681	−1.0505	−0.5108	−1.4258	−0.7141	−0.7734	−0.7376	−1.0613
29	−0.5698	−0.4724	0.6948	1.8213	0.461	−0.2123	0.8457	−0.5394	0.0274	0.2609	1.2183
30	0.6801	1.4869	0.3168	0.4771	0.9277	−0.2703	0.652	−1.223	−1.1059	−0.4979	1.1406
31	−0.3198	0.2304	−1.246	−1.0268	−0.7886	−0.503	−0.8969	−1.095	−0.5769	−0.5901	−0.4568
32	0.3051	2.1472	−0.7369	−0.1159	−0.2115	−0.4856	0.3672	−1.3903	−1.1513	−0.6593	−0.6402
33	−0.5698	−0.1718	−1.1375	−0.5536	−1.0131	0.2101	−1.0687	−0.3773	0.5774	0.6295	1.7189
34	−0.5698	−0.1122	−0.591	0.3303	−0.5716	−0.2188	−0.802	−0.3222	−0.8561	−0.1881	−0.0978
Average	0	0	0	0	0	0	0	0	0	0	0
Sigma	1	1	1	1	1	1	1	1	1	1	1
35	−0.5698	−1.3025	−0.3776	−0.2299	−1.4437	8.4422	0.6985	1.4759	0.6534	−0.218	1.0767
36	−1.0698	−1.0629	−0.2584	−0.0524	0.4301	4.6246	−1.1443	−0.8545	0.0055	11.593	0.0778
37	4.3048	−0.3981	0.0673	−0.6723	−1.7756	−0.4148	−1.6456	−1.4846	−0.6108	−0.5966	0.8396
38	1.43	0.1188	1.2765	5.8986	1.2665	−0.4004	1.6251	−0.2346	−0.9305	−0.0725	0.9327
39	−0.4448	0.4684	6.4555	5.3244	0.8854	2.0645	1.7388	−0.8467	2.4885	2.0697	0.2604

With its clear and sharp distinction, the Mahalanobis algorithm is applicable for data discrimination in correlated multiparameter systems. The sketch illustrates the principle (Fig. 7.4).

7.3 Characterization of Individual Patients in a Clinical Case Study

The following sections are based on an excerpt from a clinical case study. With 11 variables, 34 patients are assessed as medically uniform, five further patients as different. All patients are facing bypass surgery. For reasons of clarity, the physiological parameters are marked consecutively from *A* through *K* **(2–6*)** (Table 7.4).

In the first step, all data is standardized with the average value and the standard deviation (sigma) of the individuals (patients) of the unit group.

Important note: Contrary to the usual mathematical correctness, no new or altered notation is introduced for the transformed variables. To maintain uniqueness, the type of transformation is always named together with the set of variables. Since the original variables are gained with the back-transformation, the retention of the original definitions avoids confusion (Table 7.5).

Starting from the correlation matrix R from the individuals of the unit group, the inverse matrix R^{-1} is calculated (Tables 7.6 and 7.7).

The squared Mahalanobis distance D^2 of all individuals is the product of the individual parameter values (vector) with the inverse correlation matrix multiplied by the vector of the individual parameter values divided by the number of variables **(2-7*)** (Table 7.8).

The graphical representation is easier to read (Fig. 7.5).

7.4 Effectiveness of Variables

One of the advantages of using the Mahalanobis algorithm is that all input variables are converted into a metric that makes the degree of deviation from a defined unit group obvious. For practical reasons, the question arises whether the number of variables in multiparameter systems can be reduced, either to lessen effort and processing time, or to exclude variables that are expensive to collect or that occur with a time delay. The contribution of the individual parameters to the result of the squared Mahalanobis distance can be analyzed by using orthogonal arrays of suitable size. By combining both, the Mahalanobis algorithm and the Taguchi method, information on parameter effects can be obtained from existing data sets without the need for additional experiments or simulation calculations. The following assignments are used to gain the variable effects from an orthogonal array (Table 7.9).

Table 7.6 Correlation matrix R of the data of the unit group

	A	*B*	*C*	*D*	*E*	*F*	*G*	*H*	*I*	*J*	*K*
A	1	0.0427	−0.1368	−0.02	−0.2871	−0.2493	−0.2132	−0.1271	−0.3935	−0.6091	−0.0197
B	0.0427	1	0.359	0.401	0.2571	0.0127	0.3514	−0.2751	0.2035	0.0858	0.0255
C	−0.1368	0.359	1	0.7338	0.1086	−0.0629	0.1135	−0.069	0.31	0.1373	0.0402
D	−0.02	0.401	0.7338	1	0.072	−0.0443	0.0529	−0.0265	0.2947	0.0722	0.1643
E	−0.2871	0.2571	0.1086	0.072	1	0.1924	0.4864	0.2089	0.2445	0.4831	0.3516
F	−0.2493	0.0127	−0.0629	−0.0443	0.1924	1	0.1835	0.0514	0.0812	0.4923	0.0419
G	−0.2132	0.3514	0.1135	0.0529	0.4864	0.1835	1	0.2694	0.365	0.4349	0.0445
H	−0.1271	−0.2751	−0.069	−0.0265	0.2089	0.0514	0.2694	1	0.3961	0.249	0.0753
I	−0.3935	0.2035	0.31	0.2947	0.2445	0.0812	0.365	0.3961	1	0.3145	−0.0974
J	−0.6091	0.0858	0.1373	0.0722	0.4831	0.4923	0.4349	0.249	0.3145	1	0.0802
K	−0.0197	0.0255	0.0402	0.1643	0.3516	0.0419	0.0445	0.0753	−0.0974	0.0802	1

Table 7.7 Inverse correlation matrix R^{-1}

	A	*B*	*C*	*D*	*E*	*F*	*G*	*H*	*I*	*J*	*K*
A	1.9215	−0.3551	0.1874	−0.2682	0.0548	−0.1092	−0.1321	−0.384	0.6907	1.1513	0.0785
B	−0.3551	1.8545	−0.0333	−0.559	−0.3554	0.0279	−0.6237	0.8188	−0.3433	−0.0048	0.0957
C	0.1874	−0.0333	2.3779	−1.6715	−0.0847	0.218	−0.0803	0.2513	−0.1677	−0.1449	0.1842
D	−0.2682	−0.559	−1.6715	2.594	0.2594	−0.0383	0.3244	−0.1659	−0.3556	−0.1302	−0.4658
E	0.0548	−0.3554	−0.0847	0.2594	1.829	0.0686	−0.4306	−0.1425	−0.077	−0.565	−0.6074
F	−0.1092	0.0279	0.218	−0.0383	0.0686	1.3762	−0.0085	0.1459	−0.0315	−0.8266	−0.0345
G	−0.1321	−0.6237	−0.0803	0.3244	−0.4306	−0.0085	1.7682	−0.3725	−0.2444	−0.4325	0.0752
H	−0.384	0.8188	0.2513	−0.1659	−0.1425	0.1459	−0.3725	1.7121	−0.7711	−0.3412	−0.1274
I	0.6907	−0.3433	−0.1677	−0.3556	−0.077	−0.0315	−0.2444	−0.7711	1.8755	0.232	0.3489
J	1.1513	−0.0048	−0.1449	−0.1302	−0.565	−0.8266	−0.4325	−0.3412	0.232	2.5995	0.1424
K	0.0785	0.0957	0.1842	−0.4658	−0.6074	−0.0345	0.0752	−0.1274	0.3489	0.1424	1.3121

Table 7.8 Squared Mahalanobis distances D^2 for the individuals 1–34 of the unit group and the individuals 35–39 outside the unit group

Patient	1	2	3	4	5	6	7	8	9	10
D^2	0.9712	1.0268	0.9026	1.176	1.0029	0.6905	1.3324	0.9395	1.2584	0.7786
Patient	11	12	13	14	15	16	17	18	19	20
D^2	0.8767	1.7286	0.8291	0.9184	0.7444	0.5282	1.2029	0.6297	0.5595	0.8442
Patient	21	22	23	24	25	26	27	28	29	30
D^2	1.8379	2.2502	1.1193	1.9922	0.7385	0.4428	0.4548	1.3245	1.0545	0.5917
Patient	31	32	33	34						
D^2	0.4695	0.9792	1.3754	0.429						
Patient	35	36	37	38	39					
D^2	10.344	25.61	4.228	7.6107	5.9413					

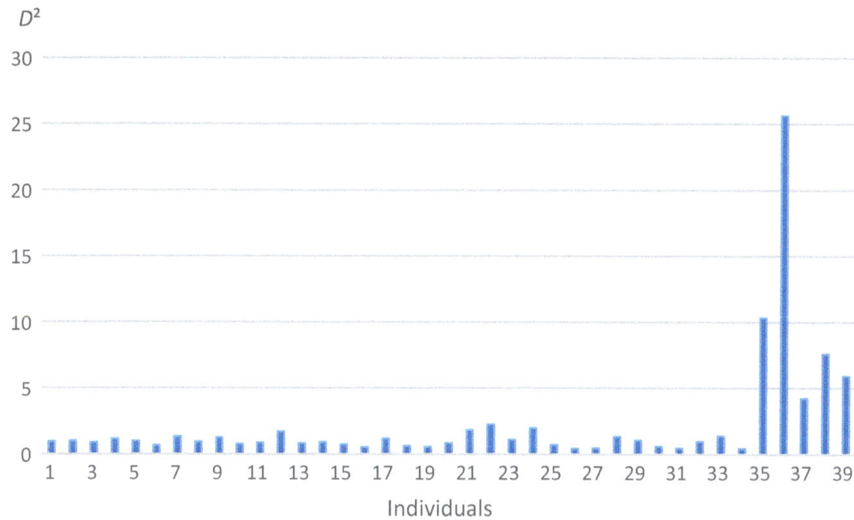

Fig. 7.5 Illustration of the squared Mahalanobis distances of the members of the unit group 1–34 and the non-associated individuals 35–39

Table 7.9 Definition of the setting values to determine the effect of the variables

Level 1	Parameter applied
Level 2	Parameter not applied

Table 7.10 Nonexistence of correlation matrices with no more than three parameters assigned to an orthogonal array L4

Test #	Parameter *A*	Parameter *B*	Parameter *C*	Correlation matrix
1	Applied	Applied	Applied	3×3
2	Applied	Not applied	Not applied	–
3	Not applied	Applied	Not applied	–
4	Not applied	Not applied	Applied	–

Table 7.11 Parameter screening according to the numerical value of their *SNR* index

$SNR_{\text{level 1}} > SNR_{\text{level 2}}$	Parameter applied
$SNR_{\text{level 1}} < SNR_{\text{level 2}}$	Parameter not applied

The number of parameters must be at least equal to four or greater, since only then correlation patterns can be calculated. For explanation, an example with too few, i.e., three parameters, is given. The smallest possible orthogonal array is an L4 consisting of three columns. The assignment of the setting values leads to the result that only test #1 can be calculated as a complete 3×3 correlation matrix. All other tests are omitted since there is no correlation with only one applied variable **(2–8*)** (Table 7.10).

The greater the distance measure D^2 of the elements that do not belong to the unit group, the clearer the data discrimination is accomplished. This can be achieved in conjunction with the Taguchi method and a suitable *SNR* index. The parameter analysis reveals which variables contribute to the squared Mahalanobis distance and which are ineffective or also weaken the distance measure D^2. The latter can be disregarded in further data analyses, so that a reduction in data and calculation effort can even be accompanied by an improvement in discrimination power. The analytical procedure of sorting out variables is also called screening (Table 7.11).

Whenever possible, the choice of a partial orthogonal array is to be preferred for parameter analysis, since interactions of parameter pairs to varying degrees may also exist when elements outside the unit group are evaluated. In the case of a purely orthogonal array, the coupling quantities of the parameter pairs appear as superposition according to the columns of the corresponding interaction table to which variables are already assigned. In the case of relatively small parameter effects, but opposite interaction relationships of parameter pairs that fall on the same column, the pure effect can be tilted, i.e., interpreted as attenuating the distance measure and sorted out. With partial orthogonal arrays, an overlapping of effect and interaction component cannot be completely ruled out, but the interactions appear only in an attenuated form due to the partial redistribution to all columns **(2–9*)**.

The risk of misinterpretation is reduced, whereby a recalculation with the reduced parameter set must be carried out to prove the results regardless of the choice of the used orthogonal array and the evaluation function. If the discrimination power between the element of the unit group with the largest distance measure and the element with the smallest distance measure outside the unit group—contrary to the results of the *SNR* indicators—is reduced, screening is not

Table 7.12 Parameters to be used to calculate the correlation matrices

Test #	*A*	*B*	*C*	*D*	*E*	*F*	*G*	*H*	*I*	*J*	*K*	Matrix
1	1	1	1	1	1	1	1	1	1	1	1	11 × 11
2	1	1	1	1	1	2	2	2	2	2	2	5 × 5
3	1	1	2	2	2	1	1	1	2	2	2	5 × 5
4	1	2	1	2	2	1	2	2	1	1	2	5 × 5
5	1	2	2	1	2	2	1	2	1	2	1	5 × 5
6	1	2	2	2	1	2	2	1	2	1	1	5 × 5
7	2	1	2	2	1	1	2	2	1	2	1	5 × 5
8	2	1	2	1	2	2	2	1	1	1	2	5 × 5
9	2	1	1	2	2	2	1	2	2	1	1	5 × 5
10	2	2	2	1	1	1	1	2	2	1	2	5 × 5
11	2	2	1	2	1	2	1	1	1	2	2	5 × 5
12	2	2	1	1	2	1	2	1	2	2	1	5 × 5

Table 7.13 Reduced standardized data set for test #2 of the orthogonal array L12

Patient	*A*	*B*	*C*	*D*	*E*
1	0.8051	−0.0886	0.5105	1.0748	−1.5251
2	1.68	−0.599	−0.612	−0.3048	−0.1112
3	−1.1947	−0.1795	1.9376	0.5055	0.7184
4	−0.3198	2.9449	0.1741	1.1845	0.378
5	0.1801	1.9371	1.3524	0.7894	1.6474
6	−1.3197	−1.3291	−0.4829	−1.2195	0.6
7	1.0551	0.7476	0.629	−1.1188	−0.024
8	−1.8197	0.144	0.3185	−0.9189	1.2689
9	−1.5697	−0.4423	0.5136	0.1055	0.0043
10	−0.8198	−0.2795	1.0334	1.3582	−0.6569
11	0.3051	0.0504	−0.7662	−0.0442	0.8854
12	1.18	1.0284	3.3069	3.6624	−0.674
13	0.3051	−0.3562	−0.4402	−0.5701	−1.8125
14	0.5551	−1.2282	−0.7596	−0.4201	0.4075
15	0.8051	−0.2276	−0.3381	−0.2261	1.6929
16	1.18	−0.5527	−0.934	−0.5953	−0.4871
17	−0.4448	0.1816	−0.5222	−0.6501	−0.3012
18	1.93	−1.2647	−1.2467	−1.1039	−1.0295
19	0.9301	−1.3661	−1.2081	−1.0049	0.1055
20	0.4301	−0.8744	−0.4772	−0.5463	0.2086
21	−1.5697	0.2379	1.1815	1.1335	1.0727
22	−0.1948	−1.0096	−0.5129	−0.5004	−1.419
23	−1.8197	−0.7961	0.13	−0.5398	0.3203
24	−0.5698	0.7872	−0.5656	0.043	2.8553
25	−0.8198	−0.5999	0.0754	−0.7008	−0.8584
26	0.3051	−0.118	−0.9995	−0.2304	−0.4206
27	0.4301	−0.7607	0.1584	0.3731	−0.5991
28	1.43	0.905	1.2441	−0.4681	−1.0505
29	−0.5698	−0.4724	0.6948	1.8213	0.461
30	0.6801	1.4869	0.3168	0.4771	0.9277
31	−0.3198	0.2304	−1.246	−1.0268	−0.7886
32	0.3051	2.1472	−0.7369	−0.1159	−0.2115
33	−0.5698	−0.1718	−1.1375	−0.5536	−1.0131
34	−0.5698	−0.1122	−0.591	0.3303	−0.5716
Average	0	0	0	0	0
Sigma	1	1	1	1	1
35	−0.5698	−1.3025	−0.3776	−0.2299	−1.4437
36	−1.0698	−1.0629	−0.2584	−0.0524	0.4301
37	4.3048	−0.3981	0.0673	−0.6723	−1.7756
38	1.43	0.1188	1.2765	5.8986	1.2665
39	−0.4448	0.4684	6.4555	5.3244	0.8854

Table 7.14 Inverse correlation matrix R^{-1} for test #2 of the orthogonal array L12

	A	*B*	*C*	*D*	*E*
A	1.1469	−0.1885	0.2867	−0.1375	0.3565
B	−0.1885	1.3114	−0.1923	−0.3637	−0.3442
C	0.2867	−0.1923	2.2694	−1.5823	−0.0008
D	−0.1375	−0.3637	−1.5823	2.2998	0.0603
E	0.3565	−0.3442	−0.0008	0.0603	1.1866

possible, and the complete set of variables must be used. The screening procedure is explained numerically step by step below.

7.5 Parameter Screening with Static Signal-to-Noise Indices

For 11 variables, the partial orthogonal array L12 is appropriate for effect analysis. The inverse correlation matrix to be calculated is determined by the setting values of the columns and their definitions according to Table 7.9. The size of the inverse correlation matrices to be calculated is displayed in Table 7.12.

Test #1 with consideration of all variables and the complete correlation matrix already exists. The further test calculations 2–12 require less effort, as some of the parameters are always omitted. Due to the diverse configurations of the remaining parameters, varied correlation patterns with different Mahalanobis distances result in the same individuals. The second test calculation is treated in detail for practice purposes, all further simulation calculations are run in the same way and are limited to the numerical result (Table 7.13).

The correlation coefficients of paired parameters are combined in the matrix form and the inverse matrix is calculated (Table 7.14).

With five variables, the squared Mahalanobis distances are calculated with (Tables 7.15, 7.16, 7.17, 7.18, 7.19, 7.20, 7.21, 7.22, 7.23, 7.24, 7.25),

$$D^2_{\text{test \#2}} = \frac{1}{5} Y R^{-1} Y^T$$

The distance measures of the individuals deviating from the unit group are summarized in the same order of the orthogonal array L12 (Table 7.26).

One possibility to evaluate the parameter effects is to interpret the squared average value of the Mahalanobis distances *D* as *SNR* index of a static system.

$$SNR = 10 \log\left(\bar{D}^2\right);$$

D is dimension-free; the average value of the unit group is one.

Table 7.15 Squared Mahalanobis distances for the individuals 1–39 for test #2 of the orthogonal array L12 with the parameters *A*, *B*, *C*, *D*, *E*

Patient	1	2	3	4	5	6	7	8	9	10
D^2	0.7943	0.7355	1.3552	2.2395	1.2802	0.9119	1.7693	1.261	0.5949	0.7764
Patient	11	12	13	14	15	16	17	18	19	20
D^2	0.4617	3.4041	0.7305	0.613	1.087	0.4245	0.2515	1.1855	0.772	0.2803
Patient	21	22	23	24	25	26	27	28	29	30
D^2	0.8361	0.5631	0.8391	1.8968	0.5959	0.3347	0.2867	2.1648	1.2819	0.6466
Patient	31	32	33	34						
D^2	0.7347	1.5784	0.7477	0.565						
Patient	35	36	37	38	39					
D^2	0.7945	0.5362	4.4052	12.867	9.7085					

Table 7.16 Squared Mahalanobis distances for the individuals 1–39 for test #3 of the orthogonal array L12 with the parameters *A*, *B*, *F*, *G*, *H*

Patient	1	2	3	4	5	6	7	8	9	10
D^2	0.7319	0.8602	0.6328	2.1444	1.0023	0.7244	1.5	0.9756	1.0366	0.2739
Patient	11	12	13	14	15	16	17	18	19	20
D^2	0.2813	0.9282	0.0785	0.9896	0.7224	0.7864	1.8245	1.1628	0.6501	0.3786
Patient	21	22	23	24	25	26	27	28	29	30
D^2	1.7102	3.7704	1.0821	3.2597	0.9821	0.1473	0.677	1.066	0.6649	0.7342
Patient	31	32	33	34						
D^2	0.4509	1.1039	0.3954	0.2714						
Patient	35	36	37	38	39					
D^2	15.385	5.2857	4.1538	1.4972	1.7175					

Table 7.17 Squared Mahalanobis distances for the individuals 1–39 for test #4 of the orthogonal array L12 with the parameters *A*, *C*, *F*, *I*, *J*

Patient	1	2	3	4	5	6	7	8	9	10
D^2	0.5001	1.2327	1.4388	0.1196	0.7699	0.561	0.5661	0.9572	1.7832	1.0199
Patient	11	12	13	14	15	16	17	18	19	20
D^2	1.185	3.0016	0.5503	0.5559	0.1832	0.4544	0.696	1.0308	0.5065	0.1623
Patient	21	22	23	24	25	26	27	28	29	30
D^2	2.5026	4.2708	1.5019	3.4797	0.5749	0.4904	0.4332	0.9241	0.1936	0.357
Patient	31	32	33	34						
D^2	0.5739	0.3654	0.5236	0.4346						
Patient	35	36	37	38	39					
D^2	20.252	40.017	5.2256	1.218	9.9911					

Table 7.18 Squared Mahalanobis distances for the individuals 1–39 for test #5 of the orthogonal array L12 with the parameters *A*, *D*, *G*, *I*, *K*

Patient	1	2	3	4	5	6	7	8	9	10
D^2	1.202	0.9195	0.7251	0.3724	0.5926	0.9174	1.5816	1.3107	0.8133	1.3392
Patient	11	12	13	14	15	16	17	18	19	20
D^2	1.0703	3.4572	1.2869	1.122	0.4319	0.5309	1.2011	1.1349	0.5724	1.0706
Patient	21	22	23	24	25	26	27	28	29	30
D^2	1.0263	0.422	1.9611	1.3632	0.9244	0.2715	0.479	0.8787	1.1368	0.8081
Patient	31	32	33	34						
D^2	0.4353	0.5988	1.5047	0.5354						
Patient	35	36	37	38	39					
D^2	0.4714	0.6263	4.8968	9.9571	6.2127					

Table 7.19 Squared Mahalanobis distances for the individuals 1–39 for test #6 of the orthogonal array L12 with the parameters A, E, H, J, K

Patient	1	2	3	4	5	6	7	8	9	10
D^2	0.6434	1.4477	0.5022	0.16	1.2572	0.7341	0.3965	1.3248	2.1141	0.4691
Patient	11	12	13	14	15	16	17	18	19	20
D^2	0.6832	0.3789	1.2491	1.4165	1.564	1.0332	1.3337	0.9294	0.3873	1.305
Patient	21	22	23	24	25	26	27	28	29	30
D^2	2.5512	0.6815	1.7774	2.9021	1.0773	0.3407	0.3613	0.726	0.4574	0.9019
Patient	31	32	33	34						
D^2	0.4476	0.506	1.713	0.2274						
Patient	35	36	37	38	39					
D^2	1.879	48.175	6.42	1.1107	1.4872					

Table 7.20 Squared Mahalanobis distances for the individuals 1–39 for test #7 of the orthogonal array L12 with the parameters *B*, *E*, *F*, *I*, *K*

Patient	1	2	3	4	5	6	7	8	9	10
D^2	0.6237	0.3365	0.2437	1.7918	1.259	1.0498	0.1946	1.0173	0.7514	1.323
Patient	11	12	13	14	15	16	17	18	19	20
D^2	0.7185	0.9684	1.5822	0.8246	0.8164	0.2841	0.4462	0.8305	0.8204	1.5646
Patient	21	22	23	24	25	26	27	28	29	30
D^2	0.2713	4.2759	1.0227	3.468	0.9298	0.2673	0.4241	0.7652	0.3819	1.1202
Patient	31	32	33	34						
D^2	0.243	1.6245	1.5791	0.1804						
Patient	35	36	37	38	39					
D^2	17.304	4.5388	1.1284	0.7359	1.9811					

Table 7.21 Squared Mahalanobis distances for the individuals 1–39 for test #8 of the orthogonal array L12 with the parameters *B*, *D*, *H*, *I*, *J*

Patient	1	2	3	4	5	6	7	8	9	10
D^2	0.8133	0.2739	0.3162	1.779	0.9403	0.9447	0.5961	1.3818	1.0931	1.393
Patient	11	12	13	14	15	16	17	18	19	20
D^2	0.8986	2.94	0.5758	0.9614	0.785	0.9684	1.547	0.8868	0.5951	0.3384
Patient	21	22	23	24	25	26	27	28	29	30
D^2	3.5218	0.4157	0.7969	1.8979	0.8403	0.2419	0.8121	0.5173	1.2223	0.877
Patient	31	32	33	34						
D^2	0.5184	1.5778	0.4035	0.2293						
Patient	35	36	37	38	39					
D^2	0.7104	32.836	0.6504	9.5472	8.1769					

Table 7.22 Squared Mahalanobis distances for the individuals 1–39 for test #9 of the orthogonal array L12 with the parameters *B*, *C*, *G*, *J*, *K*

Patient	1	2	3	4	5	6	7	8	9	10
D^2	0.9152	0.3524	1.1316	2.1615	1.2318	0.9754	1.0414	0.439	1.0855	0.4891
Patient	11	12	13	14	15	16	17	18	19	20
D^2	0.2675	2.5289	0.906	1.2744	0.3696	0.5549	1.1377	0.8403	0.638	1.0949
Patient	21	22	23	24	25	26	27	28	29	30
D^2	2.1154	0.3462	1.2066	2.4786	0.7667	0.5135	0.2871	1.2618	0.7548	0.8337
Patient	31	32	33	34						
D^2	0.6713	1.6081	1.4972	0.2237						
Patient	35	36	37	38	39					
D^2	1.0017	37.03	0.722	1.3381	9.8143					

Table 7.23 Squared Mahalanobis distances for the individuals 1–39 for test #10 of the orthogonal array L12 with the parameters *D*, *E*, *F*, *G*, *J*

Patient	1	2	3	4	5	6	7	8	9	10
D^2	1.3863	0.11	0.6188	0.3025	1.2419	0.8787	1.5024	0.6239	0.6702	0.7829
Patient	11	12	13	14	15	16	17	18	19	20
D^2	0.6796	2.9462	0.9929	1.3054	1.0486	0.3119	1.7116	0.6049	0.5807	0.1534
Patient	21	22	23	24	25	26	27	28	29	30
D^2	2.3691	4.4225	0.6459	3.6233	0.6247	0.2834	0.1884	0.4825	0.7934	0.5051
Patient	31	32	33	34						
D^2	0.4069	0.2036	0.8141	0.1845						
Patient	35	36	37	38	39					
D^2	20.163	39.797	0.9312	7.689	7.0324					

Table 7.24 Squared Mahalanobis distances for the individuals 1–39 for test #11 of the orthogonal array L12 with the parameters *C*, *E*, *G*, *H*, *I*

Patient	1	2	3	4	5	6	7	8	9	10
D^2	1.3208	0.1763	1.3595	0.277	1.2851	0.2929	1.3971	1.0659	0.8195	1.3801
Patient	11	12	13	14	15	16	17	18	19	20
D^2	1.4503	2.7695	1.4949	1.3053	1.1444	0.7232	2.3134	0.7578	0.8094	0.3335
Patient	21	22	23	24	25	26	27	28	29	30
D^2	2.0366	0.6681	0.2737	2.2744	0.5597	0.3296	0.4756	0.9519	0.3378	0.9108
Patient	31	32	33	34						
D^2	0.6915	0.7743	1.0104	0.2297						
Patient	35	36	37	38	39					
D^2	1.412	0.7382	1.0211	1.6865	8.704					

Table 7.25 Squared Mahalanobis distances for the individuals 1–39 for test #12 of the orthogonal array L12 with the parameters *C*, *D*, *F*, *H*, *K*

Patient	1	2	3	4	5	6	7	8	9	10
D^2	0.7547	0.4498	1.2289	0.6241	1.1518	0.5406	1.257	0.6594	0.5818	0.5949
Patient	11	12	13	14	15	16	17	18	19	20
D^2	0.3708	2.945	0.9078	0.7887	0.7089	0.6802	1.3177	0.4802	0.4374	1.0883
Patient	21	22	23	24	25	26	27	28	29	30
D^2	1.5547	3.0603	1.2813	2.5223	1.2745	0.5717	0.4152	1.3196	1.0623	0.6284
Patient	31	32	33	34						
D^2	0.6645	0.7395	0.9565	0.3811						
Patient	35	36	37	38	39					
D^2	14.566	4.5287	0.9531	11.106	9.7803					

Table 7.26 Summarized squared Mahalanobis distances D^2 of the screening tests of the orthogonal array L12

Patient	35	36	37	38	39
Test #	D^2	D^2	D^2	D^2	D^2
1	10.344	25.61	4.228	7.6107	5.9413
2	0.7945	0.5362	4.4052	12.867	9.7085
3	15.385	5.2857	4.1538	1.4972	1.7175
4	20.252	40.017	5.2256	1.218	9.9911
5	0.4714	0.6263	4.8968	9.9571	6.2127
6	1.879	48.175	6.42	1.1107	1.4872
7	17.304	4.5388	1.1284	0.7359	1.9811
8	0.7104	32.836	0.6504	9.5472	8.1769
9	1.0017	37.03	0.722	1.3381	9.8143
10	20.163	39.797	0.9312	7.689	7.0324
11	1.412	0.7382	1.0211	1.6865	8.704
12	14.566	4.5287	0.9531	11.106	9.7803

Table 7.27 Evaluation of the orthogonal array L12 with $SNR = 10\log(\bar{D}^2)$

Patient	35	36	37	38	39	Average	*SNR*
Test #	D	D	D	D	D	$\bar{D}$	$10\log(\bar{D}^2)$
1	3.2162	5.0607	2.0562	2.7587	2.4375	3.1059	9.8437
2	0.8914	0.7323	2.0989	3.5871	3.1158	2.0851	6.3824
3	3.9223	2.2991	2.0381	1.2236	1.3105	2.1587	6.6839
4	4.5002	6.3259	2.286	1.1036	3.1609	3.4753	10.82
5	0.6866	0.7914	2.2129	3.1555	2.4925	1.8678	5.4265
6	1.3708	6.9408	2.5338	1.0539	1.2195	2.6238	8.3785
7	4.1599	2.1305	1.0623	0.8579	1.4075	1.9236	5.6823
8	0.8428	5.7302	0.8065	3.0899	2.8595	2.6658	8.5165
9	1.0009	6.0852	0.8497	1.1568	3.1328	2.4451	7.7658
10	4.4904	6.3085	0.965	2.7729	2.6519	3.4377	10.725
11	1.1883	0.8592	1.0105	1.2986	2.9503	1.4614	3.2952
12	3.8166	2.1281	0.9763	3.3325	3.1274	2.6762	8.5502

To calculate the *SNR* index, the non-squared Mahalanobis distances must first be set up, from which the average value follows. The squared value enters the formula for the *SNR* key figure (Table 7.27).

The parameter effects are in Table 7.28 and presented in graphical form (Fig. 7.6).

Table 7.28 Parameter effect as static *SNR* index interpreted over all patients 35–39 outside the unit group

	A	*B*	*C*	*D*	*E*	*F*	*G*	*H*	*I*	*J*	*K*
Level 1	7.9225	7.4791	7.7762	8.2408	7.3846	8.7176	7.2901	7.5447	7.264	9.342	7.608
Level 2	7.4226	7.8659	7.5689	7.1043	7.9605	6.6275	8.055	7.8004	8.0811	6.0034	7.7372

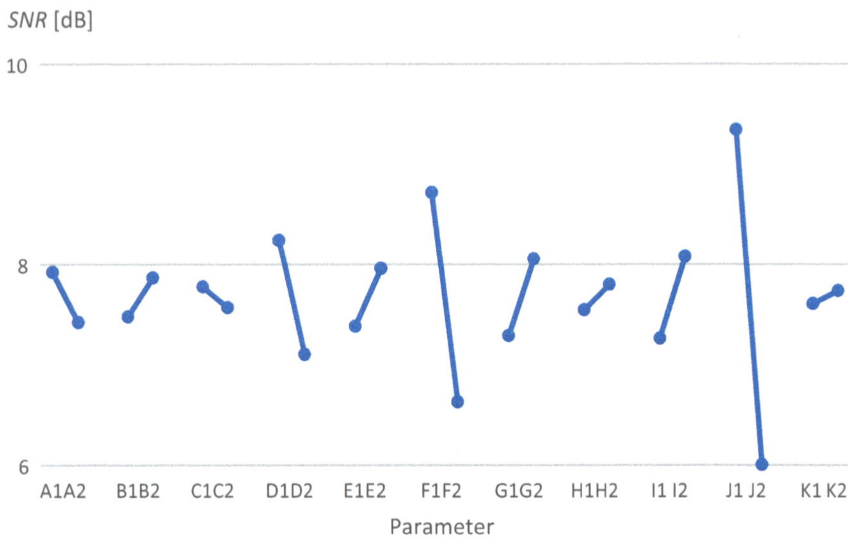

Fig. 7.6 Graphical representation of the *SNR* index of the parameter effects of the orthogonal array L12

By defining that only the parameters that increase the *SNR* index are to be included, remain the five variables *A*, *C*, *D*, *F*, and *J*. For verification, a new calculation is performed based on this selection (Table 7.29).

If the squared standard deviation is added to the squared average of the Mahalanobis distance to increase the numerical value D^2 of the individuals outside the unit group, the *SNR* index *LTB* is generated. This can cause an additional increase of the distance measures with a different variable configuration. The comparative calculation is based on the *LTB* key figure (Table 7.30).

$$SNR = LTB = 10\log\left(\frac{k}{\frac{1}{D_1^2}+\frac{1}{D_2^2}+\frac{1}{D_3^2}+\ldots+\frac{1}{D_k^2}}\right)$$

The parameter effects are presented in tabular and graphical form (Table 7.31, Fig. 7.7).

To confirm the set of variables *A*, *C*, *D*, *E*, *F*, *G*, *H*, *J*, and *K* based on the evaluation with the *LTB* index, all distance measurements are recalculated (Table 7.32).

Table 7.29 Calculation of the squared Mahalanobis distances with the variables *A*, *C*, *D*, *F*, *J* as a confirmation test

Patient	1	2	3	4	5	6	7	8	9	10
D^2	0.58	1.2156	1.2921	0.5487	0.752	0.8088	1.8143	1.1532	1.5147	0.5402
Patient	11	12	13	14	15	16	17	18	19	20
D^2	0.3682	3.2672	0.0933	0.5598	0.1539	0.4334	0.3616	0.9946	0.4697	0.1575
Patient	21	22	23	24	25	26	27	28	29	30
D^2	2.2748	4.0877	0.9167	3.369	0.7716	0.5289	0.17	1.8409	0.973	0.146
Patient	31	32	33	34						
D^2	0.5503	0.2727	0.4703	0.5491						
Patient	35	36	37	38	39					
D^2	20.046	39.489	5.6603	11.238	10.009					

Table 7.30 Evaluation of the orthogonal array L12 with the *SNR* index of type *LTB*

Patient	35	36	37	38	39	*SNR*
Test #	D^2	D^2	D^2	D^2	D^2	*LTB*
1	10.344	25.61	4.228	7.6107	5.9413	8.7164
2	0.7945	0.5362	4.4052	12.867	9.7085	1.5104
3	15.385	5.2857	4.1538	1.4972	1.7175	4.5715
4	20.252	40.017	5.2256	1.218	9.9911	6.2458
5	0.4714	0.6263	4.8968	9.9571	6.2127	0.7742
6	1.879	48.175	6.42	1.1107	1.4872	3.4076
7	17.304	4.5388	1.1284	0.7359	1.9811	2.1783
8	0.7104	32.836	0.6504	9.5472	8.1769	1.9346
9	1.0017	37.03	0.722	1.3381	9.8143	1.8581
10	20.163	39.797	0.9312	7.689	7.0324	5.4642
11	1.412	0.7382	1.0211	1.6865	8.704	1.2492
12	14.566	4.5287	0.9531	11.106	9.7803	5.14

7.6 Parameter Screening with Dynamic Signal-to-Noise Indices

A quite reliable parameter analysis for the screening of variables is in most cases achieved by a dynamic *SNR* index, since a whole range is considered. However, this presupposes that the true numerical value of the squared Mahalanobis distance is known in the form of various technical malfunctions or in the medical field from different pathological situations. Without knowledge of the real D^2 sizes, the application of a static *SNR* index is, therefore, often regarded as the simpler option. As a way out, the true but unknown D^2 values of the individuals outside the unit group can be replaced by the average value of the single Mahalanobis distance in good approximation. Averaged over every individual, a signal value is gained as an input variable, which is linearly related to the output variable. The proportionality constant is the slope, deviations around this value and in linearity are interpreted as squared standard deviation of the dynamic function. The logarithmic definition of the *SNR* key figure transforms the unit space to the origin.

$$\text{Unitspace:} \quad \bar{D} = 1$$
$$\log(1) = 0$$

Therefore, for the individuals outside the unit group, a zero-point-proportional relationship with an evaluation by the *SNR* key figure applies.

$$SNR = 10 \log\left(\frac{\beta^2}{\sigma^2} S_2\right)$$

Table 7.31 Effect size as *LTB* index of the parameters interpreted over all patients 35–39 outside the unit group

	A	*B*	*C*	*D*	*E*	*F*	*G*	*H*	*I*	*J*	*K*
Level 1	4.2043	3.4615	4.12	3.9233	3.7543	5.386	3.7723	4.1699	3.5164	4.6044	3.6791
Level 2	2.9707	3.7135	3.055	3.2517	3.4207	1.789	3.4028	3.0051	3.6586	2.5706	3.4959

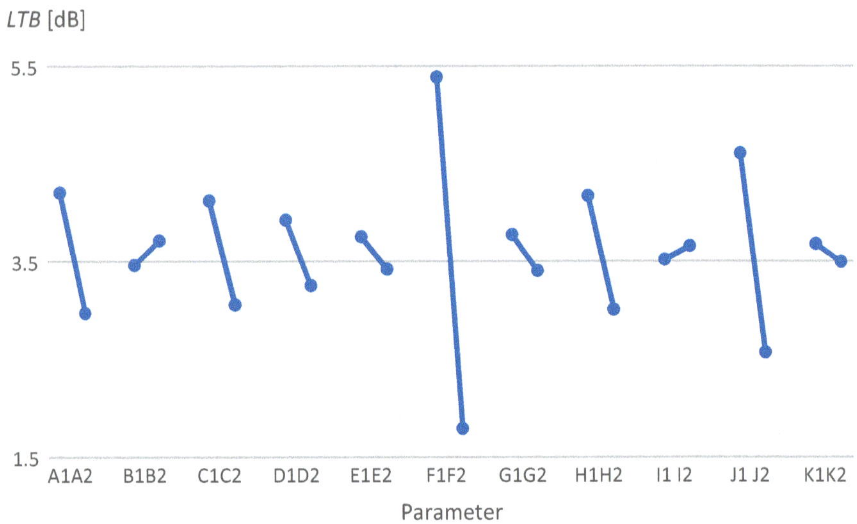

Fig. 7.7 *LTB* key figures of the parameter effects of the orthogonal array L12

Improved discrimination power can be expected if it is possible to separate the variables with the highest sensitivity coefficient (=slope β) and the smallest deviation from the variables that run in the opposite direction or have no effect at all. The table of the orthogonal array L12 of individuals outside the unit group is extended by the corresponding columns and rows to calculate the *SNR* index. For easy traceability of the results, the unsquared Mahalanobis distances are listed (Table 7.33).

The table contains the evaluation of the orthogonal array L12 with the parameter effects illustrated in the graphic (Table 7.34, Fig. 7.8)

Screening with the dynamic *SNR* key figure in this example leads to the result that each variable omitted reduces the discrimination power. In the graph, the squared Mahalanobis distances of the individuals outside the unit group are compared according to the differently defined evaluation functions (Fig. 7.9).

The results are similar, whereby the more important aspect is the highest discrimination power. For this purpose, the largest numerical value within the unit group must be compared with the smallest numerical value of the individuals outside the unit group. The greater the distance, the better the discriminability, the easier will be the decision-making for the assignment of unknown individuals (Table 7.35).

Since different correlation patterns are the consequence from differently composed sets of variables, the respective result must always be checked from the practical relevance and, if applicable, the best possible reduction of the variables is chosen, i.e., the technical expertise is paramount. If, for example, the individual with the smallest distance measure (patient no. 37) could be integrated into the unit group from a medical point of view because of minor deviations from the unit

Table 7.32 Calculation of the squared Mahalanobis distances with the variables *A*, *C*, *D*, *E*, *F*, *G*, *H*, *J*, *K* as confirmation test

Patient	1	2	3	4	5	6	7	8	9	10
D^2	0.8886	0.9342	1.0003	0.444	1.0736	0.7195	1.6246	0.9374	1.5213	0.5336
Patient	11	12	13	14	15	16	17	18	19	20
D^2	0.6168	2.0085	0.764	1.1175	0.8949	0.6224	1.3908	0.6709	0.5844	0.758
Patient	21	22	23	24	25	26	27	28	29	30
D^2	1.7094	2.6301	1.2648	2.3703	0.8534	0.5285	0.2373	1.3229	0.7985	0.5837
Patient	31	32	33	34						
D^2	0.477	0.5513	1.1917	0.3757						
Patient	35	36	37	38	39					
D^2	12.614	30.638	4.1876	6.8879	5.6845					

Table 7.33 Evaluation of the orthogonal array L12 with the dynamic *SNR* index

Patient	35	36	37	38	39	β	σ^2	*SNR*
Test #	*D*	*D*	*D*	*D*	*D*			$10 \log\left(\frac{\beta^2}{\sigma^2} S_2\right)$
1	3.2162	5.0607	2.0562	2.7587	2.4375	1.2549	0.1182	19.54
2	0.8914	0.7323	2.0989	3.5871	3.1158	0.701	2.3461	1.503
3	3.9223	2.2991	2.0381	1.2236	1.3105	0.8176	1.0961	6.1449
4	4.5002	6.3259	2.286	1.1036	3.1609	1.4524	1.1029	11.109
5	0.6866	0.7914	2.2129	3.1555	2.4925	0.6247	1.7993	1.6544
6	1.3708	6.9408	2.5338	1.0539	1.2195	1.1541	2.8246	5.028
7	4.1599	2.1305	1.0623	0.8579	1.4075	0.755	1.2901	4.7455
8	0.8428	5.7302	0.8065	3.0899	2.8595	1.1472	1.5016	7.7199
9	1.0009	6.0852	0.8497	1.1568	3.1328	1.0997	1.8191	6.5194
10	4.4904	6.3085	0.965	2.7729	2.6519	1.4552	0.8298	12.361
11	1.1883	0.8592	1.0105	1.2986	2.9503	0.5308	0.8111	3.7
12	3.8166	2.1281	0.9763	3.3325	3.1274	1.0073	1.3392	7.0868
Signal	2.5072	3.7826	1.5747	2.1159	2.4888			
S2	6.7491							

group, the evaluation with the static *SNR* index of the squared average value leads to an improved data discrimination with much less effort. It should be emphasized that each variable elimination changes the correlation pattern and as a result, new unknown individuals may no longer be correctly assigned according to the complete original pattern.

7.7 Root-Cause-Analyses

In addition to the possibility of parameter screening, the simulation calculations lead to further valuable insights. Since each element (individual, object) outside the unit group is accessible to a parameter analysis, the effect contribution of those variables can be mapped that are causally responsible for not belonging to the unit group. Each element outside the unit group differs from the others in its own way, since all conditions can be different. The value of the individual parameter analysis consists in the fact that the conspicuous parameters and their numerical values are individually identified purely by simulation calculations, i.e., without any additional experimental effort. For control functions, only these can be monitored with their critical settings. This paves the way to prevent elements from breaking out of the unit group as early as first signs develop by means of suitable countermeasures. The evaluation is based on the existing data set of the orthogonal array L12 for the parameter screening (Table 7.36).

Table 7.34 Effect amplitude as dynamic *SNR* index of parameters interpreted across all patients 35–39 outside the unit group

	A	*B*	*C*	*D*	*E*	*F*	*G*	*H*	*I*	*J*	*K*
Level 1	7.4965	7.6955	8.243	8.3109	7.813	10.165	8.32	8.2033	8.0781	10.38	7.429
Level 2	7.0221	6.8232	6.2757	6.2078	6.7057	4.3541	6.1987	6.3154	6.4405	4.1391	7.0897

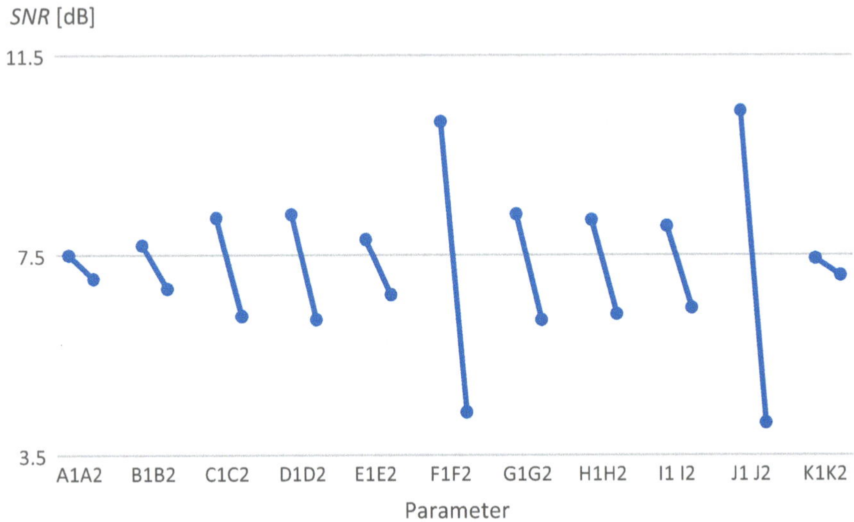

Fig. 7.8 Dynamic *SNR* ratios of the parameter effects of the orthogonal array L12

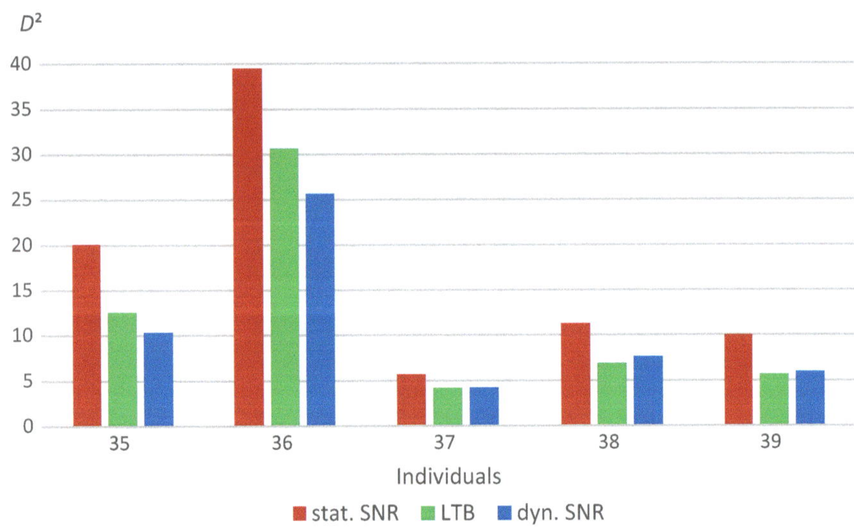

Fig. 7.9 Squared Mahalanobis distances of the individuals 35–39 outside the unit group with the evaluation functions of the static *SNR* index of the squared average of the Mahalanobis distance, the static *LTB* index, and the dynamic *SNR* index

Since exactly one set of data exists for each element outside the unit group, only the static *SNR* index of the squared Mahalanobis distance remains for evaluation (Table 7.37).

Table 7.35 Discrimination power of different *SNR* indices

Evaluation function	$SNR = 10\log(\bar{D}^2)$	$SNR = LTB$	$SNR = 10\log\left(\frac{\beta^2}{\sigma^2} S_2\right)$
Parameter	*A, C, D, F, J*	*A, C, D, E, F, G, H, J, K*	*A, B, C, D, E, F, G, H, I, J, K*
Smallest numerical value outside the unit group	5.6603	4.1876	4.228
Largest numerical value within the unit group	4.0877	2.6301	2.2502
Absolute difference	1.5726	1.5575	1.9778
Relative difference (%)	28	37	47

Table 7.36 Squared Mahalanobis Distances D^2 of the screening tests of the orthogonal array L12 of Table 7.26

Patient	35	36	37	38	39
Test #	D^2	D^2	D^2	D^2	D^2
1	10.344	25.61	4.228	7.6107	5.9413
2	0.7945	0.5362	4.4052	12.867	9.7085
3	15.385	5.2857	4.1538	1.4972	1.7175
4	20.252	40.017	5.2256	1.218	9.9911
5	0.4714	0.6263	4.8968	9.9571	6.2127
6	1.879	48.175	6.42	1.1107	1.4872
7	17.304	4.5388	1.1284	0.7359	1.9811
8	0.7104	32.836	0.6504	9.5472	8.1769
9	1.0017	37.03	0.722	1.3381	9.8143
10	20.163	39.797	0.9312	7.689	7.0324
11	1.412	0.7382	1.0211	1.6865	8.704
12	14.566	4.5287	0.9531	11.106	9.7803

Table 7.37 Static *SNR* indices of the squared Mahalanobis distances of the individuals outside the unit group of the orthogonal array L12

Patient	35	36	37	38	39
Test #	10 log (D^2)	10 log (D^2)	10 log (D^2)	10 log (D^2)	10 log (D^2)
1	10.147	14.084	6.2614	8.8142	7.7388
2	−0.9988	−2.7068	6.4397	11.095	9.8715
3	11.871	7.231	6.1844	1.7527	2.349
4	13.065	16.022	7.1814	0.8565	9.9961
5	−3.2658	−2.0324	6.8991	9.9813	7.9328
6	2.7393	16.828	8.0753	0.4559	1.7238
7	12.382	6.5695	0.5246	−1.3316	2.9691
8	−1.485	15.163	−1.8682	9.7988	9.1259
9	0.0075	15.686	−1.4148	1.265	9.9186
10	13.046	15.998	−0.3096	8.8587	8.471
11	1.4984	−1.3185	0.0906	2.2698	9.3972
12	11.633	6.5598	−0.2087	10.455	9.9035

$$SNR = 10 \cdot \log(D^2)$$

Figure 7.10 show the graphical evaluation of the parameter effects, separately for every patient outside the unit group.

In addition to the individual graphical evaluations, the effective contribution in percent of the individual parameters can be calculated quantitatively. To do this, the effect amplitude must be used as the difference between the two setting values. As in the tolerance analysis, all values must be squared to remove the sign (Table 7.38).

In the absence of an error estimation, the smallest effect can be defined as the noise background, which must be redefined for each individual. The ratio of the squared effects to the previously defined noise factor represents the relative effect. The percentage contribution of the parameters that are causally responsible for not belonging to the unit group refers to the sum of all effects, multiplied by a factor of 100. If a larger orthogonal array is used for a parameter screening, the empty columns are also evaluated and interpreted as noise factors. The effect amplitude as the difference of the underlying setting values is to be squared. The reference value for the ratio of individual parameter contributions is the sum of all squared error components including the squared parameter effects that are less than or equal to the error sum. The causal effect contribution, shown as a percentage, follows from the relative ratio of the squared effect difference to the sum of all partial effects including the total error (Table 7.39, Fig. 7.11).

Situations that essentially depend on only one variable are easy to identify. This special case can also be seen directly from the original data. The fact that patient no. 35 does not belong to the unit group is due to the exceptionally large numerical value of the variable F of 30.71 (Table 7.4). The case is similar for patient no. 36 with variable J of 136.18, for patient no. 37 with variable A of 11.6, for patient no. 38 with variable D of 164.04. This result proves not least the obviousness of the calculation effort, although alarms can be triggered when certain limit values are reached or exceeded even without simulation calculations. More frequently an appearance is observed with correlated variables such as with patient no. 39, except for variable C with the relatively large numerical value of 317.12. Individual alarms do not function if no limit values are hit. Pushing back a patient into the unit group requires an in-depth knowledge of the interaction of numerous variables for initiating suitable therapies.

7.8 Directional Recognition

The most common appearance of objects is generally used to define a characteristic unit group. To stay with clinical case studies, this can be a group of people classified as healthy. A sick state is expressed by not belonging to the unit group with squared Mahalanobis distances considerably greater than the threshold value. If the initial situation is based on such a “normal group”, deviations occur at most in one

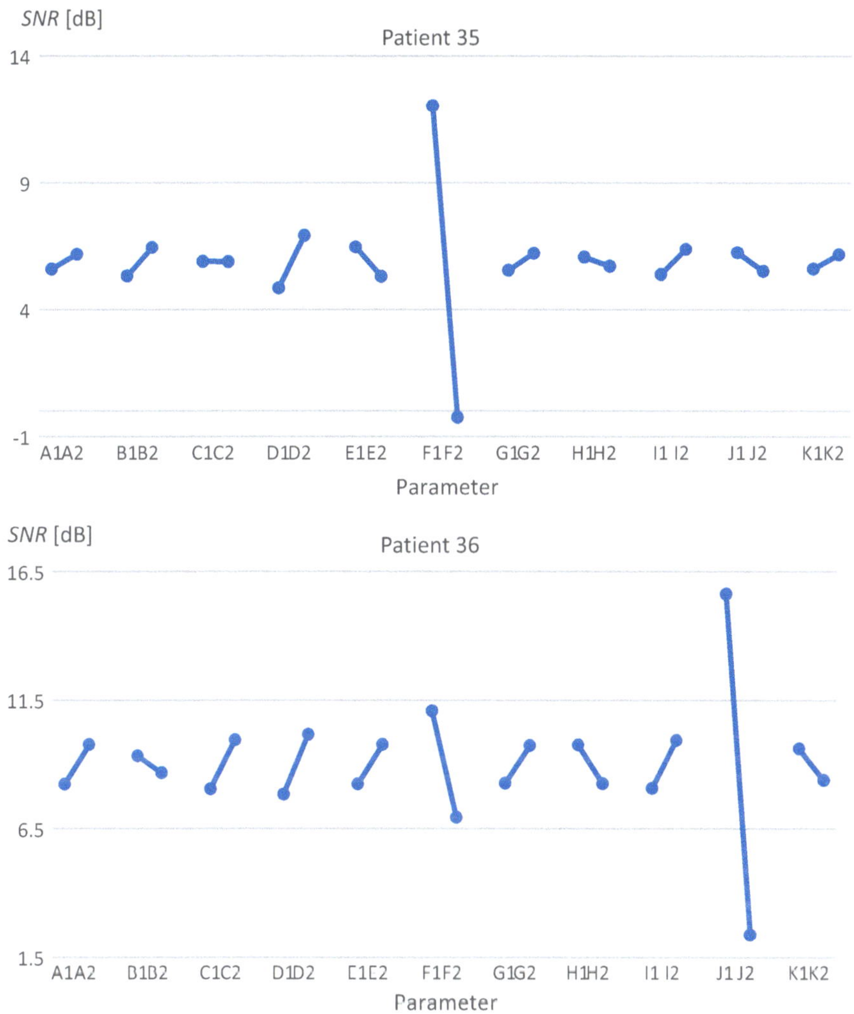

Fig. 7.10 Parameter effects of individuals 35–39 outside the unit group

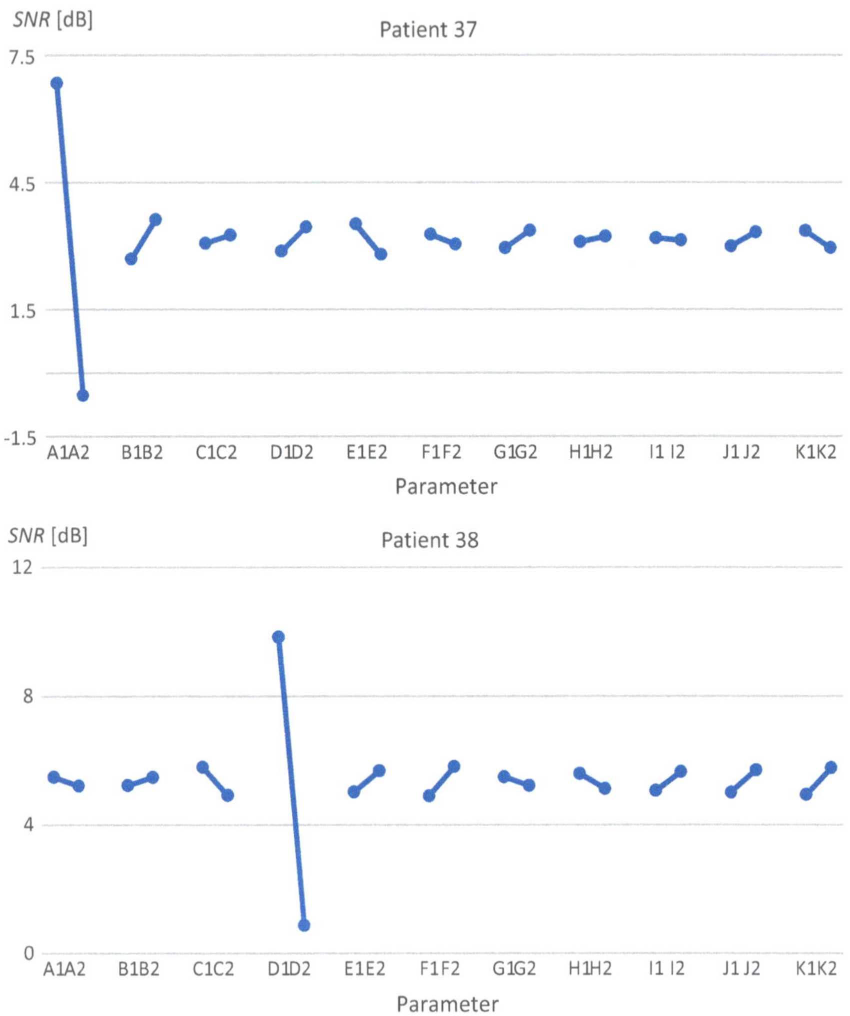

Fig. 7.10 (continued)

direction **(2–10*)**. A group of patients with similar complaints, e.g., before a surgery is performed, can also be defined as a unit group. Consequently, individuals not belonging to the unit group may be in noticeable poorer health or may be on the healthier side not needing a surgery at all. The squared Mahalanobis distance cannot make a distinction here since it only constructs a one-sided metric. The squared values alone cannot distinguish between the latter cases. The application of a therapy to approximate all patients with values $D^2 > 3$ to the unit group could, therefore, have fatal consequences just for the healthier patients. The decisive additional information is the recognition of a direction. Data standardization as a

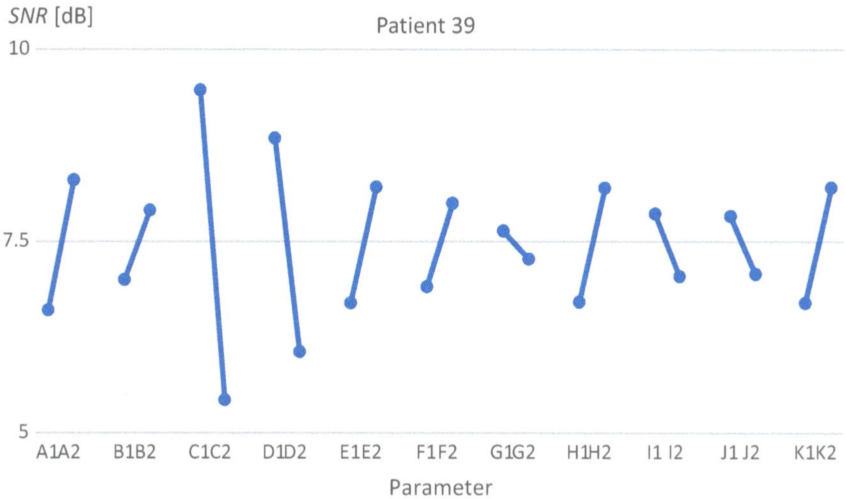

Fig. 7.10 (continued)

preliminary stage for calculating the squared Mahalanobis distances allows this important identification for such a two-sided problem. Since the transformation of standardization results in the average values of all parameters in the unit group being zero, the elements of the variables of non-associated objects are greater or less than zero, that is, they have a positive or negative sign. To make a statement about a direction, the importance of a numerical trend of the variable must be known in advance. For example, changing the parameter to a larger/smaller numerical value can represent an advantageous/disadvantageous property. Increasingly larger means a positive number, i.e., larger than the average value of

Table 7.38 Squared difference of the parameter effects of the individuals 35–39 from the evaluation of the orthogonal array L12; smallest effect is highlighted

Patient	35	36	37	38	39
Variable	Δ^2	Δ^2	Δ^2	Δ^2	Δ^2
A	0.345	2.3673	54.335	0.0746	2.8748
B	1.2816	**0.4376**	0.8714	**0.0612**	0.8256
C	**0.0001**	3.6301	0.0371	0.7627	16.341
D	4.3305	5.4068	0.3264	80.207	7.7374
E	1.3562	2.3382	0.5157	0.433	2.2774
F	150.67	17.148	0.0554	0.828	1.1889
G	0.4515	2.1454	0.1644	0.0722	**0.1366**
H	0.1307	2.2558	0.0171	0.2212	2.2106
I	0.9858	3.4273	**0.0029**	0.339	0.6731
J	0.5376	175.47	0.1116	0.4838	0.5754
K	0.3123	1.4821	0.1625	0.6919	2.2621

Table 7.39 Percentage effects of causally responsible parameters for individuals outside the unit group

Patient	35	36	37	38	39
Variable	%	%	%	%	%
A	0.22	1.10	96.00	0.09	7.75
B	0.80	0.20	1.54	0.07	2.23
C	0.00	1.68	0.07	0.91	44.04
D	2.70	2.50	0.58	95.29	20.85
E	0.85	1.08	0.91	0.51	6.14
F	93.93	7.93	0.10	0.98	3.20
G	0.28	0.99	0.29	0.09	0.37
H	0.08	1.04	0.03	0.26	5.96
I	0.61	1.59	0.01	0.40	1.81
J	0.34	81.20	0.20	0.57	1.55
K	0.19	0.69	0.29	0.82	6.10

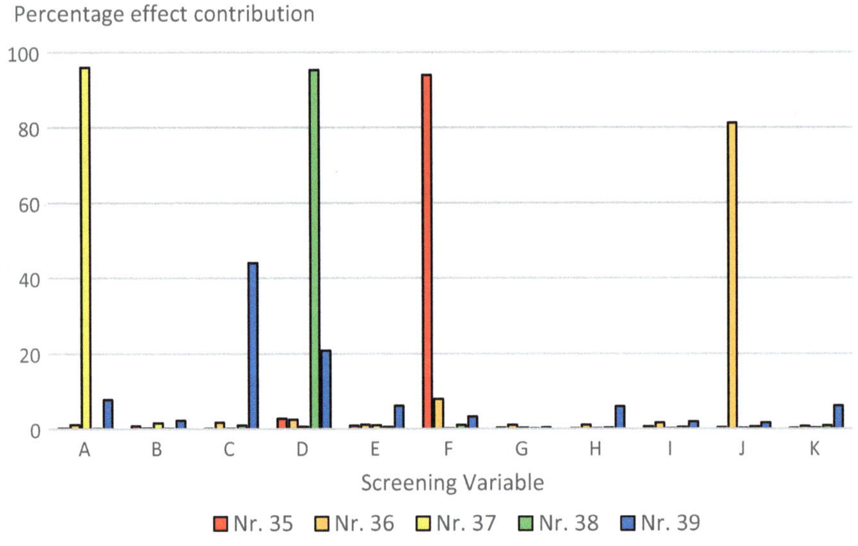

Fig. 7.11 Percentage effect contribution of the individuals 35–39 illustrated in the graphic

the elements of the unit group, decreasing smaller means a negative number (Table 7.40).

With the numerical value of the squared Mahalanobis distance in connection with the sign of the standardized variable, the individuals outside the unit group can be distinguished with respect to a directional trend (Table 7.41).

Table 7.40 Combination options for direction recognition of standardized variables

Sign of an element of standardized variables			
Positive (+)		Negative (−)	
Advantageous property	Disadvantageous property	Advantageous property	Disadvantageous property

Table 7.41 Direction assignment of objects outside the unit group

$D^2 > D_T^2$	
Outside the unit group due to a favorable appearance	Outside the unit group due to a detrimental appearance

For illustration, the clinical case study has been extended to include patients with minor impairments (minor myocardial damage) outside the unit group (Table 7.42).

The first physiological parameter *A* describes the hemoglobin concentration, which is considerably higher in patients with a mild course of the disease, i.e., values in the range of 14 g/dl. All other parameters *B*–*K* characterize the severity of the disease. The smaller the numbers for the given units, the smaller the impairments are to be classified. According to the previous explanations, it can be expected that all squared Mahalanobis distances are greater than 3, but that the direction information can additionally be read individually for each patient using the sign of the standardized data (Tables 7.43 and 7.44).

The squared Mahalanobis distances of the clinical case study extended by patients 40–42 are plotted (Fig. 7.12).

The squared Mahalanobis distances of the individuals 35–42 clearly differ from the unit group, without a further classification being possible due to the always positive signs (Fig. 7.13).

Patients 35–39 have mixed signs, for patients 40–42 the signs of parameter *A* are positive due to the better hemoglobin value, otherwise for all other parameters negative due to less complaints. This means that for the individuals not belonging to the unit group, one part shows squared Mahalanobis distances greater than 3 but due to a lighter clinical appearance compared to the original patient group. The observation of both the squared Mahalanobis distance and the directional recognition for causal differentiation contributes decisively to diagnostic certainty.

Table 7.42 Extension of patients with different clinical appearance outside the unit group

Patient	*A*	*B*	*C*	*D*	*E*	*F*	*G*	*H*	*I*	*J*	*K*
35	7.7	332.06	68.702	29.008	41.026	30.709	1622.1	348.47	21.79	30.53	255.73
36	7.3	419.8	73.038	32.917	71.163	17.917	744.03	129.35	17.67	136.18	220.74
37	11.6	663.23	84.88	19.259	35.688	1.0309	505.15	70.103	13.75	27.15	247.42
38	9.3	852.52	128.84	164.04	84.615	1.0791	2063.7	187.64	11.72	31.84	250.68
39	7.8	980.54	317.12	151.39	78.486	9.3385	2117.9	130.08	33.47	51.00	227.14
40	13.7	51.34	32.04	3.37	9.71	0.192	95.27	53.77	5.39	12.45	133.86
41	12.9	185.47	42.19	6.78	21.27	0.205	378.12	55.49	7.48	19.02	191.31
42	13.9	110.02	29.87	0.59	15.02	0.095	278.83	69.14	2.81	17.38	142.29

Table 7.43 Standardization of the variables *A*–*K*

Patient	*A*	*B*	*C*	*D*	*E*	*F*	*G*	*H*	*I*	*J*	*K*
35	−0.5698	−1.3025	−0.3776	−0.2299	−1.4437	8.4422	0.6985	1.4759	0.6534	−0.218	1.0767
36	−1.0698	−1.0629	−0.2584	−0.0524	0.4301	4.6246	−1.1443	−0.8545	0.0055	11.593	0.0778
37	4.3048	−0.3981	0.0673	−0.6723	−1.7756	−0.4148	−1.6456	−1.4846	−0.6108	−0.5966	0.8396
38	1.43	0.1188	1.2765	5.8986	1.2665	−0.4004	1.6251	−0.2346	−0.9305	−0.0725	0.9327
39	−0.4448	0.4684	6.4555	5.3244	0.8854	2.0645	1.7388	−0.8467	2.4885	2.0697	0.2604
40	6.9295	−2.0691	−1.3861	−1.3934	−3.3907	−0.6652	−2.5058	−1.6583	−1.9261	−2.2399	−2.4031
41	5.9296	−1.7028	−1.1069	−1.2386	−2.672	−0.6613	−1.9122	−1.64	−1.5975	−1.5053	−0.7626
42	7.1795	−1.9088	−1.4458	−1.5195	−3.0606	−0.6941	−2.1206	−1.4948	−2.3317	−1.6887	−2.1624

Table 7.44 Squared Mahalanobis distances D^2 for the individuals 35–42 outside the unit group

Patient #	D^2
35	10.344
36	25.61
37	4.228
38	7.6107
39	5.9413
40	7.2998
41	5.773
42	7.8697

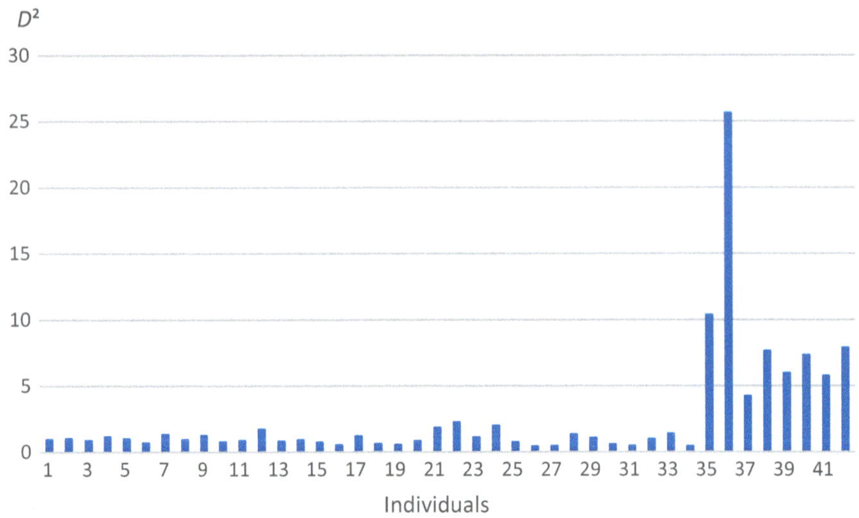

Fig. 7.12 Squared Mahalanobis distances of the patients 1–42

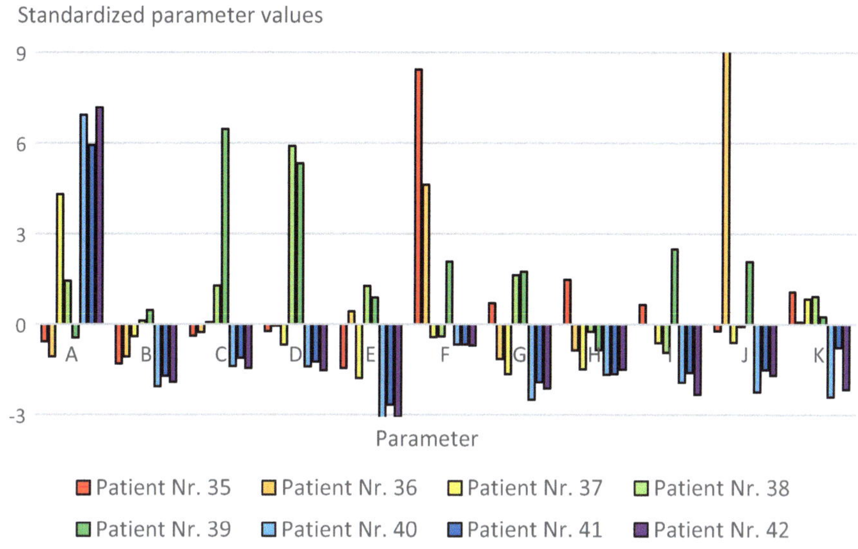

Fig. 7.13 Differentiation possibilities of individuals outside the unit group due to negative/positive signs of the standardized parameter values

Important items to memorize The most important prerequisite for the application of the Mahalanobis algorithm is the definition of the unit group, i.e., the selection and composition of its members/objects/elements.

The formalism for calculating signal-to-noise ratios refers only to objects outside the unit group.

The reliability of a diagnosis depends on the size of the *SNR* index.

Chapter 8
Alternative Numerical Procedures

Contents

Abstract To deal with the Mahalanobis algorithm, some limitations must be observed. Two further algorithms exist with identical or very close results. Both are less sensitive to the initial boundary conditions but are more calculation oriented. One algorithm makes use of an orthogonalization process for the variables. This transforms the set of data in linear, independent variables, taking advantage of the fact that the number of variables can be nearly as numerous as the number of objects defining a pattern. The other algorithm is useful if some variables are perfectly correlated. In this case, the two previous algorithms are useless, as some steps of the calculation cannot be performed. This is a crucial property in case of the third algorithm as multicollinearities can contain essential information. For comparison, screening of the variables and root-cause-analysis are performed with the same clinical case study.

8.1 Procedure of Orthogonalization According to Gram–Schmidt

Another mathematical approach to get the squared Mahalanobis distances succeeds with an orthogonal system **(2-11*)**. The transformation of a system of linear independent vectors into orthogonal vectors is known as the Gram–Schmidt orthogonalization process. The special feature of such an orthogonal system is that

© Springer Nature Switzerland AG 2019

H. Ruefer, *Living Without Mathematical Statistics*,
https://doi.org/10.1007/978-3-319-99632-5_8

the derivation of the Mahalanobis distances does not require the correlation matrix or its inverse. Here, too, the prerequisite is that the number of variables k does not exceed the number of objects or individuals n.

$$n \geq k$$

n = Objects or individuals (rows)
k = Variables (columns)

The starting point is the standardization of the measurement or observation data of the objects to obtain dimension-free quantities. For a clear explanation, only the first two standardized variables of a data set are considered as vectors X_1 and X_2, which are to be transformed into the orthogonalized vectors Y_1 and Y_2. The first vector is retained, the second vector is to be projected perpendicularly, i.e., orthogonally to Y_1. For this, X_2 is geometrically projected onto Y_1. The difference vector of X_2 and Y_1 with the projected length of X_2 is perpendicular to Y_1 and forms the orthogonalized feature Y_2 (Fig. 8.1).

Projection and the absolute value result as a pre-factor from the scalar products of the first orthogonal vector.

$$Y_1 = X_1$$

$$Y_2 = X_2 - \frac{\left(X_2^T \cdot Y_1\right)}{\left(Y_1^T \cdot Y_1\right)} Y_1$$

Each additional standardized variable is serially calculated as a vector, considering the previous orthogonalized vectors.

$$Y_3 = X_3 - \frac{\left(X_3^T \cdot Y_1\right)}{\left(Y_1^T \cdot Y_1\right)} Y_1 - \frac{\left(X_3^T \cdot Y_2\right)}{\left(Y_2^T \cdot Y_2\right)} Y_2$$

$$\cdots$$

$$Y_k = X_k - \frac{\left(X_k^T \cdot Y_1\right)}{\left(Y_1^T \cdot Y_1\right)} Y_1 - \frac{\left(X_k^T \cdot Y_2\right)}{\left(Y_2^T \cdot Y_2\right)} Y_2 - \frac{\left(X_k^T \cdot Y_3\right)}{\left(Y_3^T \cdot Y_3\right)} Y_3 - \cdots - \frac{\left(X_k^T \cdot Y_{k-1}\right)}{\left(Y_{k-1}^T \cdot Y_{k-1}\right)}$$

With the transformation of the standardized variables exists a new orthogonal data set. As can be seen from the transformation formulas, the average value of the

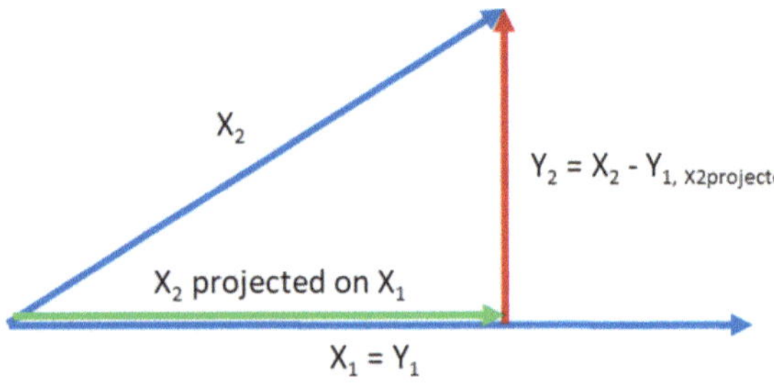

Fig. 8.1 Principle of the orthogonalization procedure according to Gram–Schmidt

vector components are kept as zero, as with the standardized variables, whereas the squared standard deviation must be recalculated for each orthogonalized variable. The squared Mahalanobis distance follows from **(2-12*)**:

$$D_i^2 = \frac{1}{k}\left(\frac{y_{1i}^2}{\sigma_1^2} + \frac{y_{2i}^2}{\sigma_2^2} + \frac{y_{3i}^2}{\sigma_3^2} + \cdots + \frac{y_{ki}^2}{\sigma_k^2}\right)$$

k = Number of variables
i = 1, 2, 3, …, n (Number of objects/individuals)

The initial example with two variables is used as numerical exercise of the Gram–Schmidt orthogonalization method. At the beginning, the standardized characteristic variables X_1 and X_2, are to be transformed into the orthogonalized variables Y_1 and Y_2. The column elements of the variables are interpreted as a vector; the column vector is transposed into a row vector.

$$Y_1 = X_1$$

$$Y_2 = X_2 - \frac{8}{10} Y_1$$

The individuals outside the unit group are orthogonalized with the same coefficient. The scalar product is useful to prove the orthogonality (Table 8.1).

Table 8.1 Gram–Schmidt orthogonalization procedure with two variables

Individuals unit group i = 1–10	Standardized weight $(x_{i,\,1})$	Standardized height $(x_{i,\,2})$	Orthogonalized variable $(y_{i,\,1})$	Orthogonalized variable $(y_{i,\,2})$
1	−1.1286	−1.7619	−1.1286	−0.8591
2	−0.5699	−0.08390	−0.5699	0.3712
3	−1.6873	−0.8030	−1.6873	0.5467
4	−0.01117	−0.4435	−0.01117	−0.4345
5	0.3240	0.1558	0.3240	−0.1034
6	−0.3464	−1.1626	−0.3464	−0.8855
7	1.1062	0.6352	1.1062	−0.2497
8	1.2180	1.2345	1.2180	0.2602
9	−0.4581	0.7551	−0.4581	1.1216
10	1.5532	1.4742	1.5532	0.2318
Average	0	0	0	0
Squared standard deviation	1	1	1	0.6001
Individuals outside the unit group i = 11–12				
11	0.2123	−1.6420	0.2123	−1.8119
12	1.3297	−0.3236	1.3297	−1.3873

$$\left(Y_1^T \cdot Y_2\right) = 0$$

The squared Mahalanobis distance for each individuum is given by, see Table 8.2

$$D_i^2 = \frac{1}{2}\left(\frac{y_{i,1}^2}{1^2} + \frac{y_{i,2}^2}{0.6001^2}\right)$$

8.2 Parameter Screening of Orthogonalized Variables with Orthogonal Array

For comparison, the same clinical case study with 11 variables is used and, based on the standardized variables, the serial orthogonalization process is performed (Table 8.3).

$$\begin{aligned} Y_1 &= X_1 \\ Y_2 &= X_2 - \frac{\left(X_2^T \cdot Y_1\right)}{\left(Y_1^T \cdot Y_1\right)} Y_1 \\ Y_3 &= X_3 - \frac{\left(X_3^T \cdot Y_1\right)}{\left(Y_1^T \cdot Y_1\right)} Y_1 - \frac{\left(X_3^T \cdot Y_2\right)}{\left(Y_2^T \cdot Y_2\right)} Y_2 \\ Y_4 &= X_4 - \frac{\left(X_4^T \cdot Y_1\right)}{\left(Y_1^T \cdot Y_1\right)} Y_1 - \frac{\left(X_4^T \cdot Y_2\right)}{\left(Y_2^T \cdot Y_2\right)} Y_2 - \frac{\left(X_4^T \cdot Y_3\right)}{\left(Y_3^T \cdot Y_3\right)} Y_3 \\ &\vdots \end{aligned}$$

$$Y_{11} = X_{11} - \frac{\left(X_{11}^T \cdot Y_1\right)}{\left(Y_1^T \cdot Y_1\right)} Y_1 - \frac{\left(X_{11}^T \cdot Y_2\right)}{\left(Y_2^T \cdot Y_2\right)} Y_2 - \frac{\left(X_{11}^T \cdot Y_3\right)}{\left(Y_3^T \cdot Y_3\right)} Y_3 - \frac{\left(X_{11}^T \cdot Y_4\right)}{\left(Y_4^T \cdot Y_4\right)} Y_4 - \cdots - \frac{\left(X_{11}^T \cdot Y_{10}\right)}{\left(Y_{10}^T \cdot Y_{10}\right)} Y_{10}$$

The squared Mahalanobis distances result from the squared value of the components of the parameters and the squared standard deviation divided by the number of variables, serially from D_1^2 to D_{39}^2, see Table 8.4.

$$D_1^2 = \frac{1}{11}\left(\begin{aligned} &\frac{0.8051^2}{1^2} + \frac{0.1229^2}{0.9991^2} + \frac{0.6656^2}{0.9208^2} \\ &+ \frac{0.6819^2}{0.6598^2} + \frac{1.2032^2}{0.918^2} + \frac{0.1688^2}{0.9534^2} \\ &+ \frac{1.3918^2}{0.826^2} + \frac{1.19^2}{0.854^2} + \frac{0.5859^2}{0.7521^2} \\ &+ \frac{0.4779^2}{0.6221^2} + \frac{0.732^2}{0.873^2} \end{aligned}\right) = 0.9712$$

Since the result is identical to the Mahalanobis algorithm, the parameters that do not change or attenuate the squared Mahalanobis distance of the individuals outside the unit group are sorted out according to the procedure described in Sect. 7.5. The orthogonal array L12 is applied in the same way. The orthogonalization process of

Table 8.2 Calculation of the squared distance of the individuals with the Mahalanobis algorithm and the Gram–Schmidt process

Individuals unit group i = 1–10	Standardized weight ($x_{i,\ 1}$)	Standardized height ($x_{i,\ 2}$)	Mahalanobis algorithm D^2	Orthogonalized variable ($y_{i,\ 1}$)	Orthogonalized variable ($y_{i,\ 2}$)	Gram–Schmidt process D^2
1	−1.1286	−1.7619	1.6617	−1.1286	−0.8591	1.6617
2	−0.5699	−0.08390	0.3545	−0.5699	0.3712	0.3545
3	−1.6873	−0.8030	1.8385	−1.6873	0.5467	1.8385
4	−0.01117	−0.4435	0.2623	−0.01117	−0.4345	0.2623
5	0.3240	0.1558	0.06735	0.3240	−0.1034	0.06735
6	−0.3464	−1.1626	1.1489	−0.3464	−0.8855	1.1489
7	1.1062	0.6352	0.6985	1.1062	−0.2497	0.6985
8	1.2180	1.2345	0.8357	1.2180	0.2602	0.8357
9	−0.4581	0.7551	1.8518	−0.4581	1.1216	1.8518
10	1.5532	1.4742	1.2808	1.5532	0.2318	1.2808
Average	0	0	1	0	0	1
Squared standard deviation	1	1		1	0.6001	
Individuals outside the unit group i = 11–12						
11	0.2123	−1.6420	4.5813	0.2123	−1.8119	4.5813
12	1.3297	−0.3236	3.5567	1.3297	−1.3873	3.5567

Table 8.3 Gram–Schmidt orthogonalization of parameters *A*–*K* of the clinical case study

Patient	*A*	*B*	*C*	*D*	*E*	*F*	*G*	*H*	*I*	*J*	*K*
1	0.8051	−0.1229	0.6656	0.6819	−1.2032	−0.1688	1.3918	−1.19	−0.5859	−0.4779	−0.732
2	1.68	−0.6708	−0.137	0.0932	0.5523	0.0263	0.2151	−0.8728	0.7	1.2243	1.1597
3	−1.1947	−0.1284	1.8211	−0.7214	0.4358	−0.4699	−1.1405	−0.408	−0.6588	−0.0529	−0.0213
4	−0.3198	2.9586	−0.9511	0.6424	−0.5123	−0.2967	−0.7226	0.6422	0.0366	−0.0739	−0.4734
5	0.1801	1.9294	0.6718	−0.4465	1.1788	0.7551	−0.3928	−0.7366	0.5369	−0.7717	0.8664
6	−1.3197	−1.2727	−0.1983	−0.5971	0.5274	−0.3808	−0.4206	−0.2873	0.4064	0.5259	−1.0683
7	1.0551	0.7025	0.5165	−1.7363	0.0189	−0.1607	1.7713	−0.3085	−0.1544	0.352	0.7252
8	−1.8197	0.2217	−0.0115	−1.0367	0.6334	−0.4947	0.2535	−0.6179	1.1929	−0.317	−0.5284
9	−1.5697	−0.3752	0.436	−0.0753	−0.3341	−0.6445	1.1332	0.7967	0.3132	−1.6532	0.567
10	−0.8198	−0.2445	1.0106	0.744	−0.7538	−0.1889	0.1513	0.4515	1.4316	0.5204	−0.0714
11	0.3051	0.0373	−0.7381	0.4554	0.9608	−0.7621	−0.8038	0.5069	1.6503	−0.4252	−0.2946
12	1.18	0.978	3.1108	1.1492	−0.4339	0.1826	−0.414	0.6751	0.7434	0.1942	−0.0745
13	0.3051	−0.3693	−0.2635	−0.2339	−1.6461	0.3433	0.8719	0.4812	1.2893	0.2261	−0.7977
14	0.5551	−1.2519	−0.2261	0.2507	0.9098	−0.5921	−1.2839	0.922	0.0066	−0.9465	0.7586
15	0.8051	−0.2619	−0.1323	−0.0134	1.9895	−0.4696	−0.1125	1.06	−0.2209	−0.5781	0.0356
16	1.18	−0.6031	−0.5522	0.0516	−0.0019	−0.1551	0.2614	1.2782	−0.3005	−0.4343	0.7166
17	−0.4448	0.2006	−0.6564	−0.2875	−0.5202	−0.5909	2.1815	1.7903	−0.0691	0.0781	0.3873
18	1.93	−1.3471	−0.4903	−0.1852	−0.138	0.0698	0.2305	−0.301	−0.3366	−0.0292	0.67
19	0.9301	−1.4058	−0.567	−0.0297	0.731	−0.31	−0.6999	0.2778	−0.2559	0.214	−0.9285
20	0.4301	−0.8928	−0.092	−0.115	0.564	−0.438	0.4821	0.26	−0.4687	−0.0194	−2.2851
21	−1.5697	0.305	0.8553	0.39	0.5888	0.2097	0.1337	1.9374	−1.7376	1.4495	−0.1925
22	−0.1948	−1.0012	−0.1736	0.0183	−1.2098	4.0107	−0.5425	0.5115	−0.7687	−0.9351	0.3334
23	−1.8197	−0.7183	0.1437	−0.386	−0.0231	−0.6849	−0.3103	−0.8006	−1.0312	0.376	1.8176
24	−0.5698	0.8115	−0.9402	0.3526	2.4584	2.8356	0.9937	−0.3784	0.495	0.694	−0.4476

(continued)

Table 8.3 (continued)

Patient	*A*	*B*	*C*	*D*	*E*	*F*	*G*	*H*	*I*	*J*	*K*
25	−0.8198	−0.5648	0.1697	−0.6067	−0.9664	1.0875	−0.6083	−0.9667	0.5309	−0.1736	−1.0486
26	0.3051	−0.1311	−0.9098	0.4553	−0.3057	−0.3768	−0.3388	−0.5037	−0.2132	0.8353	−0.6994
27	0.4301	−0.7791	0.502	0.3495	−0.2303	−0.3978	0.0325	−1.4096	−0.4616	−0.1373	−0.4312
28	1.43	0.8439	1.1313	−1.5585	−0.908	0.137	−1.2541	0.6743	−0.1829	0.4656	−0.3674
29	−0.5698	−0.448	0.7806	1.4525	0.5189	−0.3621	0.9736	−1.5032	−0.2415	−0.184	0.5294
30	0.6801	1.4578	−0.123	−0.0113	0.7248	−0.2458	0.0045	−0.9572	−0.8223	−0.3733	0.5504
31	−0.3198	0.244	−1.379	−0.1814	−1.0028	−0.6215	−0.7209	−0.4586	0.0376	−0.0609	−0.0539
32	0.3051	2.1342	−1.4752	0.0474	−0.7481	−0.5456	−0.0821	−0.514	−0.9101	−0.2639	−0.6386
33	−0.5698	−0.1475	−1.1615	0.2946	−1.1618	0.0714	−0.7192	0.0929	0.8536	0.8704	2.3698
34	−0.5698	−0.0879	−0.6369	0.793	−0.693	−0.3717	−0.5146	−0.144	−0.8044	−0.1186	−0.3328
Average	0	0	0	0	0	0	0	0	0	0	0
Sigma	1	0.9991	0.9208	0.6598	0.918	0.9534	0.826	0.854	0.7521	0.6221	0.873
35	−0.5698	−1.2781	0.0116	0.2653	−1.2485	8.4884	0.8537	1.1356	0.2117	−3.0975	1.4694
36	−1.0698	−1.0172	−0.033	0.3583	0.4145	4.3112	−1.4703	−0.8579	0.5368	9.8928	0.8754
37	4.3048	−0.582	0.869	−0.9504	−0.4009	0.8638	−0.5263	−0.6285	1.6535	2.2238	1.9717
38	1.43	0.0577	1.4511	4.9043	1.9606	−0.2819	1.9806	−2.1848	−1.4229	−0.1386	−1.6031
39	−0.4448	0.4874	6.2165	0.8366	0.8821	2.5722	1.387	−1.7568	1.0396	0.4634	−0.1118

Table 8.4 Squared Mahalanobis distances D^2 for the individuals 1–39 of the clinical case study with orthogonalized vectors

Patient	1	2	3	4	5	6	7	8	9	10
D^2	0.9712	1.0268	0.9026	1.176	1.0029	0.6905	1.3324	0.9395	1.2584	0.7786
Patient	11	12	13	14	15	16	17	18	19	20
D^2	0.8767	1.7286	0.8291	0.9184	0.7444	0.5282	1.2029	0.6297	0.5595	0.8442
Patient	21	22	23	24	25	26	27	28	29	30
D^2	1.8379	2.2502	1.1193	1.9922	0.7385	0.4428	0.4548	1.3245	1.0545	0.5917

Patient	31	32	33	34
D^2	0.4695	0.9792	1.3754	0.429

Patient	35	36	37	38	39
D^2	10.344	25.61	4.228	7.6107	5.9413

the variables must be started again for each configuration of the orthogonal array. From the squared value of the vector component divided by the squared standard deviation, added up according to the number of variables, and normalized with its number result in the squared Mahalanobis distances. The application of static or dynamic *SNR* key figures to members outside the unit group leads to the same screening results because of the same D^2 values for each *SNR* index.

8.3 Parameter Screening of Orthogonalized Variables Without Orthogonal Array

To avoid the relatively high effort to calculate all test configurations of an applicable orthogonal array, there is a further direct access to evaluate the parameter effects even without an orthogonal array. It should be noted that

- sorting of the parameters with the smallest *SNR* indices does not represent a sharp criterion
- relations between the original and the orthogonalized variables can exist to varying degrees **(2-13*)**.

To perform the screening analysis of orthogonalized parameters, only individuals that do not belong to the unit group are considered. For a better overview, Table 8.5 is transposed.

From the values of the orthogonalized parameters and the corresponding standard deviation, the individual squared Mahalanobis distances are calculated (Table 8.6).

Instead of the test number of an orthogonal array L12, the 11 orthogonalized characteristics *A*–*K* appear. By converting each line, i.e., each orthogonalized characteristic, into an *SNR* index to be determined, the effect of each variable can be read directly.

Table 8.5 Orthogonalized parameter values of individuals 35–39 outside the unit group

Patient	35	36	37	38	39	Sigma
Variable						
A	−0.5698	−1.0698	4.3048	1.43	−0.4448	1
B	−1.2781	−1.0172	−0.582	0.0577	0.4874	0.9991
C	0.0116	−0.033	0.869	1.4511	6.2165	0.9208
D	0.2653	0.3583	−0.9504	4.9043	0.8366	0.6598
E	−1.2485	0.4145	−0.4009	1.9606	0.8821	0.918
F	8.4884	4.3112	0.8638	−0.2819	2.5722	0.9534
G	0.8537	−1.4703	−0.5263	1.9806	1.387	0.826
H	1.1356	−0.8579	−0.6285	−2.1848	−1.7568	0.854
I	0.2117	0.5368	1.6535	−1.4229	1.0396	0.7521
J	−3.0975	9.8928	2.2238	−0.1386	0.4634	0.6221
K	1.4694	0.8754	1.9717	−1.6031	−0.1118	0.873

Table 8.6 Squared Mahalanobis distances of individuals 35–39 outside the unit group

Patient	35	36	37	38	39
Variable	D^2	D^2	D^2	D^2	D^2
A	0.3247	1.1444	18.531	2.045	0.1979
B	1.6366	1.0366	0.3394	0.0033	0.238
C	0.0002	0.0013	0.8906	2.4832	45.576
D	0.1616	0.2948	2.0745	55.242	1.6074
E	1.8497	0.2039	0.1907	4.5615	0.9232
F	79.272	20.449	0.8209	0.0874	7.279
G	1.0681	3.1686	0.406	5.7493	2.8194
H	1.7681	1.0091	0.5415	6.5444	4.2313
I	0.0793	0.5095	4.8339	3.5798	1.9109
J	24.792	252.89	12.779	0.0496	0.5549
K	2.8331	1.0054	5.1008	3.3718	0.0164

The same static and dynamic *SNR* indicators are applied, transferred to the orthogonalized characteristics.

Static characteristic of the squared average value:

$$SNR = 10\ \log\left(\overline{\sqrt{\frac{y_i^2}{\sigma^2}}}\right)^2; \quad i = 1, 2, 3, 4, 5$$

$$SNR = 10\ \log\left(\overline{D^2}\right)$$

Static characteristic: *LTB* (Table 8.7).

Table 8.7 Effect amplitude as *SNR* indices of the orthogonalized parameters interpreted over the patients 35–39 outside the unit group

Patient	35	36	37	38	39	*SNR*	*SNR*	β	σ^2	*SNR*
Variable	*D*	*D*	*D*	*D*	*D*	Static [dB]	*LTB* [*dB*]			Dynamic [dB]
A	0.5698	1.0698	4.3048	1.43	0.4448	3.8838	−2.8107	0.7278	2.3377	−0.4434
B	1.2793	1.0181	0.5826	0.0577	0.4878	−3.2847	−17.905	0.3528	0.1548	5.0564
C	0.0126	0.0358	0.9437	1.5758	6.751	5.4078	−31.533	0.8135	7.1536	−4.3347
D	0.402	0.543	1.4403	7.4325	1.2678	6.9158	−3.3046	1.0761	7.262	−1.9697
E	1.36	0.4515	0.4367	2.1358	0.9609	0.5793	−3.7994	0.5235	0.4537	3.8154
F	8.9035	4.522	0.906	0.2957	2.698	10.794	−4.101	1.7987	8.6904	1.7126
G	1.0335	1.78	0.6372	2.3978	1.6791	3.5538	0.7125	0.7646	0.3131	8.7158
H	1.3297	1.0045	0.7359	2.5582	2.057	3.7339	1.2006	0.751	0.5715	5.9466
I	0.2815	0.7138	2.1986	1.892	1.3824	2.2365	−4.9383	0.606	0.7193	3.084
J	4.9791	15.903	3.5748	0.2228	0.7449	14.126	−6.45	2.9104	24.462	1.398
K	1.6832	1.0027	2.2585	1.8363	0.128	2.8085	−10.994	0.6755	0.6472	4.4862
Signal		1.9849		2.5494		1.6381	1.985		1.6911	
S_2		3.9845								

$$SNR = LTB = 10\log\left(\frac{5}{\frac{1}{\frac{y_1^2}{\sigma^2}+\frac{y_2^2}{\sigma^2}+\frac{y_3^2}{\sigma^2}+\frac{y_4^2}{\sigma^2}+\frac{y_5^2}{\sigma^2}}}\right)$$

$$SNR = LTB = 10\log\left(\frac{5}{\frac{1}{D_1^2}+\frac{1}{D_2^2}+\frac{1}{D_3^2}+\frac{1}{D_4^2}+\frac{1}{D_5^2}}\right)$$

Dynamic characteristic:

$$SNR = 10\log\left(\frac{\beta^2}{\sigma^2}S_2\right)$$

For an easier comparison of the distinct *SNR* key figures, the difference with the smallest key figure is shown, so that the latter assumes the value zero (Fig. 8.2).

By sorting out the smallest *SNR* values, the parameters with the greatest effect on the individuals 35–39 remain. In the case of the static *SNR* index, at least parameter *B* could be omitted, supplemented by the smaller contributions of parameters *E*, *I*, and *K*; for the *LTB* index, parameter *C* could be omitted, extended by parameter *B*; for the dynamic *SNR* index, parameter *C* could also be omitted together with parameter *D*. The comparison with the Mahalanobis algorithm shows some similarities, but no identical screening results. For verification, the squared Mahalanobis distances of the individuals 35–39 are recalculated by omitting the above variables (Fig. 8.3).

Compared to the results with the Mahalanobis algorithm (Fig. 7.9), the above figure is similar but not identical. An important difference becomes apparent in the evaluation with the dynamic *SNR* key figure. The squared Mahalanobis distances of

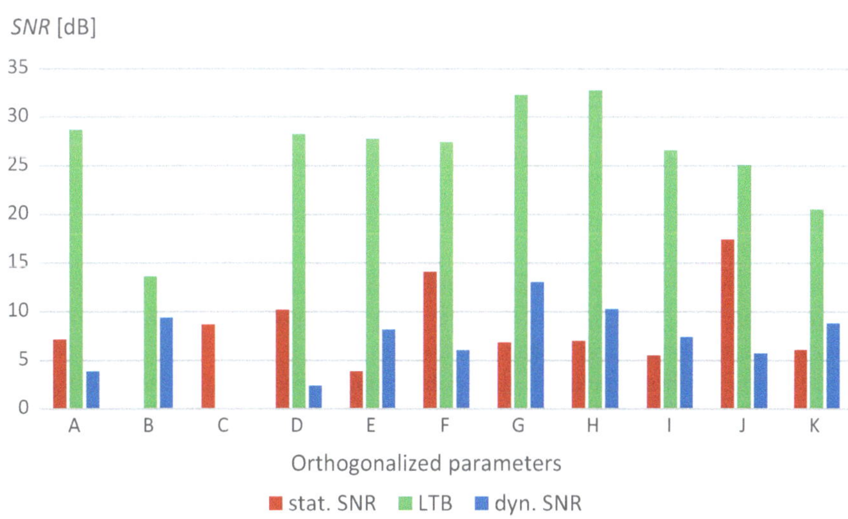

Fig. 8.2 Effect contribution of parameters *A–K* with evaluation of the static *SNR* index of the squared average value, the static *LTB* index, and the dynamic *SNR* index

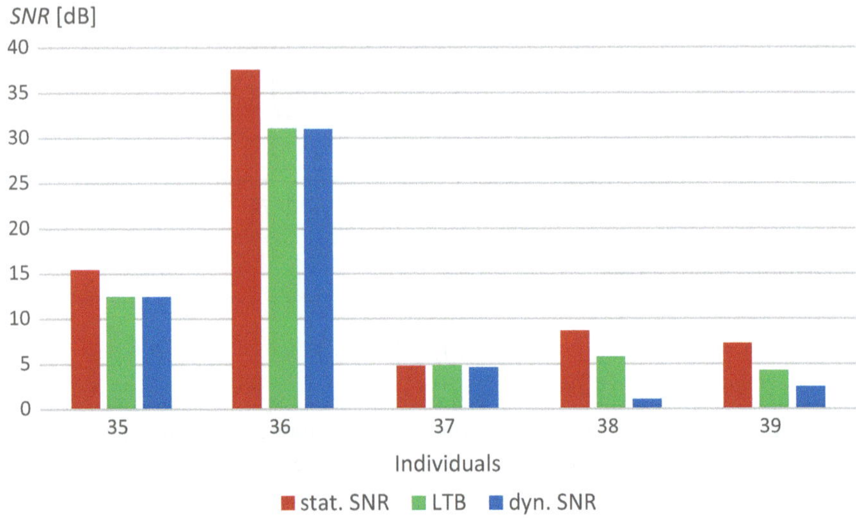

Fig. 8.3 Squared Mahalanobis distances of the orthogonalized variables of the individuals 35–39 outside the unit group with the evaluation functions: the static *SNR* index of the squared average value without *B*, *E*, *I*, and *K*, of the static *LTB* index without *B*, *C*, and of the dynamic *SNR* index without *C*, *D*

Table 8.8 Discrimination power applying different *SNR* indices based on orthogonalized variables

Evaluation function	$SNR = 10\log\left(\overline{D}^2\right)$	$SNR = LTB$	$SNR = 10\log\left(\frac{\beta^2}{\sigma^2}S_2\right)$
Parameter	*A, C, D, F, G, H, J*	*A, D, E, F, G, H, I, J, K*	*A, B, E, F, G, H, I, J, K*
Smallest numerical value outside the unit group	7.265	4.267	4.6069
Largest numerical value within the unit group	4.0877	2.6301	2.2502
Absolute difference	1.6647	1.5611	1.9017
Relative difference (%)	35	37	41

patients 38 and 39 are smaller than the threshold value $D_T = 3$ for the individuals of the unit group, i.e., both patients are to be integrated into these. The following table presents the discrimination power (Table 8.8).

Deviations in both methods, the Mahalanobis algorithm and the direct interpretation of the Gram–Schmidt orthogonalized variables, are in the first case due to the redistribution of partial interactions of the standardized variables in the orthogonal array, in the second case to partial correlations, which, if sufficient large enough, cannot be neglected.

8.4 Root-Cause-Analysis with Orthogonalized Variables

A root-cause-analysis is also carried out by taking the numerical value of the contributions of orthogonalized variables to the squared Mahalanobis distance of the individual variables.

In the previous case of the original Mahalanobis algorithm, the effect amplitudes of the parameters were determined using the orthogonal array. In squared form, the sign was omitted, so that the percentage effect contribution was determined in relation to the overall effect. Instead of transforming the squared Mahalanobis distance of the orthogonalized variables into the *SNR* index

$$SNR = 10\log\left(D^2\right)$$

with further calculation steps, the squared Mahalanobis distance can also be converted directly into its percentage effect contribution (Tables 8.9 and 8.10).

The patient's individual effect contributions are illustrated (Fig. 8.4).

The comparison of the root-cause-analysis with non-orthogonalized parameters (Fig. 7.11) provides very similar results. The strongest effects are given by parameter *F* for patient No. 35, parameter *J* for patient No. 36, parameter *A* for patient No. 37, parameter *D* for patient No. 38, and parameter *C* for patient No. 39. For both methods, the Mahalanobis algorithm and the Gram–Schmidt process, are essentially equivalent, whereby one or the other, depending on the set of data, yields more robust results.

Table 8.9 Individual squared Mahalanobis distances of the individuals 35–39 with the total effect of the orthogonalized variables

Patient	35	36	37	38	39
Variable	D^2	D^2	D^2	D^2	D^2
A	0.3247	1.1444	18.531	2.045	0.1979
B	1.6366	1.0366	0.3394	0.0033	0.238
C	0.0002	0.0013	0.8906	2.4832	45.576
D	0.1616	0.2948	2.0745	55.242	1.6074
E	1.8497	0.2039	0.1907	4.5615	0.9232
F	79.272	20.449	0.8209	0.0874	7.279
G	1.0681	3.1686	0.406	5.7493	2.8194
H	1.7681	1.0091	0.5415	6.5444	4.2313
I	0.0793	0.5095	4.8339	3.5798	1.9109
J	24.792	252.89	12.779	0.0496	0.5549
K	2.8331	1.0054	5.1008	3.3718	0.0164
Sum	113.79	281.71	46.508	83.717	65.354

Table 8.10 Percentage contribution of the orthogonalized variables for the individuals 35–39

Patient	35	36	37	38	39
Variable	%	%	%	%	%
A	0.29	0.41	39.84	2.44	0.30
B	1.44	0.37	0.73	0.00	0.36
C	0.00	0.00	1.91	2.97	69.74
D	0.14	0.10	4.46	65.99	2.46
E	1.63	0.07	0.41	5.45	1.41
F	69.67	7.26	1.77	0.10	11.14
G	0.94	1.12	0.87	6.87	4.31
H	1.55	0.36	1.16	7.82	6.47
I	0.07	0.18	10.39	4.28	2.92
J	21.79	89.77	27.48	0.06	0.85
K	2.49	0.36	10.97	4.03	0.03

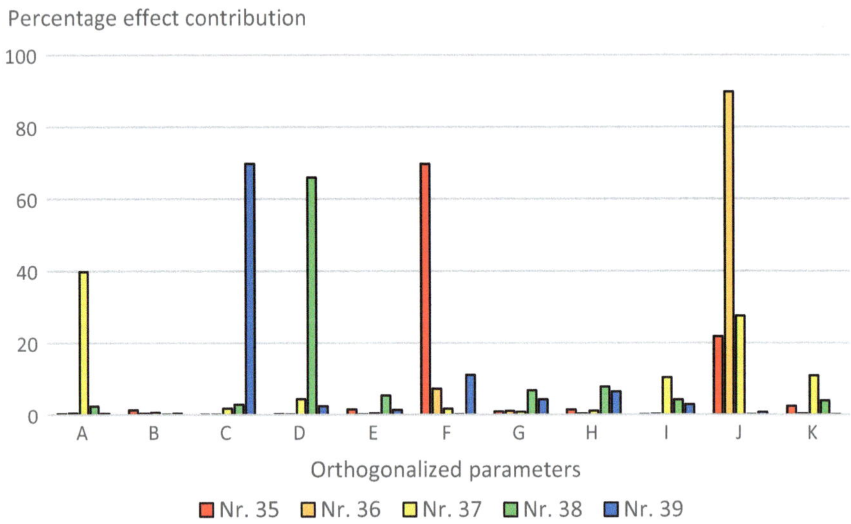

Fig. 8.4 Parameter effects of individuals 35–39

8.5 Procedure with the Adjoint Correlation Matrix

With the application of the Mahalanobis algorithm, it has been implicitly assumed that the inverse of the correlation matrix exists. The mathematical condition for this is given if the determinant of the correlation matrix is unequal zero, as is usually the case in practice. Extremely strong linear relationships between pairs of variables are expressed in a correlation coefficient close to one. In special cases, exact collinearity also occurs with a correlation coefficient identically to one. Apart from the application of the Gram–Schmidt process, one variable of the pair in question, for which the correlation coefficient is close to one or identical to one, in principle could be removed. However, this step is only allowed after parameter screening, as otherwise valuable information may be lost. An example is a control system in which two flow meters monitor the section of a pipeline. The numerical values of the variables (= sensors) are exactly synchronous if no leakage occurs. At the slightest leakage, the correlation coefficient will deviate from one that triggers an alarm signal. If one variable were ignored in advance due to perfect correlation, the entire monitoring function would be lost. Paradoxical as it sounds, multi-collinearities contain the essential information.

There is another mathematical procedure for the substitution of the inverse correlation matrix by means of its adjoint matrix.

$$R^{-1} = \frac{R_{\text{adj}}}{\text{Det } R}$$

$R^{-1} =$ Inverse matrix
$R_{\text{adj}} =$ Adjoint matrix
Det $R =$ Determinant of matrix R.

The adjoint matrix is derived from the cofactors that are to be transposed after completion **(2-14*)**. The application of the Mahalanobis algorithm results in the squared distance D^2.

$$D^2 = \frac{1}{k} Y \frac{R_{\text{adj}}}{\text{Det } R} Y^T$$

Nothing is gained with the exact equation if the determinant of the correlation matrix disappears with multicollinearity of pairwise variables. Since the determinant of the correlation matrix Det R is nothing else but a proportionality factor, the division by the determinant can be ignored.

$$D^2_{\text{adj}} = \frac{1}{k} Y\, R_{\text{adj}} Y^T \left(= D^2 \cdot (\text{Det } R)\right)$$

The calculation accuracy is maintained with the disadvantage that the average value of the squared Mahalanobis distances of the objects belonging to the unit group calculated in this way can no longer be identical to one. However, the procedure does not change, since an analog threshold value is derived from the D^2_{adjoint} values of the unit group for separating non-related individuals. The method using the adjoint matrix instead of the inverse correlation matrix is completely equivalent with the advantage that even with frequently occurring correlation coefficients close to or equal to one no interpretation problems occur.

8.6 Mahalanobis Distance with the Adjoint Matrix

The two-parameter example is again cited to dealing with the adjoint matrix. The data are standardized, and the correlation matrix is calculated (Table 8.11).

The correlation matrix of the individuals in the unit group results in

$$R = \begin{pmatrix} 1 & 0.8 \\ 0.8 & 1 \end{pmatrix}$$

with its determinant

$$\text{Det } R = 0.3601$$

The inverse of the correlation matrix follows from the adjoint correlation matrix R_{adj} and the determinant of the correlation matrix Det R.

$$R_{\text{adj}} = \begin{pmatrix} 1 & -0.8 \\ -0.8 & 1 \end{pmatrix}$$

Table 8.11 Weight and height of the unit group (individuals 1–10) and two other individuals with standardized data

Individuals unit group i = 1–10	Weight (kg) ($x_{i,\ 1}$)	Height (m) ($x_{i,\ 2}$)	Standardized weight ($y_{i,\ 1}$)	Standardized height ($y_{i,\ 2}$)
1	55	1.61	−1.1286	−1.7619
2	60	1.75	−0.5699	−0.08390
3	50	1.69	−1.6873	−0.8030
4	65	1.72	−0.01117	−0.4435
5	68	1.77	0.3240	0.1558
6	62	1.66	−0.3464	−1.1626
7	75	1.81	1.1062	0.6352
8	76	1.86	1.2180	1.2345
9	61	1.82	−0.4581	0.7551
10	79	1.88	1.5532	1.4742
Average	65.10	1.757	0	0
Standard deviation	8.95	0.08343	1	1
Individuals not belonging to the unit group i = 11–12				
11	67	1.62	0.2123	−1.6420
12	77	1.73	1.3297	−0.3236

$$R^{-1} = \frac{1}{\text{Det } R} R_{\text{adj}} = \frac{1}{0.3601} \begin{pmatrix} 1 & -0.8 \\ -0.8 & 1 \end{pmatrix} = \begin{pmatrix} 2.7783 & -2.2217 \\ -2.2217 & 2.7783 \end{pmatrix}$$

This means that the squared distances in terms of the adjoint matrix differ about the determinant Det R as a proportionality factor. The average value of the individuals in the unit group has no longer the numerical value one. Nevertheless, the individuals not belonging to the unit group are to be distinguished (Table 8.12).

The graph illustrates the equivalence of both algorithms (Fig 8.5).

The threshold value for distinguishing individuals not belonging to the unit group is calculated in analogy to the Mahalanobis formalism as three times the average value of the D^2_{adjoint} figures of the unit group. The latter is identical to the determinant of the correlation matrix.

Threshold

$$D^2_{T,\text{adjoint}} > 3 \cdot \text{Det}(R)$$

For this numerical example,

$$D^2_{T,\text{adjoint}} > 3 \cdot 0.3601 = 1.08$$

Table 8.12 Squared distances D^2 from the Mahalanobis algorithm compared with D^2_{adj} from the adjoint correlation matrix for all individuals (1–12)

Individuals unit group i = 1–10	Standardized weight ($y_{i,\,1}$)	Standardized height ($y_{i,\,2}$)	Squared Mahalanobis distance D^2	Adjoint correlation matrix D^2_{adj}
1	−1.1286	−1.7619	1.6617	0.5983
2	−0.5699	−0.08390	0.3545	0.1277
3	−1.6873	−0.8030	1.8385	0.6620
4	−0.01117	−0.4435	0.2623	0.09443
5	0.3240	0.1558	0.06735	0.02425
6	−0.3464	1.1626	1.1489	0.4137
7	1.1062	0.6352	0.6985	0.2515
8	1.2180	1.2345	0.8357	0.3009
9	−0.4581	0.7551	1.8518	0.6668
10	1.5532	1.4742	1.2808	0.4612
Average	0	0	1	0.3601
Standard deviation	1	1		
Individuals not belonging to the unit group i = 11–12				
11	0.2123	−1.6420	4.5813	1.6496
12	1.3297	−0.3236	3.5567	1.2807

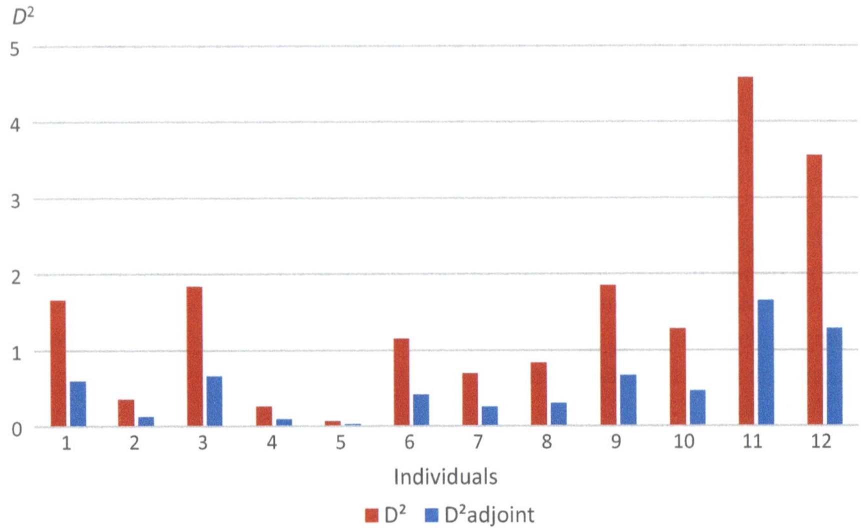

Fig. 8.5 Application of the squared distances D^2 and D^2_{adjoint} for all individuals 1–12

8.7 Clinical Case Study Revisited: Application of the Adjoint Correlation Matrix

For practice, data discrimination is performed with the same clinical case study by use of the adjoint correlation matrix. The starting point is the standardized data set in Table 7.5, and the adjoint correlation matrix gained from the cofactors (Table 8.13).

Determinant of the adjoint matrix:

$$\text{Det}\,R = 0.02343$$

The squared distances D^2_{adjoint} follows from:

$$D^2_{j,\text{adj}} = \frac{1}{k} Y_j R_{\text{adj}} Y_j^T$$

k = Number of parameters

j = 1, 2, 3…n, Number of individuals.

The threshold value for separating individuals belonging to the unit group from nonassociated individuals must be three times the numerical value of the determinant that is $D^2_{T,\text{adj}} = 0.07$ (Table 8.14; Fig. 8.6).

Table 8.13 Adjoint matrix of the correlation matrix *R*

	A	*B*	*C*	*D*	*E*	*F*	*G*	*H*	*I*	*J*	*K*
A	0.045	−0.008	0.0044	−0.006	0.0013	−0.003	−0.003	−0.009	0.0162	0.027	0.0018
B	−0.008	0.0435	−8E−04	−0.013	−0.008	0.0007	−0.015	0.0192	−0.008	−1E−04	0.0022
C	0.0044	−8E−04	0.0557	−0.039	−0.002	0.0051	−0.002	0.0059	−0.004	−0.003	0.0043
D	−0.006	−0.013	−0.039	0.0608	0.0061	−9E−04	0.0076	−0.004	−0.008	−0.003	−0.011
E	0.0013	−0.008	−0.002	0.0061	0.0429	0.0016	−0.01	−0.003	−0.002	−0.013	−0.014
F	−0.003	0.0007	0.0051	−9E−04	0.0016	0.0322	−2E−04	0.0034	−7E−04	−0.019	−8E−04
G	−0.003	−0.015	−0.002	0.0076	−0.01	−2E−04	0.0414	−0.009	−0.006	−0.01	0.0018
H	−0.009	0.0192	0.0059	−0.004	−0.003	0.0034	−0.009	0.0401	−0.018	−0.008	−0.003
I	0.0162	−0.008	−0.004	−0.008	−0.002	−7E−04	−0.006	−0.018	0.044	0.0054	0.0082
J	0.027	−1E−04	−0.003	−0.003	−0.013	−0.019	−0.01	−0.008	0.0054	0.0609	0.0033
K	0.001	0.0022	0.0043	−0.011	−0.014	−8E−04	0.0018	−0.003	0.0082	0.0033	0.0307

Table 8.14 Squared distances D^2_{adj} for the individuals of the unit group 1–34 and the nonassociated patients 35–39

Patient	1	2	3	4	5	6	7	8	9	10
D^2_{Adjoint}	0.0228	0.0241	0.0212	0.0276	0.0235	0.0162	0.0312	0.022	0.0295	0.0182
Patient	11	12	13	14	15	16	17	18	19	20
D^2_{Adjoint}	0.0205	0.0405	0.0194	0.0215	0.0174	0.0124	0.0282	0.0148	0.0131	0.0198
Patient	21	22	23	24	25	26	27	28	29	30
D^2_{Adjoint}	0.0431	0.0527	0.0262	0.0467	0.0173	0.0104	0.0107	0.031	0.0247	0.0139
Patient	31	32	33	34						
D^2_{Adjoint}	0.011	0.0229	0.0322	0.0101						
Patient	35	36	37	38	39					
D^2_{Adjoint}	0.2424	0.6002	0.0991	0.1783	0.1392					

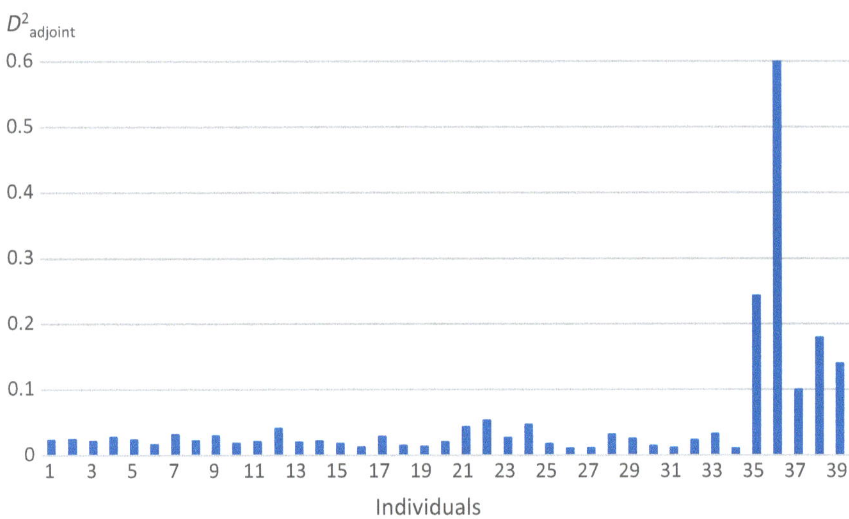

Fig. 8.6 Illustration of the squared distances D^2_{adjoint} of the members of the unit group 1–34 and the nonassociated individuals 35–39

8.8 Parameter Screening and Root-Cause-Analysis

In the presented case study with 11 parameters, the partially orthogonal array L12 is used. The calculation of the squared distances D^2_{adjoint} means that the determinant of the correlation matrix has a different numerical value for each required test calculation due to different parameter configurations. Therefore, the distances

D^2_{adjoint} of one test calculation is changed in the same way by a factor, but individually different for each test. The orthogonal array cannot be evaluated in this form. Alternatively, instead of the determinant of the correlation matrix, if it is very close to zero, another numerical value can be used to compensate the offset. This correction factor must be chosen such that for each test the average values of all individuals in the unit group assume the same numerical value.

In the case of division by the determinant, the average value would always be one, and thus identical to the value when using the Mahalanobis algorithm, otherwise a different numerical value. The numbers of the *SNR* indices of the evaluation of the orthogonal array can thus be shifted by a constant amount, the effect amplitude as difference of the setting values remains unaffected. Alternatively, a procedure without an orthogonal array can be used as an approximation method. As shown in Sect. 8.3, the square of the standardized variables is already a reliable indicator. With reference to the overall effect, the percentage contribution of each standardized variable can be displayed. Parameter screening and root-cause-analysis, therefore, lead to similar evaluations, at least for the greatest parameter effects, comparable to the parameter analysis with an orthogonal array (Fig. 8.7; Table 8.15).

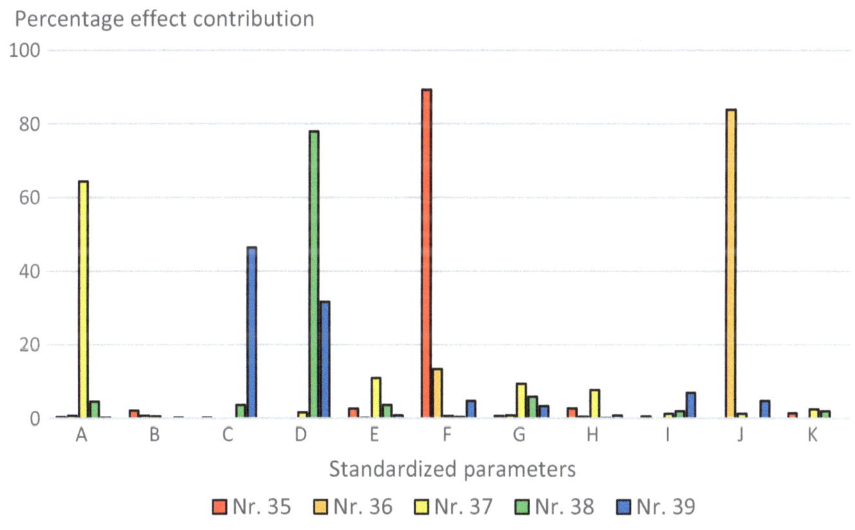

Fig. 8.7 Graphical representation of the parameter effects of the individuals 35–39

Table 8.15 Effect contribution in percent of the standardized variables for the individuals 35–39

Patient	35	36	37	38	39
Variable	%	%	%	%	%
A	0.41	0.71	64.31	4.58	0.22
B	2.12	0.70	0.55	0.03	0.24
C	0.18	0.04	0.02	3.65	46.42
D	0.07	0.00	1.57	77.87	31.58
E	2.61	0.12	10.94	3.59	0.87
F	89.23	13.34	0.60	0.36	4.75
G	0.61	0.82	9.40	5.91	3.37
H	2.73	0.46	7.65	0.12	0.80
I	0.53	0.00	1.29	1.94	6.90
J	0.06	83.81	1.24	0.01	4.77
K	1.45	0.00	2.45	1.95	0.08

Important items to memorize To derive the squared Mahalanobis distances, there exist alternative algorithms besides the Mahalanobis formalism.

The use of alternative algorithms leads to similar but not identical results of the parameter screening.

Depending on the properties of a data set, the most suitable algorithm is to be used. For small data sets, especially, if the number of objects is only slightly larger than the number of variables, the Gram–Schmidt formalism provides the most reliable values for the squared Mahalanobis distances.

Chapter 9
Methodical Extensions

Contents

Abstract There are sets of data without a clear separation of distinct patterns. This can be caused either by a high degree of complexity or insufficient homogeneity of the applied unit group. In the former situation, the discrimination threshold is adjusted in a way which ascertains that the sum of the error rates in the groups to be distinguished can be minimized. In the latter situation, the unit group is scrutinized regarding homogeneity, refined, and tried again. The latter is also used as a method to begin with if only one set of data exists to investigate if the data set is composed of hidden patterns. Thus, historic or unique data can be analyzed retrospectively to gain additional information. In the case of great amount of data, the individual Mahalanobis distances can be condensed into multiples to reduce data complexity at the risk of less accuracy. If patterns change gradually, for instance as a function of time, Mahalanobis data can be treated as a dynamic system. For therapeutic purposes, the *SNR* index is calculated individually for each patient to figure out the most efficient therapy with respect to time lapse.

9.1 Adjustments to the Threshold Value

The threshold value for distinguishing objects as belonging or not belonging to the unit group is previously described by the numerical value three from the threefold standard deviation of the individuals of the unit group. Without overlapping of the

© Springer Nature Switzerland AG 2019

H. Ruefer, *Living Without Mathematical Statistics*,
https://doi.org/10.1007/978-3-319-99632-5_9

objects belonging to the unit group or lying outside, a uniform distance between the two object groups can also be used to adjust the threshold value.

$$D_T^2 = \frac{D^2_{\text{max, within the unit group}} + D^2_{\text{min, outside the unit group}}}{2}$$

However, this does not represent a rigid limit but can be adapted to circumstances, especially if the discriminating power is insufficient and the object affiliation overlaps (Fig. 9.1).

The practical importance of the threshold value can be seen equivalent to the definition of the LD50 limit value. Objects or individuals that are very close to or have this value cannot be distinguished in terms of their affiliation. In the case of an intact technical product, this leads to rejection because it is incorrectly classified as defective, or vice versa, a defective product is used that cannot fulfill its function. If a medical diagnosis is based solely on the squared Mahalanobis distance, this means that in borderline cases, an individual is treated, although no treatment is necessary and vice versa. The same applies to sensor systems that respond at the threshold without an alarm situation or do not react, although danger circumstances are looming. Each situation is a serious mistake and entails follow-up costs.

If the costs are known or can be estimated, the threshold value can be adjusted in such a way that possible losses with the same frequency of occurrence of each type of fault are approximately equal. In the case of a sufficiently large amount of data with known affiliation of the individuals, the threshold value is to be set at the intersection of the two groups. If the economic losses of the misclassification of objects not belonging to the unit group are low compared to the incorrect valuation of a member of the unit group, the threshold value must be shifted to 4 or a larger number, for the opposite situation to values in the range 2–2.5.

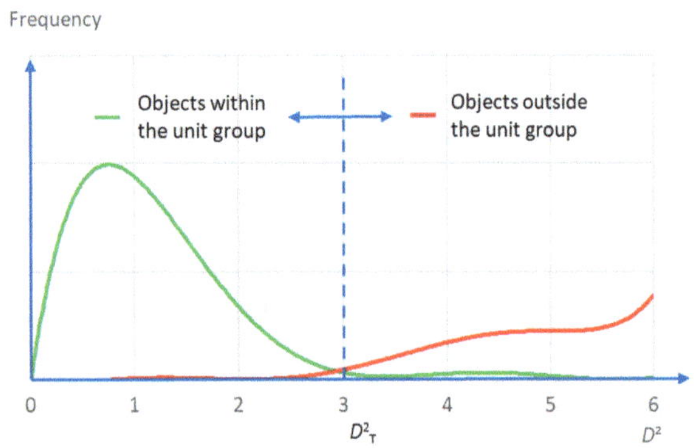

Fig. 9.1 Overlapping of object affiliation in case of insufficient discrimination power

9.2 Consideration of Sources of Noise

The integration of sources of noise is an essential prerequisite for guaranteeing robust behavior of parameter effects. The same applies to the field of pattern recognition, whereby one repetition means twice the effort. Since the number of objects considered for data discrimination based on the Mahalanobis algorithm is already relatively large, there is usually no such additional effort. If the formulation of a Mahalanobis metric with the same variables is still technically feasible, e.g., at two different locations (clinics, production facilities, etc.), the information gained consists on the one hand in ensuring the classification results and on the other hand in a robust parameter screening. Parameter effect and root-cause-analysis are based on one repetition of data under other accompanying circumstances, i.e., sources of noise, and are therefore more reliable. Data acquisition and evaluation with one repetition can be performed in several ways, whereby all situations can be converted into one another.

- Separation according to the noise levels (e.g., different locations); repetition is included in parameter screening
- Integration as variable; evaluation of the noise factor included in the parameter screening
- Integration of both noise levels; effect balanced by a common squared Mahalanobis distance; data discrimination is robust if the results in parameter screening for variable set $N1$ and variable set $N2$ are the same (Fig. 9.2).

9.3 Multiple Mahalanobis Distance

Image identification, speech recognition, and spectral analyses have become very complex regarding the number of characterizing variables. Several hundreds or thousands can come together, raising the question of an efficient reduction in complexity. To accomplish such a task, subgroups of squared Mahalanobis distances can be formed and merged into a multiple group. First, the schematic procedure is explained, followed by the example of the clinical case study, divided into subgroups (Fig. 9.3).

Each subgroup must be supplemented with the same objects outside the unit group. The number of objects or individuals must always be greater than the total number of variables, i.e., $n > g + h + i + j$. The number of variables can vary from subgroup to subgroup (Fig. 9.4).

Parameter screening is first performed at the level of the variables in the subgroups. Depending on the results, some of the subgroups may already be completely omitted. In the next step, all Mahalanobis distances of the remaining subgroups are combined, standardized, and the squared multiple Mahalanobis distances are calculated from them. In a further screening, the subgroups that do not contribute to the multiple squared distance can be sorted out.

Noise level 1

	Y_1	…	…	Y_k	D^2
1					
…					
…					
n					

Noise level 2

	Y_1	…	…	Y_k	D^2
1					
…					
…					
n					

Noise variable N

	Y_1	…	…	Y_k	N	D^2
1					1	
…					1	
…					1	
n					1	
n+1					2	
…					2	
…					2	
2n					2	

Noise levels N1, N2

	N1 Y_1	N1 …	N1 …	N1 Y_k	N2 Y_1	N2 …	N2 …	N2 Y_k	D^2
1									
…									
…									
n									

Fig. 9.2 Different approaches to consider a noise factor N at two levels

Sub group 1

	Y_1	Y_2	...	Y_g	D^2_1
1					D^2_{11}
2					D^2_{12}
3					D^2_{13}
...					...
n					D^2_{1n}

Sub group 2

	Y_{g+1}	Y_{g+2}	...	Y_{g+h}	D^2_2
1					D^2_{21}
2					D^2_{22}
3					D^2_{23}
...					...
n					D^2_{2n}

Sub group 3

	Y_{g+h+1}	Y_{g+h+2}	...	Y_{g+h+i}	D^2_3
1					D^2_{31}
2					D^2_{32}
3					D^2_{33}
...					...
n					D^2_{3n}

Sub group 4

	$Y_{g+h+i+1}$	$Y_{g+h+i+2}$	...	$Y_{g+h+i+j}$	D^2_4
1					D^2_{41}
2					D^2_{42}
3					D^2_{43}
...					...
n					D^2_{4n}

Fig. 9.3 Example for a separation into four extendable unit subgroups from a set of $g + h + i + j$ variables

Fig. 9.4 Merging of the non-squared Mahalanobis distances of the subgroups as a new variable to determine the squared multiple Mahalanobis distances

	D_1	D_2	D_3	D_4	...	D_k	MD^2
1	D_{11}	D_{21}	D_{31}	D_{41}	...	D_{k1}	MD^2_1
2	D_{12}	D_{22}	D_{32}	D_{42}	...	D_{k2}	MD^2_2
3	D_{13}	D_{23}	D_{33}	D_{43}	...	D_{k3}	MD^2_3
...	...	...	...	...	...	...	...
n	D_{1n}	D_{2n}	D_{3n}	D_{4n}	...	D_{kn}	MD^2_n

Depending on the application, not only many variables can be assigned to subgroups but numerous subgroups can also be formed based on a few variables (at least 2). The latter option is advantageous for the recognition of letters and numbers (known as character recognition), so that each symbol defines its own unit space.

9.4 Clinical Case Study Revisited: Application of Squared Multiple Mahalanobis Distance

The starting point is the standardized patient data from Sect. 7.3. For example, 11 parameters can be divided into 2 subgroups with 6 and 5 variables. The squared Mahalanobis distances of each subgroup are determined (Tables 9.1 and 9.2).

Table 9.1 Squared Mahalanobis distances of variables *A–F* of the first subgroup

Patient	*A*	*B*	*C*	*D*	*E*	*F*	D_1^2
1	0.8051	−0.0886	0.5105	1.0748	−1.5251	−0.5927	0.6672
2	1.6800	−0.5990	−0.6120	−0.3048	−0.1112	−0.3151	0.6130
3	−1.1947	−0.1795	1.9376	0.5055	0.7184	−0.3672	1.1698
4	−0.3198	2.9449	0.1741	1.1845	0.3780	−0.0765	1.8824
5	0.1801	1.9371	1.3524	0.7894	1.6474	0.8136	1.1714
6	−1.3197	−1.3291	−0.4829	−1.2195	0.6000	−0.0062	0.7865
7	1.0551	0.7476	0.6290	−1.1188	−0.0240	−0.5248	1.4792
8	−1.8197	0.1440	0.3185	−0.9189	1.2689	0.0162	1.0957
9	−1.5697	−0.4423	0.5136	0.1055	0.0043	−0.3629	0.5720
10	−0.8198	−0.2795	1.0334	1.3582	−0.6569	−0.1919	0.6535
11	0.3051	0.0504	−0.7662	−0.0442	0.8854	−0.6036	0.4913
12	1.1800	1.0284	3.3069	3.6624	−0.6740	−0.4964	2.8428
13	0.3051	−0.3562	−0.4402	−0.5701	−1.8125	0.0659	0.6304
14	0.5551	−1.2282	−0.7596	−0.4201	0.4075	−0.6031	0.5751
15	0.8051	−0.2276	−0.3381	−0.2261	1.6929	−0.3969	0.9462
16	1.1800	−0.5527	−0.9340	−0.5953	−0.4871	−0.3933	0.3582
17	−0.4448	0.1816	−0.5222	−0.6501	−0.3012	−0.4717	0.2736
18	1.9300	−1.2647	−1.2467	−1.1039	−1.0295	−0.4061	0.9888
19	0.9301	−1.3661	−1.2081	−1.0049	0.1055	−0.4083	0.6610
20	0.4301	−0.8744	−0.4772	−0.5463	0.2086	−0.4837	0.2688
21	−1.5697	0.2379	1.1815	1.1335	1.0727	0.5922	0.7048
22	−0.1948	−1.0096	−0.5129	−0.5004	−1.4190	3.8977	3.4187
23	−1.8197	−0.7961	0.1300	−0.5398	0.3203	−0.2814	0.7853
24	−0.5698	0.7872	−0.5656	0.0430	2.8553	3.4506	3.0551
25	−0.8198	−0.5999	0.0754	−0.7008	−0.8584	1.1101	0.7134
26	0.3051	−0.1180	−0.9995	−0.2304	−0.4206	−0.3686	0.3050
27	0.4301	−0.7607	0.1584	0.3731	−0.5991	−0.6050	0.2680
28	1.4300	0.9050	1.2441	−0.4681	−1.0505	−0.5108	1.8075
29	−0.5698	−0.4724	0.6948	1.8213	0.4610	−0.2123	1.0923
30	0.6801	1.4869	0.3168	0.4771	0.9277	−0.2703	0.5499
31	−0.3198	0.2304	−1.2460	−1.0268	−0.7886	−0.5030	0.6831
32	0.3051	2.1472	−0.7369	−0.1159	−0.2115	−0.4856	1.3699
33	−0.5698	−0.1718	−1.1375	−0.5536	−1.0131	0.2101	0.6240
34	−0.5698	−0.1122	−0.5910	0.3303	−0.5716	−0.2188	0.4962
Average	0	0	0	0	0	0	1
Sigma	1	1	1	1	1	1	
35	−0.5698	−1.3025	−0.3776	−0.2299	−1.4437	8.4422	13.8742
36	−1.0698	−1.0629	−0.2584	−0.0524	0.4301	4.6246	3.8550
37	4.3048	−0.3981	0.0673	−0.6723	−1.7756	−0.4148	3.8078
38	1.4300	0.1188	1.2765	5.8986	1.2665	−0.4004	10.7370
39	−0.4448	0.4684	6.4555	5.3244	0.8854	2.0645	9.3035

Table 9.2 Squared Mahalanobis distances of variables *A–F* of the second subgroup

Patient	*G*	*H*	*I*	*J*	*K*	D_2^2
1	0.5634	−0.9705	−0.8729	−1.2134	−0.9480	1.0084
2	−0.1774	−0.6480	−0.3587	0.2222	1.0393	0.3496
3	−0.8001	−0.6082	−0.4043	0.4523	0.3388	0.3713
4	0.1521	−0.4453	0.7419	0.0950	−0.3596	0.2469
5	0.8079	−1.1493	0.6478	−0.2085	1.2812	1.2358
6	−0.3297	0.1254	0.3206	1.0828	−1.1517	0.6982
7	2.0196	−0.2375	−0.2357	0.0557	0.0150	1.1788
8	1.0759	−0.3295	1.6633	0.7476	−0.8303	0.9535
9	1.1376	1.4233	1.5427	−0.7819	0.3985	1.3618
10	−0.2324	0.5932	2.1954	0.6717	−0.5598	1.2659
11	−0.5756	0.5110	1.5489	−0.7811	−0.0752	0.9818
12	−0.8059	0.2082	1.4532	−0.5061	0.0257	0.9209
13	0.1199	0.3375	1.2661	−0.3176	−2.0442	1.1257
14	−1.5787	1.0199	−0.3836	−1.4852	1.3981	1.6212
15	0.3544	1.5746	−0.0291	−0.5013	0.9392	0.8823
16	−0.1982	1.3905	−0.4121	−1.0834	0.7438	1.0064
17	2.1792	2.4057	1.2581	0.5960	−0.1993	1.7883
18	−0.6599	−0.1745	−1.5985	−1.3465	0.5348	0.7911
19	−1.1152	0.4827	−1.0645	−0.4722	−0.5884	0.7134
20	0.2699	0.7778	−0.5723	−0.1883	−2.0575	1.3988
21	0.7166	2.3531	0.0664	3.0666	0.4696	3.0307
22	−0.9871	0.2120	−1.0702	0.0029	0.1200	0.5005
23	−0.1989	−0.5537	−0.9146	1.1047	1.9965	1.3299
24	2.5864	0.5947	1.0907	3.0006	0.3299	2.2459
25	−0.8386	−1.3299	−0.0150	0.0459	−1.6654	1.0262
26	−0.6154	−0.6600	−0.7926	0.3012	−0.7056	0.4242
27	−0.5411	−1.2420	−1.2124	−0.7773	−0.2672	0.4668
28	−1.4258	−0.7141	−0.7734	−0.7376	−1.0613	0.6422
29	0.8457	−0.5394	0.0274	0.2609	1.2183	0.5824
30	0.6520	−1.2230	−1.1059	−0.4979	1.1406	1.0150
31	−0.8969	−1.0950	−0.5769	−0.5901	−0.4568	0.3448
32	0.3672	−1.3903	−1.1513	−0.6593	−0.6402	0.8235
33	−1.0687	−0.3773	0.5774	0.6295	1.7189	1.4411
34	−0.8020	−0.3222	−0.8561	−0.1881	−0.0978	0.2263
Average	0	0	0	0	0	1
Sigma	1	1	1	1	1	
35	0.6985	1.4759	0.6534	−0.2180	1.0767	0.7984
36	−1.1443	−0.8545	0.0055	11.5934	0.0778	38.2054
37	−1.6456	−1.4846	−0.6108	−0.5966	0.8396	1.0310
38	1.6251	−0.2346	−0.9305	−0.0725	0.9327	1.2404
39	1.7388	−0.8467	2.4885	2.0697	0.2604	2.7371

Each subgroup can be subjected to a parameter screening. Since this is already known from Sect. 7.6, only the subsequent steps are dealt with. For this purpose, the results D^2 are returned to the unsquared Mahalanobis distance and standardized again. This means that each subgroup contributes to the multiple group with exactly one variable. From this reduced set of variables, the squared Mahalanobis distance, now called squared multiple Mahalanobis distance MD^2, is calculated (Table 9.3).

Even if, as in this example, the number of variables is drastically reduced from 11 to 2, as can be seen in the graph, the non-related individuals 35–39 are distinguished from the unit group (Fig. 9.5).

Table 9.3 Subgroups as new standardized variables for calculating the squared multiple Mahalanobis distance

Patient	D_1	D_2	$D_{1,\text{ standardized}}$	$D_{2,\text{ standardized}}$	MD^2
1	0.8168	1.0042	−0.3805	0.1560	0.0816
2	0.7829	0.5912	−0.4824	−1.3378	1.0484
3	1.0816	0.6094	0.4165	−1.2722	0.8704
4	1.3720	0.4969	1.2908	−1.6791	2.1340
5	1.0823	1.1117	0.4187	0.5449	0.2489
6	0.8869	0.8356	−0.1696	−0.4538	0.1218
7	1.2162	1.0857	0.8218	0.4512	0.4604
8	1.0468	0.9765	0.3117	0.0559	0.0512
9	0.7563	1.1670	−0.5627	0.7450	0.4148
10	0.8084	1.1251	−0.4058	0.5936	0.2464
11	0.7009	0.9909	−0.7294	0.1079	0.2684
12	1.6861	0.9596	2.2362	−0.0051	2.5067
13	0.7940	1.0610	−0.4492	0.3617	0.1582
14	0.7584	1.2733	−0.5564	1.1296	0.7616
15	0.9728	0.9393	0.0889	−0.0786	0.0067
16	0.5985	1.0032	−1.0377	0.1525	0.5432
17	0.5230	1.3373	−1.2648	1.3611	1.6397
18	0.9944	0.8895	0.1541	−0.2590	0.0434
19	0.8130	0.8446	−0.3919	−0.4212	0.1747
20	0.5184	1.1827	−1.2787	0.8020	1.0879
21	0.8395	1.7409	−0.3121	2.8214	3.9933
22	1.8490	0.7075	2.7265	−0.9173	4.0166
23	0.8861	1.1532	−0.1718	0.6953	0.2509
24	1.7479	1.4986	2.4222	1.9449	5.0886
25	0.8446	1.0130	−0.2967	0.1880	0.0589
26	0.5522	0.6513	−1.1769	−1.1204	1.3939
27	0.5177	0.6833	−1.2810	−1.0049	1.3975
28	1.3444	0.8014	1.2077	−0.5775	0.8616

(continued)

Table 9.3 (continued)

Patient	D_1	D_2	$D_{1,\ \text{standardized}}$	$D_{2,\ \text{standardized}}$	MD^2
29	1.0451	0.7631	0.3068	−0.7160	0.2925
30	0.7416	1.0075	−0.6069	0.1680	0.1934
31	0.8265	0.5872	−0.3514	−1.3525	1.0044
32	1.1704	0.9075	0.6840	−0.1939	0.2464
33	0.7900	1.2004	−0.4613	0.8661	0.4616
34	0.7044	0.4757	−0.7188	−1.7558	1.8719
Average	0.9432	0.9610	0	0	1
Sigma	0.3322	0.2764	1	1	
35	3.7248	0.8935	8.3732	−0.2443	35.0748
36	1.9634	6.1811	3.0710	18.8848	186.6261
37	1.9514	1.0154	3.0348	0.1966	4.6689
38	3.2767	1.1137	7.0244	0.5524	25.0994
39	3.0502	1.6544	6.3424	2.5085	24.1691

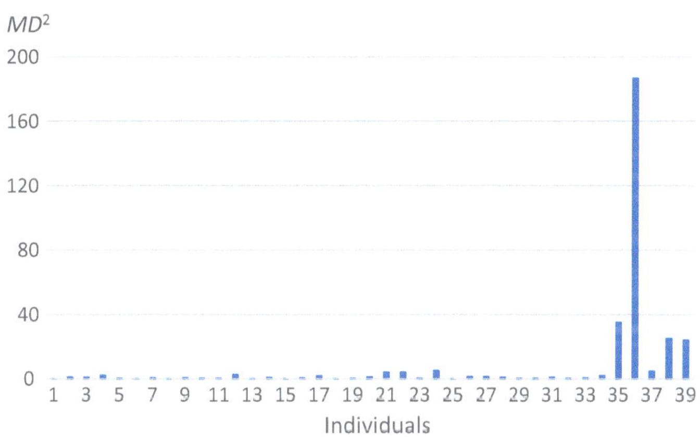

Fig. 9.5 Graphical representation of the squared multiple Mahalanobis distances of the individuals 1–39

9.5 Evaluations of Therapies

Apart from a purely medically based examination of different therapies by a physician, the time-based development of the squared Mahalanobis distance can be used as an accompanying measure for an objective evaluation. As a matter of fact, on the clinical side, the appearance of a patient is described as better or worse because of the large natural variability of the values of most physiological variables. Consequently, the effectiveness of new drugs has to be evaluated for statistical reasons for larger groups of patients up to cohort size even at the expense of clinical relevance. Thus, a certain therapy will be advantageous for the majority of the groups, others, however, have no benefit or in extreme cases, their condition can even worsen. Another aspect is the efficiency of a therapy, i.e., to emphasize the effectiveness as a function of time. In certain cases, a slower healing process with less side effects, and, therefore, more beneficial to the patient, is overlooked. Typically, placebo therapies fall into this category, which—in case of patients sensitive to placebo effects—may reach an identical state of health, but over a longer time span. As a matter of fact, medical therapies are evaluated on their efficiency, and side-effects are tolerated up to a level still considered acceptable. For hospitals, this means that time is the characteristic and decisive factor, i.e., when a patient can be moved from the intensive care unit (ICU) or discharged from stationary care. By comparison, the coverage of the individual patient yields more details.

The definition of a healthy group of people is of utmost importance because the values of physiological and other factors are used to create the unit space. The unit space is the basis for characterizing individuals. Patients who need a therapy are distinguished from the unit group if the value of the squared Mahalanobis distance is

$$D^2 \gg 3$$

In case this distance diminishes continuously towards the healthy group, the applied therapy is successful. The advantage of applying the Mahalanobis algorithm is based on only one objective, continuously varying figure instead of a categorical, qualitative evaluation. For describing the progress of a therapy which means the change of the squared Mahalanobis distance with time, an exponential function can be assumed.

$$D^2 = D_0^2 e^{-\beta t}$$

D^2 : squared Mahalanobis distance applied to medical treatment for the time t
D_0^2 : squared Mahalanobis distance at the starting point of the medical therapy
t : time (unit: days [d])
β : diminishing constant [1/d]

Provided a therapy is successful the best fit coefficients for the exponential function based on some serial measurements are used to calculate the squared Mahalanobis distance. When the extrapolated function hits the intersection for $D^2 = 3$, the length of time can be concluded, i.e., when the results concerning the patient are close enough to those of the (healthy) unit group, and the patient needs no further care (Fig. 9.6).

In addition to individual patient monitoring, different therapies or, more commonly, differences in smaller test and control groups can be discriminated

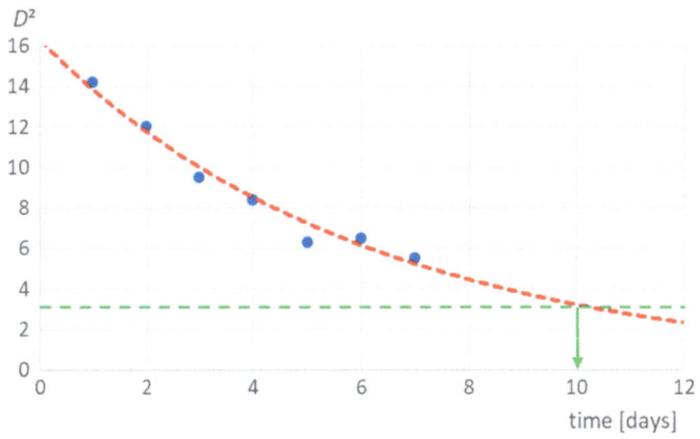

Fig. 9.6 Development of a patient's successful therapy as a function of time

quantitatively. The appropriate characteristic dynamic *SNR* index is derived of the linearized exponential function.

$$\frac{D^2}{D_0^2} = e^{-\beta t}$$

$$\ln\left(\frac{D^2}{D_0^2}\right) = -\beta t$$

$$y = \ln\left(\frac{D^2}{D_0^2}\right)$$

$$M = t$$

$$y = -\beta M$$

$$SNR = 10\log\left(\frac{\beta^2}{\sigma^2} S_2\right)$$

For example, different pain therapies after a surgery are displayed in the figure below (Fig. 9.7).

The slope of the dynamic system behavior indicates the effectiveness of the respective therapy (Table 9.4).

The time-based recording of the squared Mahalanobis distance is useful for continuously monitoring patients. Instead of controlling single physiological parameter values which trigger an alarm if certain lower or upper limits are exceeded, the squared Mahalanobis distance considers all measured variables including their correlation pattern. Based on expert knowledge, the squared

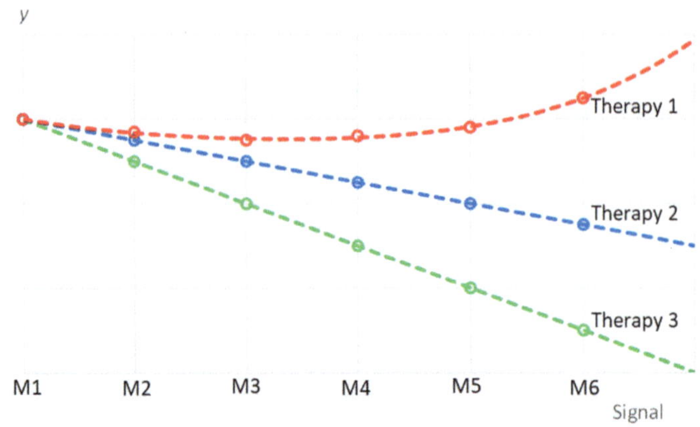

Fig. 9.7 Comparison of different painkilling therapies

Table 9.4 Medical therapies as characterized by their dynamic *SNR* indices

Type	Characteristic dynamic *SNR* values $SNR_{Therapie1} < SNR_{Therapie2} < SNR_{Therapie3}$	Result
Therapy 1	Slope changes sign	Therapy must be stopped immediately
Therapy 2	$Slope_{Therapy2} < Slope_{Therapy3}$	Patient matches unit group later in time
Therapy 3	$Slope_{Therapy3} > Slope_{Therapy2}$	Patient matches unit group sooner in time

Mahalanobis distance outside the unit group can be scaled for categorization the degree of deviation. Thus, with the definition of a unit group composed of healthy people or at least those in an uncritical state, a single control limit can be established such as

$$D^2 \geq 50$$

to set up an effective and transparent online monitoring system in real time.

9.6 Data Discrimination Without Unit Group

A proof of the consistency of an object classification is based on a proposal from P. C. Mahalanobis. According to him, data discrimination is repeatedly carried out with one-half of the unit group. If the composition of the objects in the unit group is relatively homogeneous, there will be little change in the result. This thought can be pushed to its extreme that, reduced to single elements, only one object at a time is removed from the unit group for test purposes and assigned to the nonassociated objects **(2–15*)**. If the numerical value of the squared Mahalanobis distance of an element in the unit group is close to 2 or smaller, it will be below the threshold value if interchanged outside the unit group. If, on the other hand, the numerical value is 3 or greater, the threshold can be exceeded. If the objects of the unit group are selected according to conspicuously large Mahalanobis distances, which are already close to the threshold value, the test result can by far exceed the threshold value. In this case, the object, which was previously assigned to the unit group, is better removed from it.

Note that, each element or object taken from the unit group changes the correlation pattern. Therefore, the extraction of objects from a total set of data must be done rarely and carefully, and assuming appropriate expertise, must be justified by conspicuous parameter values or value combinations. Even without knowledge of a unit group, the individual test offers a starting point for separating conspicuous objects from the correlation pattern of the majority of data from a larger data set. This can be useful for retrospective detailed analyses such as interpreting historical data that neither can be reproduced nor verified.

The following example serves as an illustration. A data set consisting of 33 objects with 7 variables must be scanned for conspicuous objects. If any are to be identified, they are removed from the unit group. The remaining objects can be defined as a slightly modified unit group and the unrelated objects become available for a more detailed examination. After standardization of the available data, the Mahalanobis algorithm is to be applied (Tables 9.5 and 9.6).

The graphic illustration of the squared Mahalanobis distances makes interpretation easier (Fig. 9.8).

The squared Mahalanobis distance of patient no. 9 has a relatively high numerical value. This is removed from the entire group and the Mahalanobis algorithm is performed again with a preliminary unit group consisting now of 32 patients. Patient No. 9 is considered outside the unit group and the associated squared distance is calculated based on the modified unit group. The result of the squared Mahalanobis distances is shown in the graph (Fig. 9.9).

To verify the decision to remove patient no. 9, the squared Mahalanobis distances of the entire group must be compared with the new group consisting of 32 individuals. The correlation pattern is valid if the D^2 values of the individuals are essentially the same, that is, patient no. 9 possesses indeed different characteristic values and does not match the correlation pattern (Fig. 9.10).

Except for individual no. 24, the squared Mahalanobis distances of the members of the unity group are essentially unchanged. Patient 24 with a value of 3.8 is another conspicuous candidate that can be screened out for test purposes. This reduces the size of the unit group to 31 individuals. The squared Mahalanobis distances, excluding individual no. 9 and no. 24 from the unit group, are shown graphically (Fig. 9.11).

Table 9.5 Historical clinical case study consisting of 33 objects with 7 variables

Variable	*A*	*B*	*C*	*D*	*E*	*F*	*G*
Patient							
1	219	12.3	1	17.5	11.2	0.2	472
2	194	19.8	1.4	9	18.3	0.4	353
3	194	39.9	1.3	11.8	16.6	0.3	237
4	436	22.9	1.3	13.9	17.3	0.5	335
5	413	40.4	1	15.8	19.6	1.4	514
6	78	15.7	1.1	1.8	18.4	0.6	274
7	327	31.8	1.1	2.6	18.2	0.2	680
8	187	20.4	1.2	3	17.1	0.6	393
9	87	18	1.2	7.6	9.6	7.8	425
10	159	26.3	0.9	14.4	12.4	0.4	297
11	235	15.5	1.2	9.4	19.7	0.1	270
12	313	53.5	1.2	30.3	14.9	0.2	239
13	190	18.8	1.1	5.7	9.3	0.7	377
14	97	14.8	0.7	6.7	17.7	0.1	145
15	209	20.2	1	8	23.6	0.3	401
16	165	14.3	1.1	5.7	15.4	0.3	325
17	211	15.1	1.1	4.7	13.9	0.2	554
18	110	11.8	1.1	3.1	13.4	0.3	310
19	94	11.9	1.2	3.6	18.2	0.3	216
20	131	19.2	1.3	6.5	16.9	0.3	380
21	155	22.4	1.6	4.9	11.5	1.4	234
22	206	12.4	1.4	7.8	14.6	0.3	252
23	193	24.7	1.4	5.6	9.6	0.3	147
24	252	81.5	1.7	38	19.7	4.2	521
25	148	22.4	1	11.8	15.4	0.1	262
26	333	36	1.3	6.7	13.4	0.2	172
27	180	30	1.2	21	17.7	0.4	405
28	425	29.5	1.1	14	20.9	0.4	472
29	243	10.1	1.1	3.4	19.8	0.2	268
30	410	14.3	1	8.1	20.2	0.2	369
31	191	9.9	1.4	5.6	11.7	0.8	188
32	182	14.2	0.8	9.8	15.8	0.4	215
33	237	34.4	1.2	43.8	23.1	0.3	551
Average	218.303	23.770	1.173	10.958	16.215	0.739	341.000
Standard deviation	96.962	96.962	0.203	9.634	3.722	1.439	128.705

Table 9.6 Standardization of the data and calculation of the squared Mahalanobis distance with the inverse correlation matrix

Variable	*A*	*B*	*C*	*D*	*E*	*F*	*G*	D^2
Patient								
1	0.0072	−0.8010	−0.8491	0.6791	−1.3476	−0.3747	1.0178	1.4220
2	−0.2506	−0.2772	1.1173	−0.2032	0.5602	−0.2358	0.0932	0.4462
3	−0.2506	1.1264	0.6257	0.0874	0.1034	−0.3052	−0.8080	0.5898
4	2.2452	−0.0607	0.6257	0.3054	0.2915	−0.1663	−0.0466	1.1670
5	2.0080	1.1613	−0.8491	0.5026	0.9095	0.4589	1.3442	1.0314
6	−1.4470	−0.5635	−0.3575	−0.9506	0.5871	−0.0968	−0.5206	0.6512
7	1.1210	0.5608	−0.3575	−0.8675	0.5333	−0.3747	2.6339	2.0077
8	−0.3228	−0.2353	0.1341	−0.8260	0.2378	−0.0968	0.4040	0.2807
9	−1.3542	−0.4029	0.1341	−0.3485	−1.7775	4.9050	0.6527	4.0239
10	−0.6116	0.1767	−1.3407	0.3573	−1.0251	−0.2358	−0.3419	0.7839
11	0.1722	−0.5775	0.1341	−0.1617	0.9364	−0.4442	−0.5516	0.3789
12	0.9766	2.0762	0.1341	2.0078	−0.3534	−0.3747	−0.7925	1.4376
13	−0.2919	−0.3470	−0.3575	−0.5457	−1.8581	−0.0274	0.2797	0.7591
14	−1.2510	−0.6264	−2.3240	−0.4419	0.3990	−0.4442	−1.5229	1.4615
15	−0.0959	−0.2493	−0.8491	−0.3070	1.9843	−0.3052	0.4662	0.8025
16	−0.5497	−0.6613	−0.3575	−0.5457	−0.2190	−0.3052	−0.1243	0.1003
17	−0.0753	−0.6054	−0.3575	−0.6495	−0.6221	−0.3747	1.6549	0.9591
18	−1.1170	−0.8359	−0.3575	−0.8156	−0.7564	−0.3052	−0.2409	0.3060
19	−1.2820	−0.8289	0.1341	−0.7637	0.5333	−0.3052	−0.9712	0.5690
20	−0.9004	−0.3191	0.6257	−0.4627	0.1840	−0.3052	0.3030	0.4338
21	−0.6529	−0.0957	2.1005	−0.6288	−1.2670	0.4589	−0.8314	0.8840
22	−0.1269	−0.7940	1.1173	−0.3278	−0.4340	−0.3052	−0.6915	0.5409
23	−0.2610	0.0650	1.1173	−0.5561	−1.7775	−0.3052	−1.5073	0.8343
24	0.3475	4.0315	2.5921	2.8070	0.9364	2.4041	1.3985	2.8807
25	−0.7251	−0.0957	−0.8491	0.0874	−0.2190	−0.4442	−0.6138	0.2970
26	1.1829	0.8541	0.6257	−0.4419	−0.7564	−0.3747	−1.3131	1.0556
27	−0.3950	0.4351	0.1341	1.0424	0.3990	−0.2358	0.4973	0.3261
28	2.1317	0.4002	−0.3575	0.3158	1.2588	−0.2358	1.0178	0.7662
29	0.2547	−0.9546	−0.3575	−0.7845	0.9633	−0.3747	−0.5672	0.4969
30	1.9770	−0.6613	−0.8491	−0.2966	1.0707	−0.3747	0.2176	1.0565
31	−0.2816	−0.9686	1.1173	−0.5561	−1.2132	0.0421	−1.1888	0.7459
32	−0.3744	−0.6683	−1.8324	−0.1202	−0.1116	−0.2358	−0.9790	0.7266
33	0.1928	0.7423	0.1341	3.4091	1.8500	−0.3052	1.6316	2.7779
Average	0	0	0	0	0	0	0	1
Standard deviation	1	1	1	1	1	1	1	

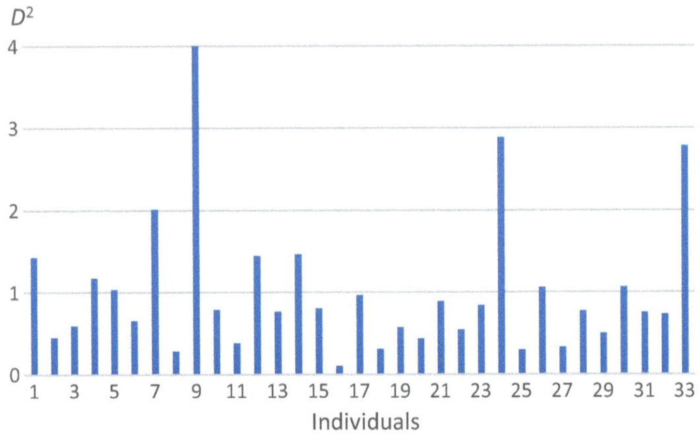

Fig. 9.8 Squared Mahalanobis distances of the whole group

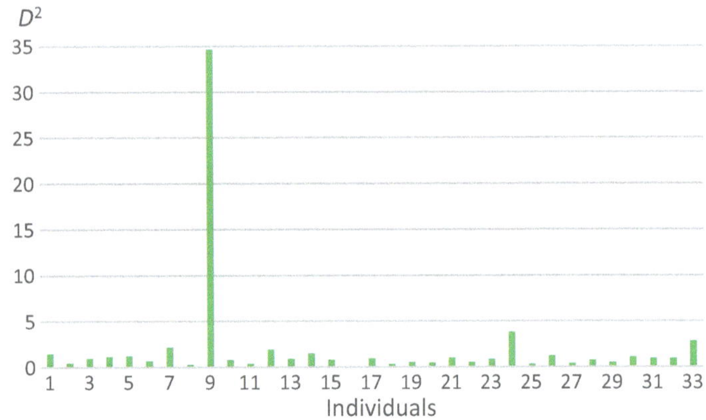

Fig. 9.9 Formation of a modified unit group; individual no. 9 is outside the unit group

A comparison of the squared Mahalanobis distances of the entire group consisting of 33 members as well as the first unit group with 32 and the second with 31 members is appropriate for scrutinizing the correlation pattern (Fig. 9.12).

Individual no. 5 and no. 21 may appear conspicuous, but all others show a uniform correlation pattern. Since the squared Mahalanobis distances of all 31 members are smaller than three, the testing with further individuals can be terminated. As a result, there is a unit group against which unknown individuals can be discriminated. As an additional confirmation, the two individuals who do not belong to the group can be subjected to a parameter screening to identify the decisive differences in characteristics. In both cases, the variable *F* has an exceptional numerical value, i.e., the separation from the whole group appears justified.

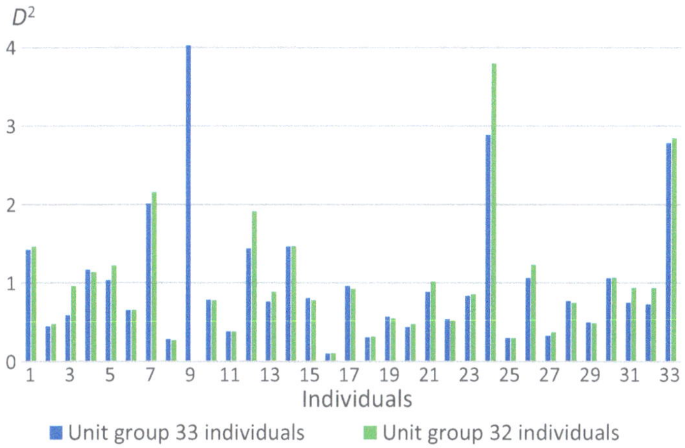

Fig. 9.10 Comparison of the squared Mahalanobis distances of the entire group consisting of 33 individuals with the newly formed unit group of 32 members

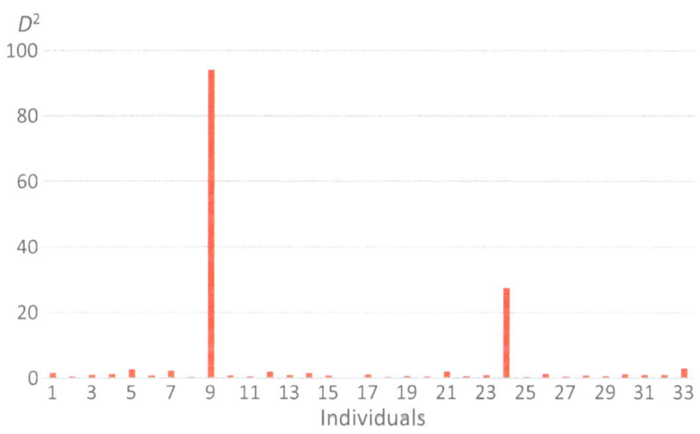

Fig. 9.11 Mahalanobis distances with arrangement of individuals no. 9 and no. 24 outside the unit group

9.7 Scope of the Mathematical Procedures and the Re-normalization

The prerequisite for the construction of a Mahalanobis space is a much larger number of individuals or objects than the number of variables. If the correlation coefficients are different from zero, there is a certain correlation pattern of the objects that form the unit group. The squared Mahalanobis distance distinguishes between objects of the unit group that belong to the correlation pattern and objects of the unit group that do not. If the number of objects n is reduced compared to the

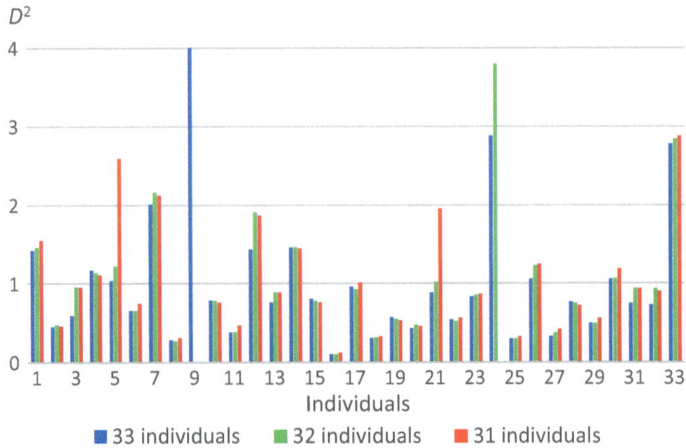

Fig. 9.12 Comparison of the squared Mahalanobis distances of the entire group consisting of 33 individuals with the unit groups with 32 and 31 members

number of variables k, the squared Mahalanobis distances of the objects of the unit group approach the value one, whereby the objects outside the unit group are increasingly overrated. In the limiting case

$$n = k + 1$$

all objects in the unit group reach the value

$$D^2 = 1$$

and the objects outside of the unit group have D^2 values which are exaggerated by orders of magnitudes, depending on their correlation coefficients **(2–16*)**. Since the trend of a rapid increase of the squared Mahalanobis distances for objects outside the unit group is already apparent before the mathematical boundary condition has been reached, reliable data discrimination raises the question of the minimum size of a unit group. Since all cases are different, it is only possible to indicate a minimum condition on an empirical basis.

$$n > 4k$$

n = Number of objects or individuals
k = Number of variables

A unit group with several objects clearly above this reference value is always advantageous. If, for reasons of data availability, the number of objects in the unit group to the number of variables falls below this ratio, the squared Mahalanobis distances will appear conspicuously large. A direct evaluation of the figures is no longer permitted, as each object by far exceeds the threshold value. The counter-measure consists in a re-normalization to reset the D^2 values of the discriminated objects. If the true value of one object is known, it serves as a reference point for all others.

$$\text{Ratio number} = \frac{D^2_{\text{true object}}}{D^2_{\text{calculated value of object from too small unit group}}}$$

All objects outside the unit group must be multiplied by this ratio and thus can be interpreted. If no reference point is known, the ratio is to be calculated from the D^2 value of the object to be classified and the object with the smallest numerical value for D^2, multiplied by the threshold value, commonly by 3.

$$D^2_{\text{re-normalized object}} = \frac{D^2_{\text{object}}}{D^2_{\text{minimum}}} \cdot 3$$

The quotient of the smallest number with itself is one. Extended by the threshold value, this object is not assigned unambiguously. However, all other objects approximate a more accurate numerical value and can be assessed based on their position relative to the threshold value.

The use of the adjoint matrix to calculate the squared Mahalanobis distances in case of low object numbers in the unit group is not an alternative, since the correlation matrix is required. Only the Gram–Schmidt process bypasses the correlation matrix. As the number of objects in the unit group decreases, the D^2 values increase, but to a lesser extent. The Gram–Schmidt method is, therefore, characterized by a relatively higher insensitivity, even if the number of objects n approaches the variable number k. To avoid exaggerated interpretations due to excessive D^2 values, it is preferable in any case, both when using the Mahalanobis and the Gram–Schmidt algorithm, to apply the re-normalization for evaluating the objects.

9.8 Dealing with Categorical Data

For the Mahalanobis algorithm, data in analog form, from digital acquisition or of categorical origin are permitted. Irrespective of their origin, the first step is to standardize the data, get the correlation matrix, calculate the inverse correlation matrix and from this the squared Mahalanobis distances. Without restriction of the data origin, i.e., the measurement acquisition systems or variable assignments, this contributes to the universal applicability of the method, since digital and analog data are often collected and processed simultaneously. Typically, clinical studies can be considered to include not only analog data of physiological parameters but also information such as

- Gender: male/female,
- Taking medication: yes/no,
- Emergency situation: yes/no etc.

If two discrete states are present, they must be coded numerically with 0/1 or any other pair of numbers. Category data appear as classes, which are also assigned for instance as consecutive numbers such as 1, 2, 3, and 4 in case of four different classes. This can be a different dosage of medication, or in a completely different area, e.g., a classification of real estate in the meaning: 1-room apartment, 2-room apartment, 3-room apartment with/without garden, etc. The determination of whether one or the other falls out of the range that means does not belong to the unit group, is immediately visible if the associated squared Mahalanobis distance is larger than the threshold value. The parameter screening also indicates the contribution to the squared Mahalanobis distance or discloses the omission of one or the other coded parameter to improve the selectivity. In the case of categorical parameters, if a variable with the number of classes is taken as setting values, the average overall effect remains; the information of the individual classes is not accessible. If the latter is important, any class with a digital character of 0/1 can be defined as a single variable, assigned to single columns. This increases the number of variables, but the influence of each class can be assessed separately in the screening process. Thus, the effect of digital data can be further explored, provided that the screening result contributes to the squared Mahalanobis distance and each condition is considered separately. This situation is by no means far-fetched but belongs among other things to the repertoire of pharmaceutical research and clinical studies to considering possible gender-specific differences in the development of drugs or therapies. For this purpose, an existing unit group is divided into two groups, one consisting only of male test persons and the other only of females. The gender-specific variable is omitted because the dispersion is zero and cannot be standardized. Only the comparison of independent test persons with both subgroups shows similar or deviating squared Mahalanobis distances that stand for differences in parameter screening which are due to causally responsible variables.

Important items to memorize An extreme compression of the variables is achieved by use of the multiple Mahalanobis distance method.

If the squared Mahalanobis distance is a function of time (or any other variable), its online recording can be used as a monitoring or all-clear indicator.

An unclassified set of data can be split retrospectively into a unit group and a group of objects that do not belong to it. Parameter screening reveals the causes of their non-affiliation.

Chapter 10
Areas of Application

Contents

Abstract Applications of pattern recognition followed by root-cause-analysis are not limited to clinical data. In any field where a decision is required to act upon a critical situation is of interest, such as activating a fire alarm, a collision warning, authentication, the monitoring of patients, character recognition, and more. The examples illustrate the main intent: minute differences in patterns must be detected and mathematically amplified to enable an unambiguous decision characterized by either spending a lot of money to act precautionary or to risk damage which might cause even higher losses. In addition, unresolved problems in applied sciences can be revisited. With more than enough data, attempts can be made to reveal hidden patterns so far undetected to improve the understanding of nature's laws.

10.1 Structured Approach

Practical examples from a wide variety of disciplines are often associated with the recognition of critical situations. Typical applications are listed in Table 10.1.

Irrespective of the existing system, the same principle must always be applied for the most reliable data discrimination. Observable and/or measurable characteristics must be defined for the objects/individuals/situations to be classified. The number of variables should not be limited a priori that means to collect them as completely and extensively as possible based on professional judgement.

© Springer Nature Switzerland AG 2019

H. Ruefer, *Living Without Mathematical Statistics*,
https://doi.org/10.1007/978-3-319-99632-5_10

Table 10.1 Discrimination of data using the squared Mahalanobis distance

Normal or frequent appearance	Unusual development or situation which rarely occurs
Monitoring of technical processes and products **(2-3)**	Defects
Support of medical diagnosis, general appearance of patients **(2-4)**, **(2-5)**	Classifications
Aerial photo interpretation, near-surface structures with infrared remote reconnaissance **(2-6)**	Potential archaeological sites

The objects that characterize a normal situation form the unit group, the unusual or exceptional situations are outside the unit group. The numerical values of the unit group members are to be standardized, whereby the average value and standard deviation are used to standardize the nonassociated objects. The Mahalanobis algorithm provides the squared distance of all objects. The appropriate definition of the threshold value for data discrimination allows the separation into objects belonging or rather not belonging to the unit group. After successful testing of a sufficiently large number of known unusual or exceptional situations, parameter screening can be used to reduce the number of characteristics to a minimum—while maintaining or improving discrimination power. From this verified stage onwards, object data can be read in (or scanned) automatically and continuously. If they are assigned to the unit group, no action is required; if they are outside the unit group, however, a special immediate action is to be taken depending on the task of the system in question. Since the definition of the required characteristics usually entails a relatively high amount of data, the following examples for deepening the application possibilities are limited to the basic principles.

10.2 Pattern Recognition

The automatic sorting of letters by postcode often requires reading manually written numbers with suitable sensors and subsequent correct allocation. More generally, this also includes the recognition of handwritten words or even texts, the identification of certain forms or objects as well as the voice analysis from their spectrum **(2-7)**. If, for example, the handwritten letter “a” should be recognized, a variety of people must deliver writing patterns in order to capture as many variants as possible. Patterns are characterized by means of a horizontal/vertical scanning method, coded in digital form and defined as a unit group. When reading or recognizing unknown handwritten characters, the calculation of the squared Mahalanobis distance reveals the belonging to the same unit group, that means identification of the letter “a” or something else outside the unit group (Table 10.2).

Table 10.2 Discrimination against known/unknown signs

Squared Mahalanobis distance	Result
$D^2 \leq$ Threshold	Known sign (as in unit group)
$D^2 >$ Threshold	Unknown sign

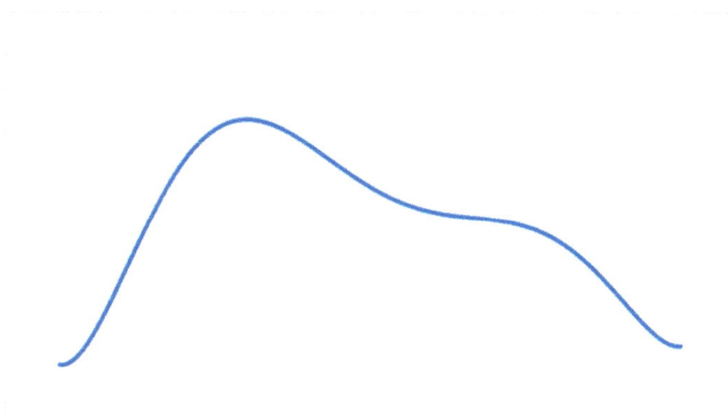

Fig. 10.1 Curve progression as part of a spectrum, object or sign is to be characterized

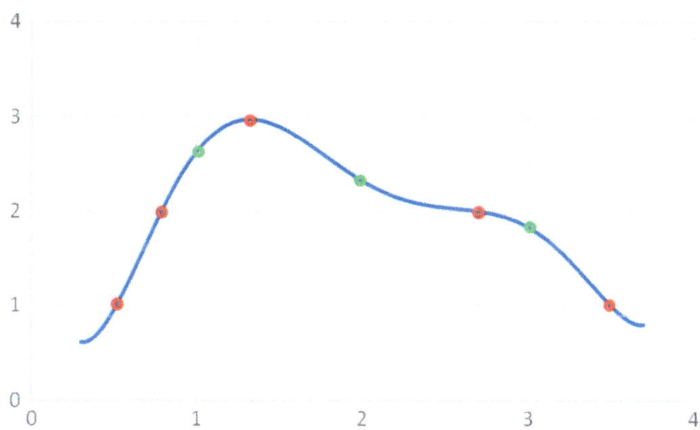

Line number	0	1	2	3	4
Horizontal intersection	0	2	2	1	0
Vertical intersection	0	1	1	1	0

Fig. 10.2 Counting the intersections from line grid with the curve

A line by line horizontal and vertical screening is suitable as raster scanning. The raster width depends on the form, i.e. a few lines for simple objects, higher resolution for finer features to be distinguished. The basic procedure is explained in the following figures and tables (Figs. 10.1 and 10.2).

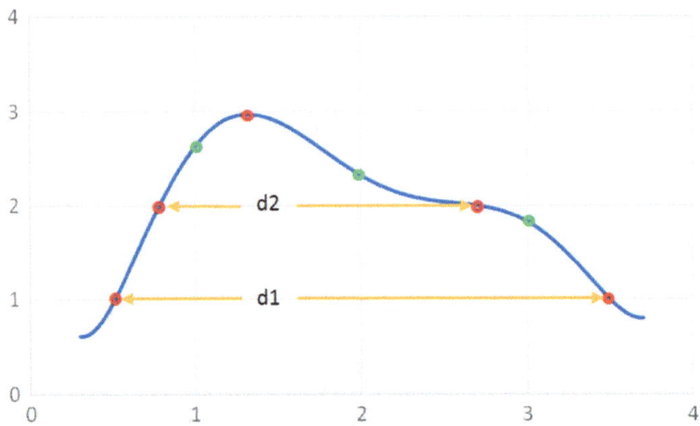

Line number	0	1	2	3	4
Horizontal distance	0	d1	d2	0	0
Vertical distance	0	0	0	0	0

Fig. 10.3 Recording of all distances between adjacent points

Table 10.3 Data set of a raster scanning method

Characteristic features	Differential value horizontal	Differential value vertical	Integral value horizontal	Integral value vertical
Line grid:	0 1 2 3 4	0 1 2 3 4	0 1 2 3 4	0 1 2 3 4
Parameter:	1 2 3 4 5	6 7 8 9 10	11 12 13 14 15	16 17 18 19 20
Pattern #				
1	0 2 2 1 0	0 1 1 1 0	0 d1 d2 0 0	0 0 0 0 0
2	…	…	…	…
3	…	…	…	…
…	…	…	…	…
…	…	…	…	…
n	…	…	…	…
Average				
Standard deviation				

The recorded intersections in horizontal and vertical direction are also called differential characteristics. Each line represents one parameter set (Fig. 10.3).

The integral characteristic is the distances between adjacent points of the horizontal and vertical row grid. With this method, each horizontal and vertical grid line represents two parameters, the differential and the integral measured value. All other patterns are coded in the same way and arranged line by line (Table 10.3).

With the average value and the standard deviation, all data must be standardized. For this, the inverse correlation matrix is used to calculate the squared Mahalanobis distances of the patterns of the unit group. Each new, initially unknown character is raster scanned according to the same procedure, standardized with the average value and the standard deviation of the members of the unit group, and their squared Mahalanobis distances are displayed. If the numerical value is below the threshold value, it is identified as a known character, otherwise, it is classified as unknown. The total number of parameters is relatively large but can be screened with a sufficient number of characters to be tested. Mechanical (fingerprints), optical (iris scanner) and phonetic (voice analysis) authentication or identification systems work in the same way.

10.3 Alarm Systems

Smoke and temperature detectors are used to monitor rooms for activating a sprinkler system when an alarm is triggered and to start the siren to prompt evacuation of the building. This may sound relatively simple, but it is linked to the occurrence of a clear situation. Nothing is more annoying than false alarms, which then induce people to ignore the corresponding signals. This applies in principle to such systems, regardless of whether it is the failure of an air conditioning system or a coolant failure in a nuclear power plant. Decisive for an unambiguous functionality is to simulate in advance as many different situations as possible, which a sensor responds to and to select the number of detectors, their sensitivity, the arrangement etc. in such a way that no false alarms can occur. For fire monitoring in a building, there are usually several different sensors distributed around the building. Every few seconds the measured values are read out, combined, and processed. The result must lead to a simple and unambiguous interpretation, i.e., to be able to distinguish a real alarm situation from a similar but meaningless state with high selectivity. The crucial aspect is to simulate the most complete error-prone conditions possible and to enter them into one unit group. A complete investigation with the smallest possible test effort is achieved with an orthogonal array **(2-8)** (Table 10.4).

More parameters are imaginable that can be recorded with the orthogonal array L12 in 12 test configurations. Depending on the conditions and technical possibilities, parallel data sets of the sensors can be supplemented with modified positions (window, door, ceiling, wall) and setting sensitivities (low, high). Further sensor-specific properties can be assigned in an outer orthogonal array of corresponding size. By starting a controlled fire, the sensor signals are recorded for the exceptional real state as a function of time. The squared Mahalanobis distances resulting from the previous 12 different test conditions are calculated for every recorded point in time. If the situation is recognized as such the data set of the sensor signals is subjected to a parameter screening. In the case of several data sets, the one with the highest selectivity must be selected after screening. The result is a reduced data set with unambiguous data discrimination of the actual alarm situation (Fig. 10.4).

Table 10.4 Definition of variables for simulation of common situations

Parameter	Level 1	Level 2
A: Room temperature	Cold	Warm
B: Four-burner gas stove (electric cooker)	Off	All burners on
C: Flat iron	Off	On
D: Cigarette smoke	Without	With
E: Toaster	Off	On
F: Match, lighter, candle	Off	On
G: Incense sticks, insect spiral	Off	On
H: Oven, barbecue	Off	On
I: Windows, doors	Open	Closed

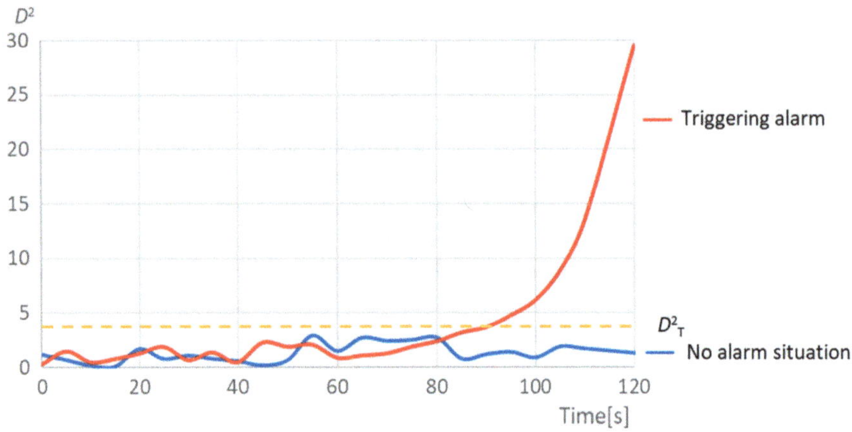

Fig. 10.4 Development of the squared Mahalanobis distances as a function of time of a reliably functioning alarm system

The fire condition is detected, i.e., an alarm is triggered as soon as the squared Mahalanobis distance exceeds the threshold value at any time.

10.4 Driving Assistants

Just as the captain of an aircraft can switch on the automatic pilot, assistance systems are becoming increasingly available to the driver. In particular, great importance is attached to safety systems. The airbag is a well-known one. In the event of a collision, acceleration sensors must be able to distinguish an ordinary vibration pattern, such as high-speed bumps or impact with the curbside from a dangerous accident. At the instant of the collision, an explosive charge is ignited whose gases can inflate the airbags in a few milliseconds. More advanced are systems that respond long enough before an accident is expected to occur. Such a collision protection system has several functions:

- alerting the driver
- intervention in the steering system
- actuation of the brakes.

This requires extensive all-round sensor technology for measuring speeds and distances as well as image recognition using video cameras. Measured values from a wide variety of situations must be recorded for reliable operation such as

- single-lane, multilane roads
- intersections
- speed of the own car, cars on the same lane, vehicles from the opposite direction
- obstacles such as construction sites and signs single-track detours
- tunnel passages and more.

With the data and their further processing for calculating trajectories including video image evaluation, promising approaches for constructing a unit space already exist **(2-9)**. Special parameter configurations are recognized by the Mahalanobis algorithm and automatically address the driving assistant. Although some potentially dangerous maneuvers are anticipated in advance, sensor and data technology improvements are needed before a serial introduction can be considered.

10.5 Monitoring of Patients

Patients in critical condition, mainly after severe medical surgery, are under special supervision in the intensive care unit (ICU). This applies to cases that can be clearly diagnosed. An unambiguous diagnosis, for discharge from the hospital, depends on the professional medical assessment based on the individual physiological and surgical parameters. The Mahalanobis algorithm is useful for diagnostic support, on the one hand, to detect endangered patients more reliably and on the other hand not to extend the patient's length of stay beyond the necessary extent. In the present case study, patients undergoing open heart surgery were subjected to a retrospective analysis according to their physiological parameter values and surgical conditions **(2-10)**. Of a total of 780 patients, 29 variables were recorded. All patients without severe postoperative problems are collected to form a unit group consisting of 758 patients. Patients who did not survive the surgery or showed conspicuous postoperative problems, here 22 individuals, are placed outside the unit group. Based on the defined unit group, all squared Mahalanobis distances were calculated, which clearly discriminates the majority of high-risk patients with their D^2 values from the patients of the unit group (Fig. 10.5).

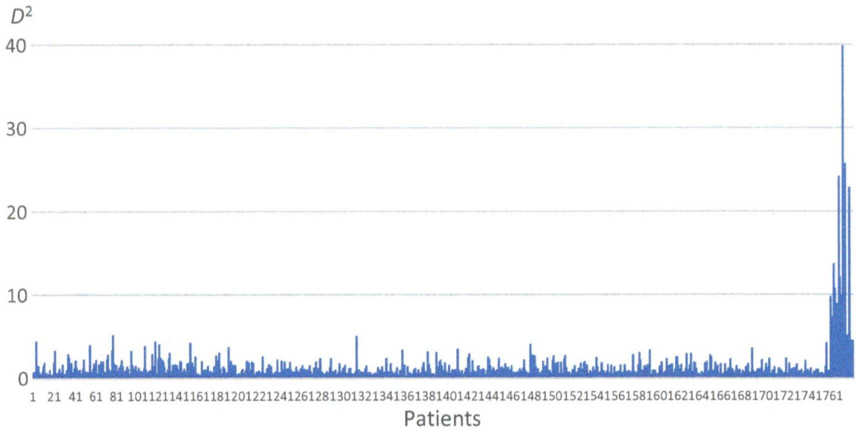

Fig. 10.5 Squared Mahalanobis distances of the unit group composed of the individuals 1–758, and the high-risk group with the remaining individuals 759–780

The incorrect classification of the patient status in this study has been in the range of 0.3%. If more parameters are included, it can be assumed that the percentage of error can be further reduced. Thus, the considered procedure appears promising as a method to support medical diagnosis and the time-dependent monitoring of the individual patient's condition. Used prospectively, there is a chance of reducing patient mortality even more in case of severe surgery by means of the earliest possible therapeutic measures. Costs can also be reduced by continuous patient monitoring (Sect. 9.5) to indicate the earliest possible point in time for relocation from the ICU or discharge from the hospital.

10.6 Original and Fake

Original, imitation or falsification is identical to the question of authentication. The first terms usually refer to objects such as works of art, identity cards or banknotes, the latter concerns, e.g., access rights for persons. The crucial task is always to find a way for a meaningful, purpose-adapted characterization. The effort in the form of numerous variables is initially acceptable, since it can be reduced later with parameter screening. In banknotes there are obvious, superficial features such as metal filaments, fluorescent patterns when exposed to UV light, hologram-like structures that are relatively easy to check if present or not. However, the situation is more comparable to the arms race in cryptography between coders and decoders. The methods are becoming more and more sophisticated with the use of software. The same applies to counterfeiters through the availability of modern color printers and additional special treatments.

Pattern deviations as such are difficult to detect. Instead of screening notes and reading them with differential and integral features, it can be advantageous to screen and mathematically transform only certain details with high resolution (**2-11**). For periodic structures, Fourier transformation is recommended. A local period appears as a signal at a certain frequency. For example, if the falsification lacks precise repetition due to mechanical tolerances of the color copiers or printers, the frequency range in the Fourier spectrum widens accordingly. A line at a fixed frequency becomes a broadened line. From the position (frequency), curve width and height (amplitude) and other characteristics such as paper thickness and consistency, a practicable set of variables are created. With old and new banknotes, crumpled and soiled, coming out of different presses, suitable objects are available for the unit group. Known forgery patterns are to be characterized with the same set of variables. The calculation of the squared Mahalanobis distance shows whether sufficient selectivity has been achieved for unambiguous identification of counterfeits or imitations. If there is an overlap, the set of variables must be revisited and, if necessary, extended until the discrimination power is large enough to perform an error free classification.

10.7 Proposal for Research Project in the Field of Biology

Biology meets medicine quite often when processes within cells are investigated. As part of its task, the cell is instructed by the genetic material to produce certain proteins. For control purposes, there are the so-called DNA methyl-transferases, which switch off genes by attaching methyl groups to the genetic material and thus decide which proteins are produced and which are not. Among other things, this leads to the conclusion that some therapies are only successful in the long term if certain genes are blocked by methylation (**2-12**). Since some types of therapies, e.g., in connection with glioblastomas, are not risk-free, progress in medicine appears to be feasible to determine in advance whether a patient will respond to a therapy or not. For this purpose, the methylation patterns of various pathogenic cells in which the disease has been successfully treated should be characterized with the variables of the cell types and docking sites that form the unit group. Those cases in which the therapy failed must be equally parameterized regarding their methylation pattern. Calculation of the squared Mahalanobis distance shows whether the discrimination power is sufficient to distinguish between a long-term successful or unsuccessful therapy. Since the methyl groups are relatively tightly bound to the DNA, the pathogenic cell types could be treated successfully. Further applications of this type are the early detection of pathogenic cells, with the consequence to initiate an appropriate therapy before the disease is irreversibly released.

10.8 Proposal for Research Project in the Field of Geophysics

Triggered by the tsunami catastrophes of 2004 (Thailand) and 2011 (Japan), a reliable early warning system appeared again urgently needed. The facts are quickly listed. If a severe seaquake occurs, the epicenter of which is located on the deep-sea floor by means of seismic wave propagation measurements and model calculations, a tsunami can spark off and spread out over the ocean or nothing happens at all. The causal detailed mechanical mechanisms are less well known. Ultimately, it can be assumed that a relatively rapid vertical displacement of fracture slabs causes a rapid change in the overburdening water column above it, which flows off to all sides as a disturbance and spreads out at a speed of approximately 800 km/h above the deep-sea floor. Certainly, refined models will be able to better describe reality, but first, the need for accurate information remains, which must be passed on extremely quickly to potentially affected coastal areas because of the enormous wave velocity. The seismic event monitored by seismographs, data from sea buoys linked to the Global Positioning System (GPS) and certain models have so far been incorporated into early warning systems. It cannot be ruled out that the decisive mechanical motion pattern is encoded in the seismic signal itself. If it is possible to characterize the earthquake waves using different methods, whereby the Fourier transformation could be worth trying, resulting parameters with their values enter the Mahalanobis algorithm. All quake events worldwide are recorded **(2-13)**. Of these, the earthquake wave patterns with all polarizations of all seaquakes with a magnitude greater than 6 with epicenter on the sea floor without the formation of a tsunami form the unit group. All equivalent seaquakes with tsunami formation are outside the unit group. If the interpretation of seismic signals is improved, it appears to be not only faster but also more reliable to distinguish between these situations with sufficient selectivity. In principle, the hazard warning of a tsunami is thus clearly available immediately after the earthquake waves have reached the seismometers.

10.9 Proposal for Research Project in the Field of Astronomy

Today's research institutions are often overwhelmed by a flood of data, which sometimes add only a modest increase in knowledge. In astronomy, there are many theories apart from many stars, especially if the nature of the objects is mysterious. Such objects are, for example, quasi-stellar objects, which have been coined in literature as quasars. In contrast to black holes, they can be observed, measured spectrally and thus attributed special properties. Because of certain assumptions, they are pushed to the limits of today's universe, which means that they must have an unimaginably high luminosity, otherwise, they would not be accessible for observation. In a survey of the sky, numerous quasars whose data are accessible have been detected to this day **(2-14)**. It remains to be seen whether the theoretical model developed so far is correct. Maybe some objects do not match the assumed pattern, and a more detailed classification could be a way for a better understanding. For the time being, no obvious unit group is known. Here, the jackknife method is a suitable method to find possibilities for a scientifically based classification (Sect. 9.6). If this is the case, there may be starting points for checking theoretical statements regarding the detected objects outside the unit group. The mysterious objects will not reveal their secrets very soon but if one or the other theoretical assumptions must be given up, it would be easier to get alternative scientific elucidation.

Important items to memorize

Due to its universal character, the Mahalanobis algorithm can be applied to almost any area and any type of data.

If the discrimination power is unsatisfactory, additional variables (features) must be detected to extend the parameter space.

Parameter screening can be used to reduce the effort and to investigate the causal parameter effects.

Part III
Prognoses

Chapter 11
Methodical Approaches

Contents

Abstract Orthogonal arrays with a size to assign almost any number of variables, followed by the procedure to calculate the *SNR* indices, combined with the ability for efficient pattern recognition opens a way to inverse the formulation. That means, if in a well-known set of data composed of variables and their respective characteristic outcome of objects, the values of the variables will change, then the corresponding new characteristic outcome can be calculated with reasonable accuracy. Consequently, experiments or simulation calculations are unnecessary. The generic term prognosis is appropriate as it is valid regarding almost any number of variables, more or less correlated, digital or continuous, with or without deviation. This inverse application makes use of the *SNR* indices and defines its numerical values as weight coefficients for prognosticating unknown characteristic outcome values. Incidentally, the number of variables can by far exceed the number of variables. Another approach makes use of the Mahalanobis algorithm that can provide more accurate results in case of nonlinearities. Both procedures are applicable and can be evaluated to select the superior one. Depending on the results, combinations can be used as well to even further improve the accuracy of a prognosis.

© Springer Nature Switzerland AG 2019

H. Ruefer, *Living Without Mathematical Statistics*,
https://doi.org/10.1007/978-3-319-99632-5_11

11.1 Issues in Retrospect

A review of the core contents of the two preceding parts appears to be appropriate both for recognizing the similarities of methodological approaches and for understanding the extended problems for the third and last part. In the first part, which focuses on data analysis, the effects of independently adjustable variables on a system response (a technically and scientifically measured quantity) are treated under unavoidable noise conditions. The purpose of an analytical examination is to be able to guarantee a reliable and reproducible result in the form of hitting advantageous target values. To get there, the response of different parameter configurations is usually recorded as a minimum data set using an orthogonal array. From at least two measured values per setting, which represent the influence of noise, the information of dispersion can be obtained in addition to the average value of the target under consideration. The combination of the two calculated figures—average and deviation—into one characteristic figure, the signal-to-noise index (*SNR*), stands for the reliability of the system under consideration. The higher the numerical value, the greater is the accuracy for hitting reproducibly a given target with the lowest deviation. More technically speaking, the use of nonlinear transfer functions of the variables for largely neutralizing noise conditions is expressed by the value of the *SNR* index. If an area is to be covered instead of one specific target value, dynamic system behavior becomes effective. The range extension stands for operating states at different levels, such as different amplification factors, zoom settings, speed ranges of motors, or also size classes within a product family while maintaining the parameter settings. Different operating states are called signal levels *M*. Also, for the dynamic extension, the same objective is to achieve reproducible results with minimal dispersion over the entire bandwidth of all signal values. A data analysis must, therefore, be extended to every signal value. Since the measurement data is generated several times at different signal levels, possible interfering noise is already implicitly contained, so that additional noise parameters can be omitted. The deciding factor for system evaluation is the corresponding dimensionless *SNR* key figure, not referred to an average response value of a static system, but to the slope that results from the responses of all signal levels. The numerical value of the *SNR* index stands and falls with the signal amplification, i.e., the system sensitivity, the linearity of the response behavior and the deviation of the measured values around the straight-line slope. The prerequisite is that a linear system response as a function of the signal values is approximately given. If this assumption is not valid, the data must be transformed in such a way that a linear relationship is established. Otherwise, the algorithm for nonlinear system behavior must be used. After modeling the parameter effects and confirming a prediction derived from them, the corresponding system response can be determined for any given signal values. It should be noted that only dynamic systems exist. Merely in a limited view, does a system appear to be static, combined with a superficially simplified data acquisition with limited information content. The superiority of a dynamic approach is achieved by analyzing the sensitivity of the system, which in

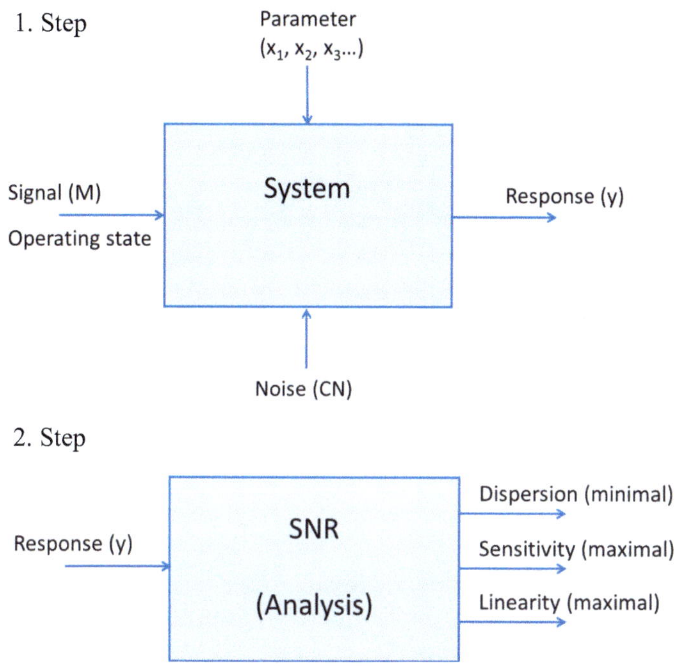

Fig. 11.1 Schematic display of a system analysis

terms of application technology leads to the maximum energy efficiency or yield with the greatest possible signal amplification. The parameter analysis as such allows the effects to be used both technically and economically in accordance with their contribution to the *SNR* key figure. Advancing creatively with new parameters and innovatively with an extended parameter range to enter unknown territory is to be seized as an opportunity to find new and better ways of solving problems and further applications. The sequential execution of a dynamic system analysis is sketched schematically (Fig. 11.1).

For reasons of clarity, the following Table 11.1 for data entry is used to determine the *SNR* key figures.

The evaluation of a performed test can be represented geometrically with the slope β of the regression line through the origin. Deviations of the measuring points from the straight line result in the associated standard deviation σ. Each slope represents one completed experiment with their characteristic properties. The test results of an orthogonal array L4 dynamically applied with three signal values are shown in the graph (Fig. 11.2).

Part 2 deals with parameters, which can be pairwise correlated. This is usually not the case for technical processes because interactions can lead to unwanted and unintentional complications. If deviations occur in a process sequence, they must be corrected unambiguously and reliably by intervention or control mechanisms to guarantee functional reliability. This is always successful if coupling of parameters

Table 11.1 Arrangement of measurement data for evaluating dynamic systems

Test #	Parameter $x_1, x_2, x_3, \ldots$	Signal $M1, M2, M3, \ldots, Mk$					Slope β	Squared standard deviation σ^2	*SNR* $10 \log\left(\frac{\beta^2}{\sigma^2} S_2\right)$ S_2 = Dimension factor
1	Orthogonal array	y_{11}	y_{12}	y_{13}	…	y_{1k}			
2		y_{21}	y_{22}	y_{23}		y_{2k}			
3		y_{31}	y_{32}	y_{33}		y_{3k}			
⋮		⋮	⋮	⋮		⋮			
n		y_{n1}	y_{n2}	y_{n3}		y_{nk}			

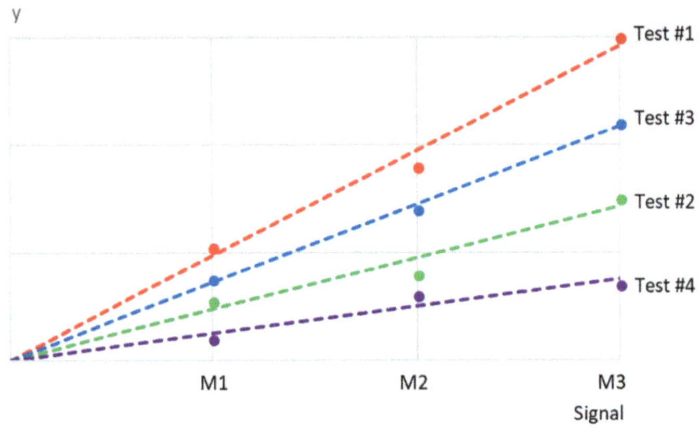

Fig. 11.2 Graphical representation for the geometric interpretation of an orthogonal array, e.g., an L4 with three signal levels

is avoided in process development. In contrast to this are biological processes. Almost all physiological parameters seem to be linked to each other to fulfill their tasks, ultimately preserving the life function. In addition, biological diversity and variance cannot just be considered as narrowly tolerated. Situations of this kind are typically common in the medical field, where unavoidable and sometimes unknown interactions are more the rule than the exception. A direct and unambiguous parameter analysis with the approaches from Part 1 is, therefore, doomed to failure. The most important point for approaching this topic is to make a comparison to classify individuals or objects with known parameter values but an unknown appearance. In everyday medical practice, it is important to clearly and unambiguously distinguish a sick person from a healthy state or—in technical terms—to make a reliable diagnosis. Apart from pathological habitual patterns, the deviation range of possible values agreed upon, e.g., for concentrations, pressure, temperature, etc., is known from many individual data. As a rule, these ranges of values are usually referred to as the normal range, assuming a healthy appearance.

The frequency distribution of the physiological parameter values does not necessarily follow a Gaussian bell curve **(1–3*)**, i.e., reaching an average value is only one out of the many possibilities. If the range of values is exceeded or falls short, a disease pattern can result, in extreme cases the vital function is ceased. A particularly striking example is the body temperature with the beginning of the death zone from 42 °C upwards. If the value of a physiological parameter changes, e.g., under physical exertion, all other parameter values must follow to a certain extent to preserve the vital function. Due to parameter coupling, however, compensation mechanisms can become effective, which are within the limits of normal defined ranges for perfectly healthy individuals.

Conversely, in the case of disturbed parameter coupling, all physiological numerical values can still lie in the normal range, although a disease pattern is unmistakable. This means that the decisive criteria are not necessarily obvious measured values but are expressed in the correlation pattern of the variables. Only if the underlying correlation pattern is considered together with the measured values, a clear diagnosis can be made. By combining the parameter values with all pairwise interactions, a dimensionless positive number can be derived for the execution of a comparison. Named after its inventor, it is called Mahalanobis distance (MD = Mahalanobis distance, D^2 = squared Mahalanobis distance). A precise and reliable knowledge of a reference group with a homogeneous appearance and the corresponding parameter values is a prerequisite for diagnostic purposes. Since such a reference group in medical practice usually refers to a healthy group of people, it is also known as a normal group. From a mathematical point of view, the average value of the squared Mahalanobis distance over all individuals of a normal group is always equal to 1 with appropriate scaling, from which the definition of the unit space is derived. For reasons of general validity, the term "unit group" is preferred. The comparison is, therefore, a task of classifying an individual or object according to their affiliation to the unit group or not. In the absence of the true key indicator (state), the squared Mahalanobis distance is applied as a substitution to classify a situation or characteristic value. A threshold value D_T^2 is used for discrimination as a criterion for group membership. The schematic sequence is shown in the illustration (Fig. 11.3).

To evaluate a data set, the parameters are preferably arranged like in Table 11.2.

After classification, the algorithms known from Part 1 can be applied to individuals or objects that do not belong to the unit group. For this purpose, the parameters are assigned to an (preferably partial) orthogonal array and their contribution to discrimination against the unit group is evaluated with an *SNR* index. After performing the parameter analysis, those who are causally responsible for not belonging to the unit group become visible (Fig. 11.4).

The knowledge of the individual contributions of the parameters allows sorting out those with low or even opposite amplitude. The resulting improvements in the discrimination power are displayed with a greater distance from the threshold value, i.e., an increase in diagnostic reliability is achieved. In addition, the effort of data acquisition is reduced if some parameters can be omitted (Fig. 11.5).

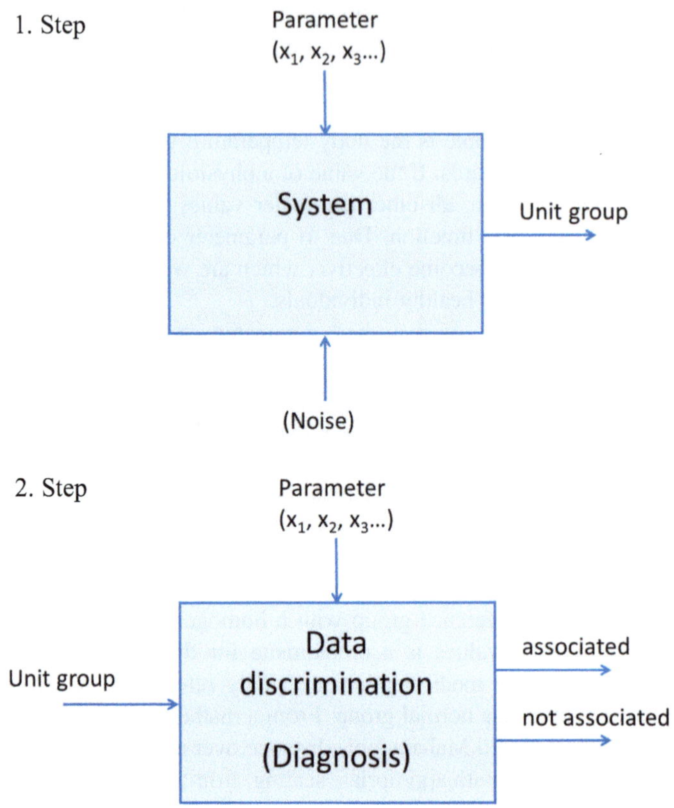

Fig. 11.3 Classification of individuals or objects according to group membership

Apart from medical issues, the same situation of decision making exists for tasks such as surveillance systems to trigger or suppress an alarm, or access rights of authorized persons, to name a few examples.

In the following **Part 3**, an inverted question is examined. The starting point is known or measured parameter values to which an observed result or evaluation is assigned. Depending on the subject area, the terms outcome parameters, score, i.e., ranking according to points (data collection from questionnaires) or similar can also be found. An existing reference group of this kind represents the data basis for forecasting unknown results for different parameter settings. If the functional relationship is known, it goes without saying that forecasting methods are not necessary, since the result of any parameter values can be derived mathematically. This is not the case, however, if there is no model available or due to mathematical complexity. Instead,values of sufficient accuracy are more easily generated. If the initial characteristic parameters have new, i.e., different numerical values, a statement is problematic, not only because of unclear relations but also because of unknown interactions of the parameters with each other or even incomplete

Table 11.2 Determination of the unit group and classification of individuals or objects of unknown affiliation with the squared Mahalanobis distance

Object	Variables $x_1, x_2, x_3, \ldots, x_k$ $(k < m)$	Squared Mahalanobis distance D^2	Properties (D_T^2 = Threshold)
1	$x_{11}, x_{12}, x_{13}, \ldots, x_{1k}$	D_1^2	Unit group $D^2 < D_T^2$
2	$x_{21}, x_{22}, x_{23}, \ldots, x_{2k}$	D_2^2	
3	$x_{31}, x_{32}, x_{33}, \ldots, x_{3k}$	D_3^2	
⋮	…	⋮	
m	$x_{m1}, x_{m2}, x_{m3}, \ldots x_{mk}$	D_m^2	
$m + 1$	$x_{m+1,1}, x_{m+1,2}, x_{m+1,3}, \ldots, x_{m+1,k}$	D_{m+1}^2	Outside the unit group if $D^2 > D_T^2$ (advantageous/disadvantageous compared to objects of the unit group)
$m + 2$	$x_{m+2,1}, x_{m+2,2}, x_{m+2,3}, \ldots, x_{m+2,k}$	D_{m+2}^2	
$m + 3$	$x_{m+3,1}, x_{m+3,2}, x_{m+3,3}, \ldots, x_{m+3,k}$	D_{m+3}^2	
⋮	…	⋮	
n	$x_{n1}, x_{n2}, x_{n3}, \ldots, x_{nk}$	D_n^2	

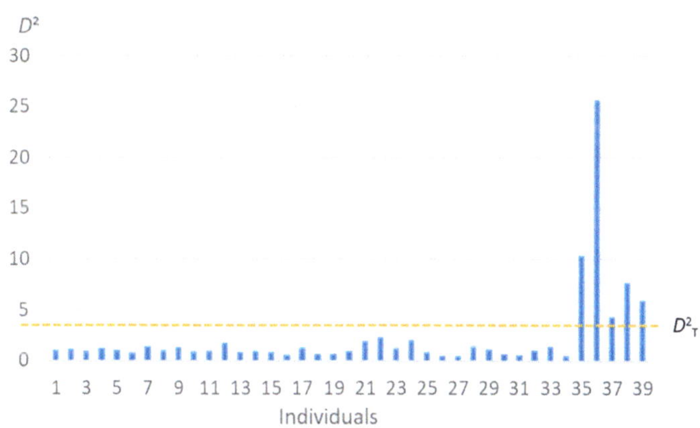

Fig. 11.4 Analysis of the parameter effects with regard to their discriminatory impact contribution

parameter sets. The result of a forecasting approach thus appears to be model-dependent and is not entirely free of speculation.

With the emergence of numerous mathematical forecasting algorithms, the terms training set for the reference group and test set for events to be predicted of an

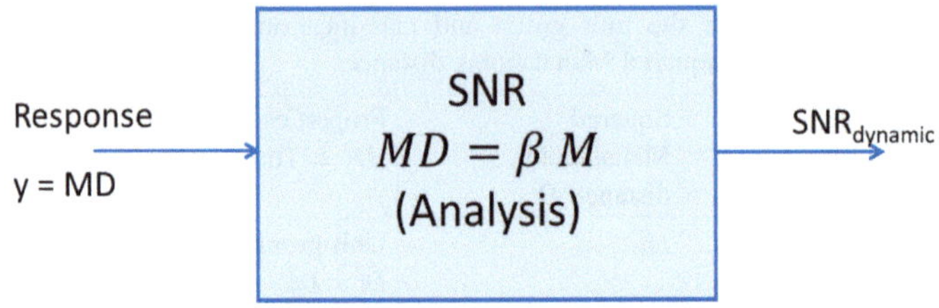

Fig. 11.5 Classification of individuals according to group membership with $D_T^2 = 3$ as threshold value

identical parameter group but with different characteristic values have become established. As the following methodological enhancements are based on the explanations in Part 1 and Part 2, the terms reference group and group of unknown assignment or test group are retained.

Dynamic system behavior is characterized by different operating states. If unambiguous operating states, i.e., signal values, can be determined from many variables with their characteristic values, then, according to the inversion of the thought, it must be possible to predict the associated signals for different parameter values within the same considered system without any experimental effort. Required prerequisite is a reference group or, in extreme cases, at least one reference point for signal calibration. Typical examples are predictions of yields in a manufacturing process, natural events such as volcanic eruptions based on many sensory measurements, the water level of a spring tide, life expectancy in case of certain diseases and many other things or events. The characteristic outcome is a continuous figure, which, as far as it is known from a given set of variables, is referred to as true value (=reference group), otherwise as calculated value (=test group). The challenge is not to derive a dimensionless key indicator as in Part 2 and to use it for classification and diagnosis purposes, but to derive a calculated characteristic value with the best possible agreement with the true value.

Approximately, this appears to be attainable if the relationship in which individual characteristics stand for the result is unknown but there exists a direct or indirect relation. This results in the feasibility to execute predictions or prognoses from a known set of parameters, but with an unknown overall effect—assuming similar relations. In a reference group, each data set represents a true result or event. In accordance with the nomenclature used so far, the result or event variables are interpreted as signal values, since they represent different operating conditions from a technical point of view. If it is possible to determine the effect of the individual parameters for generating the signal value regarding their size and precision, unknown signal values from different parameter settings can be predicted. For this purpose, the individual contributions of the parameters are characterized with their *S/N* ratios. The higher the numerical value of the *S/N* ratio turns out to be, the

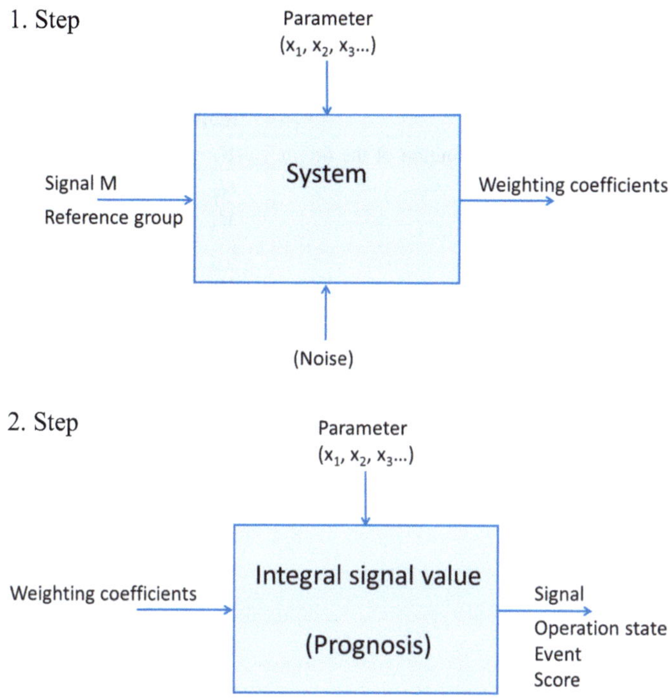

Fig. 11.6 Assignment of signal values from the effect of all available parameters without knowledge of functional relationships

stronger is the impact of the parameter to the prognosis. Dimensionless *S/N* ratios, thus, assume the task of weighting coefficients for determining the integral signal value referred to as the prognosticated result. As with the procedures of the previous parts, the result is produced in two sequential steps (Fig. 11.6).

The data scheme to be used is also in tabular form for a practical reason (Table 11.3).

From a geometric point of view, the reliability of a prognosis can be seen in the dispersion range of the calculated values relative to the true values of a reference group (Fig. 11.7).

If an acceptable agreement between the true and calculated values is achieved, the conclusion is that the predicted values also reflect the true result with appropriate accuracy. The decisive prerequisite is not only to compare the true and calculated numerical values but also to evaluate the approximate solution objectively to assess the scope of the predicted values, i.e., the meaningfulness. The accuracy of the predicted estimated values is quantified by the *SNR* key figure. If the numerical value of the *SNR* index is sufficiently large, an additional modeling approach appears to be justified. For this purpose, corresponding characteristic data is assigned to an orthogonal array and the prognosis results are used as a response. After the evaluation, there is a chance to design a model at least within a given

Table 11.3 Summary of the true values M and the calculated values $\widehat{M}$ of the reference group with prognosticated values of an unknown group

Object	Variables $x_1, x_2, x_3, \ldots, x_k$	True value result	Calculated value result
1	Reference group for definition of the origin	M_1	
2		M_2	
3		M_3	
⋮		⋮	
l		M_l	
$l + 1$	Reference group for derivation of weighting coefficients	M_{l+1}	$\widehat{M}_{l+1}$
$l + 2$		M_{l+2}	$\widehat{M}_{l+2}$
$l + 3$		M_{l+3}	$\widehat{M}_{l+3}$
⋮		⋮	⋮
m		M_m	$\widehat{M}_m$
$m + 1$	Known settings with unknown results		$\widehat{M}_{m+1}$
$m + 2$			$\widehat{M}_{m+2}$
$m + 3$			$\widehat{M}_{m+3}$
⋮			⋮
n			$\widehat{M}_n$

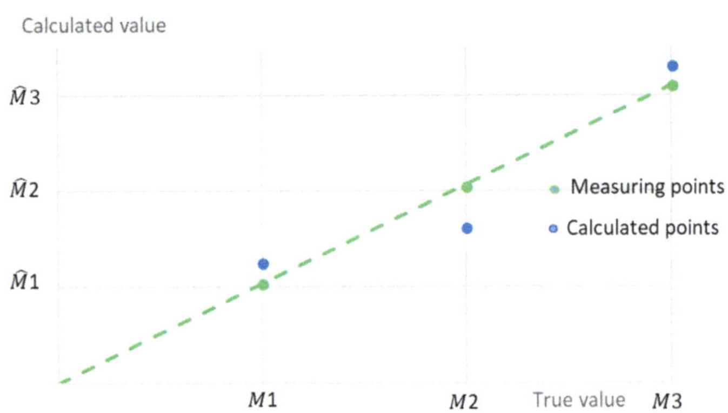

Fig. 11.7 Comparison of prognosticated values $\widehat{M}$ with the true values M of a reference group to evaluate the accuracy of prognosis

range of values, with which causal relations can be revealed. In this way, parameter analysis is seamlessly linked to Part 1, thus closing the circle of applications (Fig. 11.8).

Not subject to the mathematical treatments in Part 3 are the usual and widespread forecasts associated with the following:

- the extension (extrapolation) of a trend due to regular processes in the past

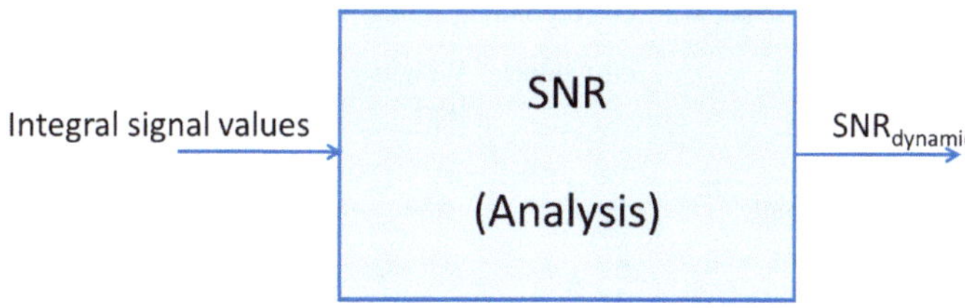

Fig. 11.8 Investigation of parameter effects on the result of prognosis

- regularities of natural growth processes
- periodic patterns in nature, technology, and economy (Kondratjew cycles).

These topics are referred to **(3–1)**.

11.2 Prognosis Algorithms with Linear Equation

Just as it is important in critical situations to be able to make a decision with the consequence of a great impact, the value of a prediction stands and falls with the extent of the agreement of the actual values or occurring events at a later point in time.

In a first attempt, a linear input–output relation provides a reliable approach for an objective evaluation with *SNR* key figures, provided that the subsequent result justifies this. This requires the definition of a zero point around which known reference data are arranged as positive and/or negative numerical values. Using the origin and derived metrics from the application of a zero-point-proportional equation, the prognosticated results for the reference group should be approximated as closely as possible to their true values in a first step as a preceding test. The consideration is that if the outcome calculated for the reference group corresponds satisfactorily with the true values, the same should apply to other parameter values whose outcome is unknown. The starting point for specifying an origin are parameter values for common, frequently occurring outcomes with an approximately homogeneous appearance. Their average values serve to determine the zero point by subtracting these from all parameter values and outcomes or rather the values of characteristic features—to the extent available. Thus, the variables and outcomes are normalized with respect to the average values and have positive and negative signs. An example composed of two parameters, weight and height, which are related to a characteristic feature (outcome, appearance, rating number, or score), serves as an illustration. The task is to derive missing outcomes from the existing parameter values (Table 11.4).

The first group is used to determine the zero point by subtracting the average value from all corresponding characteristics. By normalizing the data, the average of the first group defines the origin. In analogy to the Mahalanobis algorithm, this is

Table 11.4 Weight and height with corresponding rating number

Groups	Individuals	Weight [kg]	Height [m]	Characteristic feature, rating number
1. Group: 1–10				
Definition of the zero point	1	55	1.61	21.22
	2	60	1.75	19.59
	3	50	1.69	17.51
	4	65	1.72	21.97
	5	68	1.77	21.71
	6	62	1.66	22.50
	7	75	1.81	22.89
	8	76	1.86	21.97
	9	61	1.82	18.42
	10	79	1.88	22.35
	Average	65.10	1.757	21.012
2. Group: 11–15				
Known reference group or training group	11	73	1.68	25.86
	12	48	1.95	12.62
	13	98	1.70	33.91
	14	58	1.88	16.41
	15	81	1.65	29.75
3. Group: 16–20				
Group with unknown rating number or test group	16	59	2.00	
	17	79	1.69	
	18	46	1.98	
	19	105	1.75	
	20	62	1.85	

also called a unit group. The second group contains a complete data set, also known as a reference or training group. For the third group, the parameters are to be determined (Table 11.5).

In the case of unknown system behavior, which is also beyond the scope of practical experimentation, the existing or measured parameter values are available instead of the system response, assigned to the characteristic features (observed phenomena, events, outcomes) as an empirical response. A transposed and rearranged representation shows the similarity to the analytical procedure from Part 1, whereby the previous empirical response is to be understood as a characteristic operating state, i.e., as a signal (Table 11.6).

Formulated in a generally valid manner and extended by an evaluation using a *S/N* ratio, the familiar context of data analysis evaluation becomes apparent (Table 11.7).

Thus, for each characteristic of slope and dispersion, the corresponding *S/N* ratio of dynamic systems can be determined numerically or graphically (Fig. 11.9).

Table 11.5 Normalization of the data

Individuals	Weight less average	Height less average	Rating number less average
1–10	0	0	0
11	7.9	−0.077	4.8524
12	−17.1	0.193	−8.3888
13	32.9	−0.057	12.8979
14	−7.1	0.123	−4.6020
15	15.9	−0.107	8.7400
16	−6.1	0.243	
17	13.9	−0.067	
18	−19.1	0.223	
19	39.9	−0.007	
20	−3.1	0.093	

Table 11.6 Transposed presentation of the individuals 11–15 of the known reference group

Individuals	11	12	13	14	15
Signal M normalized rating number	4.8524	−8.3888	12.8979	−4.6020	8.7400
Parameter x_1 normalized weight	7.9	−17.1	32.9	−7.1	15.9
Parameter x_2 normalized height	−0.077	0.193	−0.057	0.123	−0.107

Table 11.7 Determination of dynamic *S/N* indicators with signal values as empirical response

Parameter	Signal (empirical response, characteristic feature) $M1, M2, M3, \ldots, Mn$	Slope β	Squared standard deviation σ^2	*S/N* $\frac{\beta^2}{\sigma^2} S_2$
x_1	$x_{11}, x_{12}, x_{13}, \ldots, x_{1n}$			
x_2	$x_{21}, x_{22}, x_{23}, \ldots, x_{2n}$			
x_3	$x_{31}, x_{32}, x_{33}, \ldots, x_{3n}$			
$\vdots$	$\vdots$			
x_k	$x_{k1}, x_{k2}, x_{k3}, \ldots, x_{kn}$			

By assigning the normalized results as signal values, all calculation steps in connection with dynamic systems with zero-point-proportional behavior can be applied identically. The objective evaluation of a dynamic system is carried out with the *S/N* ratio. Geometrically, the slope of each variable is determined by the origin, considering the range of deviation.

S/N index of dynamic systems

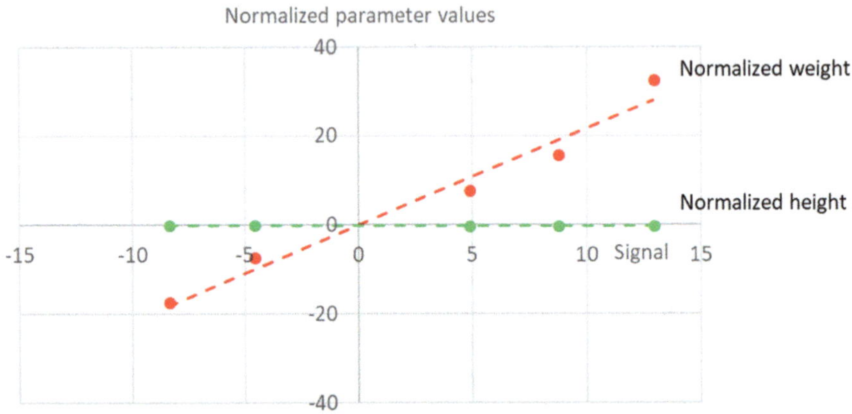

Fig. 11.9 Graphical illustration of the normalized characteristic values as a function of the normalized signal values

$$\frac{S}{N} = \frac{\beta^2}{\sigma^2} S_2$$

Slopes of the parameters as a function of the signal values

$$\beta_1 = \frac{x_{11} \cdot M_1 + x_{12} \cdot M_2 + x_{13} \cdot M_3 + x_{14} \cdot M_4 + x_{15} \cdot M_5}{M_1^2 + M_2^2 + M_3^2 + M_4^2 + M_5^2}$$

$$\beta_2 = \frac{x_{21} \cdot M_1 + x_{22} \cdot M_2 + x_{23} \cdot M_3 + x_{24} \cdot M_4 + x_{25} \cdot M_5}{M_1^2 + M_2^2 + M_3^2 + M_4^2 + M_5^2}$$

in general

$$\beta_i = \frac{x_{i1} \cdot M_1 + x_{i2} \cdot M_2 + x_{i3} \cdot M_3 + \cdots + x_{in} \cdot M_n}{M_1^2 + M_2^2 + M_3^2 + \cdots + M_n^2}$$

$$\text{with } i = 1, 2, 3, \ldots, k$$

$$\text{and } k < n$$

The squared standard deviation of the characteristics follows from

$$\sigma_1^2 = \frac{1}{5}\left[(x_{11} - \beta_1 \cdot M_1)^2 + (x_{12} - \beta_1 \cdot M_2)^2 + (x_{13} - \beta_1 \cdot M_3)^2 + \cdots + (x_{15} - \beta_1 \cdot M_5)^2\right]$$

$$\sigma_2^2 = \frac{1}{5}\left[(x_{21} - \beta_2 \cdot M_1)^2 + (x_{22} - \beta_2 \cdot M_2)^2 + (x_{23} - \beta_2 \cdot M_3)^2 + \cdots + (x_{25} - \beta_2 \cdot M_5)^2\right]$$

in general

$$\sigma_i^2 = \frac{1}{n}\left[(x_{i1} - \beta_i \cdot M_1)^2 + (x_{i2} - \beta_i \cdot M_2)^2 + (x_{i3} - \beta_i \cdot M_3)^2 + \cdots + (x_{in} - \beta_i \cdot M_n)^2\right]$$

with $i = 1, 2, 3, \ldots, k$

and $k < n$

with the dimension factor

$$S_2 = \frac{1}{5}\left(M_1^2 + M_2^2 + M_3^2 + M_4^2 + M_5^2\right)$$

in general

$$S_2 = \frac{1}{n}\left(M_1^2 + M_2^2 + M_3^2 + \cdots + M_n^2\right)$$

Table 11.8 shows the corresponding values for the two-parameter numerical example.

From the zero-point-proportional relationship

$$x = \beta \cdot M$$

the signal value can be calculated if the parameter value is known

$$M = \frac{x}{\beta}$$

where for each variable a corresponding calculated signal value $\widehat{M}$ results.

$$\widehat{M}_1 = \frac{x_{11}}{\beta_1} \quad \widehat{M}_2 = \frac{x_{12}}{\beta_1} \quad \widehat{M}_3 = \frac{x_{13}}{\beta_1} \quad \widehat{M}_4 = \frac{x_{14}}{\beta_1} \quad \widehat{M}_5 = \frac{x_{15}}{\beta_1}$$

$$\widehat{M}_1 = \frac{x_{21}}{\beta_2} \quad \widehat{M}_2 = \frac{x_{22}}{\beta_2} \quad \widehat{M}_3 = \frac{x_{23}}{\beta_2} \quad \widehat{M}_4 = \frac{x_{24}}{\beta_2} \quad \widehat{M}_5 = \frac{x_{25}}{\beta_2}$$

As a simple approximation to the true signal value, the average value seems obvious.

Table 11.8 Characteristics of the numerical example

Normalized parameters	Slope β	Squared standard deviation σ^2	*S/N* index $S_2 = 71.5680$
1 (Weight)	$\beta_1 = 2.1735$	$\sigma_1^2 = 9.9962$	$S/N_1 = 33.8221$
2 (Height)	$\beta_2 = -0.0118$	$\sigma_2^2 = 0.0046$	$S/N_2 = 2.1708$

Table 11.9 Signal values as average of two parameters

Individuals	11	12	13	14	15
True signal normalized rating number	4.8524	−8.3888	12.8979	−4.6020	8.7400
Signal value derived from parameter 1	3.6347	−7.8675	15.1369	−3.2666	7.3154
Signal value derived from parameter 2	6.5153	−16.3305	4.8230	−10.4075	9.0537
Calculated signal as average of parameters 1 and 2	5.0750	−12.0990	9.9800	−6.8371	8.1846

As Table 11.8 shows, the *S/N* ratio as a measure of precision is clearly greater for the first parameter than for the second. The larger the *S/N* figure, the better the calculated value of the signal will match the true value. As a result, the signal value

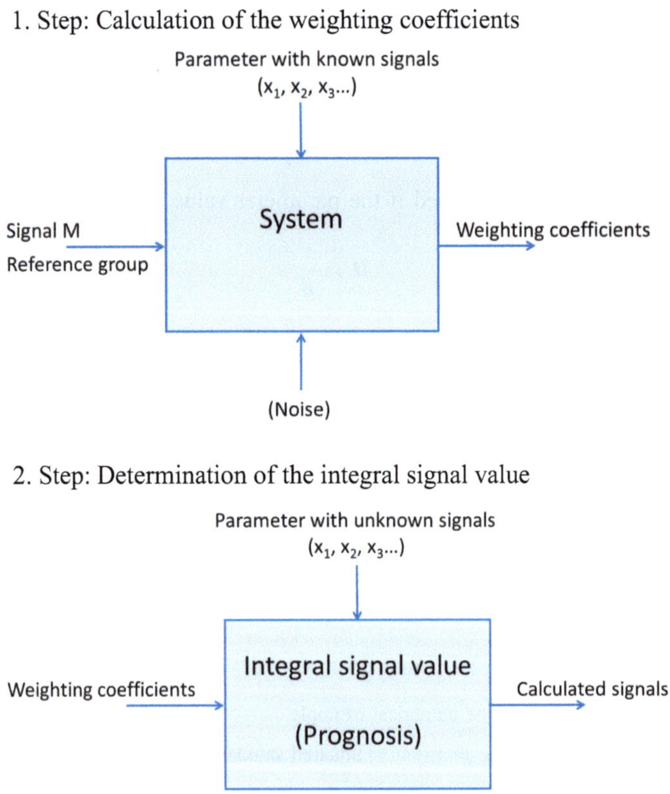

Fig. 11.10 Schematic representation of the determination of prognosticated values

of higher precision is upgraded accordingly considering the S/N ratio or, in the case of lower precision, devalued accordingly. This leads to the idea of taking all parameter contributions for a prediction, but to weight them according to their S/N ratio. Since the S/N indicators are different for all characteristics, a weighting with the respective S/N numerical value leads to the most accurate signal value. If the S/N key figures are added as weighting coefficients, the quotient of the sum of the S/N key figures must be formed for compensation. This results in a so-called integral value for the signals. One or more known operating states, i.e., the existence of at least one reference point or a reference group, are prerequisites for obtaining the weighting coefficients. Therefore, a prognosis takes place in two steps (Fig. 11.10; Table 11.9).

For the given example of two parameters and five signals, the corresponding integral signal values are calculated as follows.

$$\widehat{M}_i = \frac{S/N_1 \cdot \frac{x_{1i}}{\beta_1} + S/N_2 \cdot \frac{x_{2i}}{\beta_2}}{S/N_1 + S/N_2} \qquad i = 1, 2, 3, 4, 5$$

The integral value for k parameters and n signals follows from **(3–1*)**.

$$\widehat{M}_i = \frac{\frac{S/N_1}{\beta_1} \cdot x_{1i} + \frac{S/N_2}{\beta_2} \cdot x_{2i} + \frac{S/N_3}{\beta_3} \cdot y_{3i} + \cdots + \frac{S/N_k}{\beta_k} \cdot x_{ki}}{S/N_1 + S/N_2 + S/N_3 + \cdots + S/N_k} \qquad i = 1, 2, 3, \ldots, n$$

Table 11.10 compares the integral signal values with the true signal values.

The decisive idea for a reliable predicted integral value is to come as close as possible to the true value or event. The accuracy of the integral values, related to the true signals, is evaluated with the numerical value of the SNR index to be determined from the data of the known reference group (Table 11.11).

$$SNR = 10 \log\left(\frac{\beta^2}{\sigma^2} S_2\right)$$

Table 11.10 True signal values and their integral value from the weighted normalized parameters

Signal	M_1	M_2	M_3	M_4	M_5
True values	4.8524	−8.3888	12.8979	−4.6020	8.7400
Integral values	3.8084	−8.3779	14.5149	−3.6973	7.4202

Table 11.11 Dynamic SNR index as an accuracy measure for the integral values

Slope β	Squared standard deviation σ^2	Dimension factor S_2	SNR
$\beta = 1$	$\sigma^2 = 1.2529$	$S_2 = 71.5680$	$SNR = 17.5680\,[\text{dB}]$

Since the coefficients are calculated from the true known signal values, the numerical value of the slope must always be one. The larger the *SNR* index, the closer the integral values are to the straight line with slope one **(3–2*)**. Deviations are usually permitted with three times the standard deviation with respect to the slope with the numerical value one. This results in the limit lines with the smallest and largest slope, within which the integral values are to be assumed (Sect. 2.10) (Table 11.12).

$$\beta_{\min/\max} = \beta \pm \frac{3\sigma}{\sqrt{S_2}}$$

The tolerance range results from the zero-point-proportional behavior, which can also be determined graphically (Fig. 11.11).

$$\text{Tolerance line with largest slope } x_{\max} = \beta_{\max} \cdot M$$

$$\text{Tolerance line with smallest slope } x_{\min} = \beta_{\min} \cdot M$$

The result of the two-parameter numerical example is shown in the Fig. 11.12.

Table 11.12 Smallest and largest slopes as boundary lines of integral signal values

Slope β	Squared standard deviation σ^2	Dimension factor S_2	Dispersion of the slope
$\beta = 1$	$\sigma^2 = 1.2529$	$S_2 = 71.5680$	$\frac{3\sigma}{\sqrt{S_2}} = 0.3969$
$\beta_{\min} = 0.6$			
$\beta_{\max} = 1.4$			

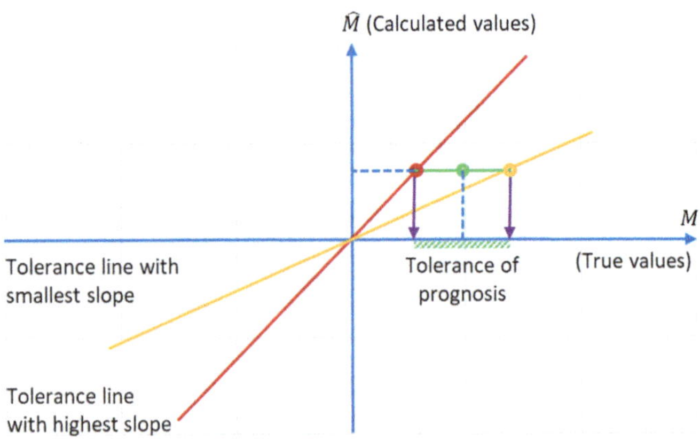

Fig. 11.11 Prognosticated signal value and associated true value tolerance range

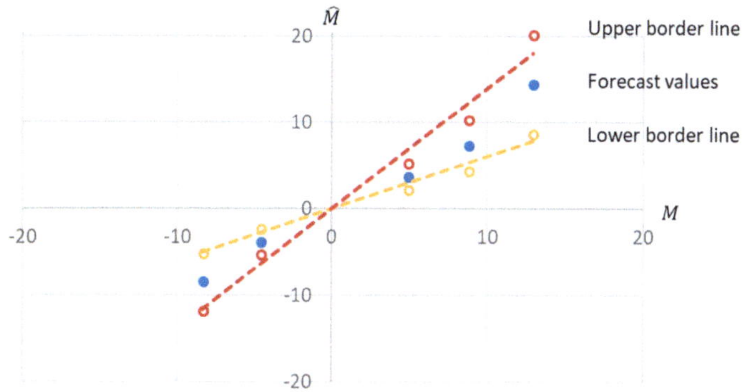

Fig. 11.12 Graphic of the integral values as a function of the true signal values

Table 11.13 Integral values of the test group from their parameter values

Individuals	Weight less average	Height less average	Rating number less average	Integral values
1–10	0	0	0	
11	7.9	−0.077	4.8524	3.8084
12	−17.1	0.193	−8.3888	−8.3779
13	32.9	−0.057	12.8979	14.5149
14	−7.1	0.123	−4.6020	−3.6973
15	15.9	−0.107	8.7400	7.4202
16	−6.1	0.243		−3.8774
17	13.9	−0.067		6.3514
18	−19.1	0.223		−9.3957
19	39.9	−0.007		17.2861
20	−3.1	0.093		−1.8149

With the determination of the weighting coefficients, the rating number (characteristic outcome) can be prognosticated from data of the test group of the same system without knowledge of the underlying functional relationships. The calculation is made with the coefficients derived from the data of the known reference group (Table 11.13).

$$\widehat{M}_i = \frac{S/N_1 \cdot \frac{x_{1i}}{\beta_1} + S/N_2 \cdot \frac{x_{2i}}{\beta_2}}{S/N_1 + S/N_2}$$

$$\widehat{M}_i = \frac{15.5612 \cdot x_{1i} + (-183.6790) \cdot x_{2i}}{33.8221 + 2.1708}; \quad i = 1, 2, 3, 4, 5$$

If the integral values are of sufficient accuracy, the tolerance range in which the actual values are found can be completed.

Table 11.14 Integral values with tolerances

Individuals	Rating number less average	Integral values	Lower tolerance limit	Upper tolerance limit
11	4.8524	3.8084	2.2967	5.32013
12	−8.3888	−8.3779	−5.0524	−11.7034
13	12.8979	14.5149	8.7534	20.2763
14	−4.6020	−3.6973	−2.2297	−5.1649
15	8.7400	7.4202	4.4749	10.3656
16		−3.8774	−2.3382	−5.4164
17		6.3514	3.8303	8.8725
18		−9.3957	−5.6662	−13.1252
19		17.2861	10.4246	24.1475
20		−1.8149	−1.0945	−2.5352

Table 11.15 Calculated result of the non-normalized rating number with tolerances

Individuals	Rating number	Integral values	Lower tolerance limit	Upper tolerance limit
11	25.86	24.82	23.31	26.33
12	12.62	12.63	15.96	9.31
13	33.91	35.53	29.77	41.29
14	16.41	17.31	18.78	15.85
15	29.75	28.43	25.49	31.38
16		17.13	18.67	15.60
17		27.36	24.84	29.88
18		11.62	15.35	7.89
19		38.30	31.44	45.16
20		19.20	19.92	18.48

$$x_{\text{lower limit}} = \beta_{\min} \cdot \widehat{M}$$

$$x_{\text{upper limit}} = \beta_{\max} \cdot \widehat{M}$$

Table 11.14 is extended accordingly.

To obtain the original valuation number, the scaling must be reversed, i.e., the previously determined average value must be added to all data (Table 11.15).

$$\text{Calculated results}_i = \widehat{M}_i + \text{Average} \quad i = 1, 2, 3, \ldots, n$$

The true rating numbers of the known reference group are compared with the prognosticated values (Fig. 11.13).

Fig. 11.13 Comparison of the original outcome values with the integral values of the rating numbers

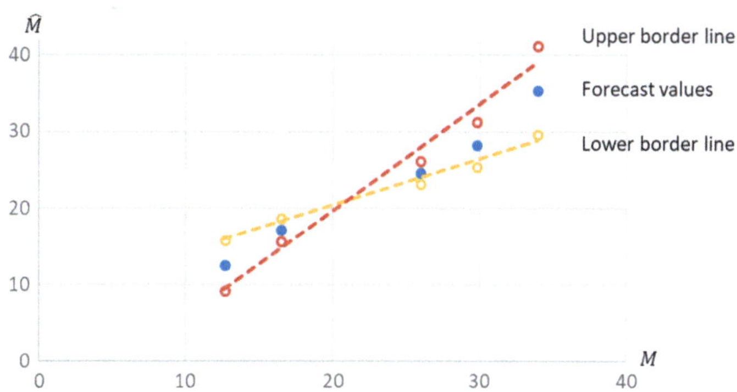

Fig. 11.14 Integral values of the reference group as a function of the original rating number with tolerances

In addition, the integral values with the tolerance limits as a function of the original rating number are shown. In case of a perfect match, all data are on a straight line through the zero point with the slope one (Fig. 11.14).

11.3 Selecting a Zero Point

In the previous subchapter, the unit group of the introductory example is used to define the origin. If the question arises as to which number of objects is to be used to determine the zero point, then, due to the transformation into an origin to be carried out, one object is already completely sufficient. The normalization process is the same, i.e., the subtraction of the characteristic data of the selected object from all other corresponding data. Without taking data correlation into account, there is also no need for a unit group to exist. This allows values for the origin to be determined directly from the reference group. Instead of the average value, the median or an individual object of the reference group can also be selected as a reference point. Various objects can also be tested for their suitability as reference points to determine and apply the one with the smallest tolerances for prognostications. For this purpose, the associated *SNR* key figures for the integral and true values must be compared and the object that displays the largest numerical value for the *SNR* index must be retained for the zero-point transformation. If the smallest (or largest) result in number is used as the reference point, this corresponds to a one-sided consideration, starting from this object as the origin. It should be noted that with a position of the origin outside the reference group, the *SNR* index as a measure of accuracy continuously increases with increasing distance. This is because the viewing angle to the deviation values of the group appears smaller when viewed from a more distant origin. However, this does not lead to any real improvement in precision, as the initial state is restored by the retransformation. For clarification, the same example, but without a unit group, with a one-sided zero-point position is selected (Table 11.16).

The smallest rating number shows individual no. 12, which is defined as the zero point. Depending on the parameter values, it cannot be ruled out in advance that the values to be determined for the group without a known rating number may also be exclusively one-sided (Table 11.17).

The *S/N* ratios of the reference group resulting from this zero-point follow from the respective parameter slopes and their squared standard deviations (Table 11.18).

The integral signal values weighted with the *S/N* indicators must be compared with the true signal values (Tables 11.19, 11.20, 11.21 and 11.22).

Table 11.16 Weight and height with corresponding rating number without unit group

Groups	Individuals	Weight [kg]	Height [m]	Characteristic feature, rating number
Reference group	11	73	1.68	25.86
	12	48	1.95	12.62
	13	98	1.70	33.91
	14	58	1.88	16.41
	15	81	1.65	29.75
Group without rating number	16	59	2.00	
	17	79	1.69	
	18	46	1.98	
	19	105	1.75	
	20	62	1.85	

Table 11.17 Use of a reference group member for normalization

Individuals	Weight less value of individual #12	Height less value of individual #12	Rating number less value of individual #12
11	25	−0.27	13.2412
12	0	0	0
13	50	−0.25	21.2868
14	10	−0.07	3.7869
15	33	−0.3	17.1288
16	11	0.05	
17	31	−0.26	
18	−2	0.03	
19	57	−0.2	
20	14	−0.1	

Table 11.18 Characteristic features for reference point (=individual #12) as origin

Normalized parameters	Slope β	Squared standard deviation σ^2	*S/N* ratio $S_2 = 243.0481$
1 (Weight)	$\beta_1 = 2.1347$	$\sigma_1^2 = 11.9582$	$S/N_1 = 89.1891$
2 (Height)	$\beta_2 = -0.0153$	$\sigma_2^2 = 0.00296$	$S/N_2 = 18.4246$

Table 11.19 True signal values and their integral value with individual #12 as reference point

Signal	M_1	M_2	M_3	M_4	M_5
True values	13.2412	0	21.2868	3.7869	17.1288
Integral values	12.7324	0	22.2144	4.6671	16.1747

Table 11.20 Dynamic *SNR* key figure of integral values

Slope β	Squared standard deviation σ^2	Dimension factor S_2	*SNR* index
$\beta = 1$	$\sigma^2 = 0.7011$	$S_2 = 243.0481$	$SNR = 25.2350$ [dB]

Table 11.21 Tolerance range of integral values

Slope β	Squared standard deviation σ^2	Dimension factor S_2	Dispersion of the slope
$\beta = 1$	$\sigma^2 = 0.7011$	$S_2 = 243.0481$	$\frac{3\sigma}{\sqrt{S_2}} = 0.1642$
$\beta_{min} = 0.8358$			
$\beta_{max} = 1.1642$			

Table 11.22 Re-transformed results and integral values with tolerance limits

Individuals	Rating number	Integral values	Lower tolerance limit	Upper tolerance limit
11	25.86	25.36	23.27	27.45
12	12.62	12.62	12.62	12.62
13	33.91	34.84	31.19	38.49
14	16.41	17.29	16.52	18.06
15	29.75	28.80	26.14	31.45
16		16.33	15.72	16.94
17		27.57	25.12	30.03
18		11.51	11.69	11.33
19		37.00	32.99	41.00
20		19.18	18.10	20.26

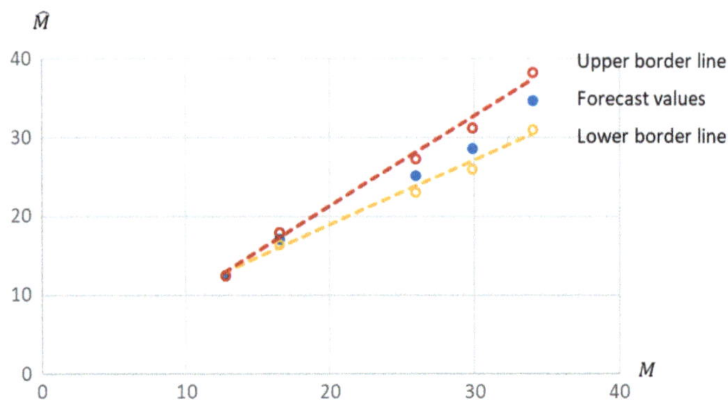

Fig. 11.15 Graphic view of the integral values as a function of the true signal values

For a graphic illustration of the one-sided zero-point position, the limit slopes are also plotted (Fig. 11.15).

11.4 Size of the Reference Group

Once the origin has been defined, the question arises as to the minimum size of the reference group. The importance of the reference group lies in the determination of the weighting coefficients for calculating the integral values. The weighting coefficients stand for the precision of the prognosis of the individual characteristics as *S/N* ratio. Due to the prerequisite of zero-point-proportional behavior, two reference points are sufficient for determining the slope and deviation for deriving the numerical value of the *S/N* ratio, whereby additional reference points contribute to increasing data certainty. Using the same example, the prognosis reliability is demonstrated with decreasing number of objects. To ensure comparability, the previous group (individuals #16–20) of unknown rating numbers is supplemented with true values. For simplicity's sake, individual #12 serves as the origin for one-sided representation. This leaves four reference points for calculating the weighting coefficients with the already existing result in Sect. 11.3, supplemented by the true values of the test group (Table 11.23).

Table 11.24 shows the parameters based on the reference points 11, 13, 14 (Tables 11.25 and 11.26).

The results of the reduction to two reference points, 11 and 13, are listed below (Tables 11.27, 11.28 and 11.29).

Table 11.23 Integral values with the reference points 11, 13, 14, 15

Groups	Individuals	Characteristic feature rating number	Integral values
Reference point	11	25.86	25.36
Zero point	12	12.62	12.62
Reference point	13	33.91	34.84
Reference point	14	16.41	17.29
Reference point	15	29.75	28.80
Test group	16	14.75	16.33
	17	27.66	27.57
	18	11.73	11.51
	19	34.29	37.00
	20	18.12	19.18

Table 11.24 Characteristic features for three reference points

Normalized parameter	Slope β	Squared standard deviation σ^2	*S/N* index
1 (Weight)	$\beta_1 = 2.2297$	$\sigma_1^2 = 9.7749$	$S/N_1 = 108.975$
2 (Height)	$\beta_2 = -0.0143$	$\sigma_2^2 = 0.00324$	$S/N_2 = 13.4436$

Table 11.25 Dynamic *SNR* index standing for the accuracy and tolerance range for three reference points

Slope β	Squared standard deviation σ^2	Dimension factor S_2	*SNR* index
$\beta = 1$	$\sigma^2 = 0.7696$	$S_2 = 214.2656$	$SNR = 24.4468$ [dB]
Tolerance of the slope	Minimum slope	Maximum slope	
$\frac{3\sigma}{\sqrt{S_2}} = 0.1798$	0.8202	1.1798	

Table 11.26 Re-transformed integral values with the reference points 11, 13, 14

Groups	Individuals	Characteristic feature rating number	Integral values
Reference point	11	25.86	24.68
Zero point	12	12.62	12.62
Reference point	13	33.91	34.51
Reference point	14	16.41	17.16
Test group	16	14.75	16.63
	17	27.66	27.00
	18	11.73	11.59
	19	34.29	36.92
	20	18.12	18.98

Table 11.27 Characteristic features for two reference points

Normalized parameter	Slope β	Squared standard deviation σ^2	*S/N* index $S_2 = 314.2283$
1 (Weight)	$\beta_1 = 2.2203$	$\sigma_1^2 = 13.4235$	$S/N_1 = 115.400$
2 (Height)	$\beta_2 = -0.0142$	$\sigma_2^2 = 0.00473$	$S/N_2 = 13.3265$

Table 11.28 Precision and tolerances with two reference points

Slope β	Squared standard deviation σ^2	Dimension factor S_2	*SNR* index
$\beta = 1$	$\sigma^2 = 0.9537$	$S_2 = 314.2283$	$SNR = 25.1781$ [dB]
Tolerance of the slope	Minimum slope	Maximum slope	
$\frac{3\sigma}{\sqrt{S_2}} = 0.1653$	0.8347	1.1653	

Table 11.29 Integral values with the reference points 11 and 13

Groups	Individuals	Characteristic feature, rating number	Integral values
Reference point	11	25.86	24.69
Zero point	12	12.62	12.62
Reference point	13	33.91	34.64
Test group	16	14.75	16.70
	17	27.66	27.04
	18	11.73	11.60
	19	34.29	37.10
	20	18.12	19.01

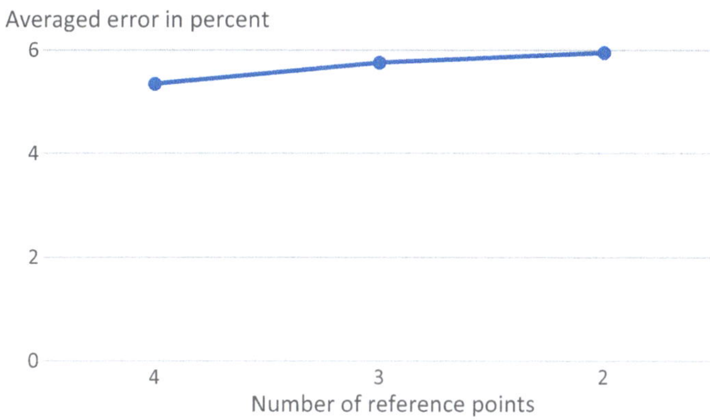

Fig. 11.16 Average relative error of integral values calculated from four, three, and two reference points

To evaluate the approaches of reference groups of different sizes, the average relative error of the integral values, related to the true values of the test group, is plotted (Fig. 11.16).

Depending on the data situation, the reliability of the prognosis is subject to deviations with increasing amplitude and increasing error with decreasing size of the reference group. The larger the reference group, the higher the precision of the prognosis. If only a few objects of the reference group can be considered representative, two or three known objects are already completely sufficient for prognosticating purposes. This reveals the decisive advantage of this method since for practical applications the number of variables can be even greater than the number of known objects or events for deriving the weighting coefficients. Even though this method does not allow to obtain exact solutions of a system with more variables than objects, but still allows the derivation of integral values with small tolerances. This describes the situation of a system of equations dealing with more unknowns than existing equations as mentioned in the introduction.

11.5 Transformation of Data

The simple normalization process, i.e., deduction of the average value or reference point from all other data, is not dimensionless. A representation of characteristic data without dimension is possible with mathematical standardization or the formation of a quotient. Since a multiplicative constant does not change the *SNR* index, the calculated integral values also remain unchanged. Thus, the procedure of including a zero point into the data field by means of the normalization process is at the same time the simplest data transformation.

Another variant is the orthogonalization of the characteristic data. For this purpose, all data must be standardized, and the transformation equations must be derived from the reference group or, if available, from the unit group to get linear independent variables. With the retransformation of the standardization of the integral values, the predicted signal values are available on the original scale.

One more possible transformation is the formation of a quotient and its logarithmic calculation, provided that all values have a positive sign. The reference point on the logarithmic scale becomes the zero point, as follows:

$$\log\left(\frac{x}{x_{\text{ref}}}\right) = 0 \quad \text{for}\, x = x_{\text{ref}}$$

The retransformation with inverse logarithm and multiplication of the reference point leads back to the dimension of the original values. Both types of transformation, if applicable, can result in a more advantageous *SNR* index of a prognosis to further narrow down the tolerance interval. The procedures are summarized in Table 11.30.

The first two procedures are already described in Sect. 11.4. For reasons of comparison, the same data is used for the additional variants. Orthogonalization is carried out as described in Sect. 8.1, whereby the standardization $\left(\frac{x-\bar{x}}{\sigma}\right)$ of all characteristics and signal values must be performed first (Tables 11.31 and 11.32).

The parameter-specific slopes and *S/N* key figures are derived from the reference group to determine the weighting coefficients (Table 11.33).

The integral values are taken from Sect. 11.2.

$$\widehat{M}_i = \frac{S/N_1 \cdot \frac{x_{1i}}{\beta_1} + S/N_2 \cdot \frac{x_{2i}}{\beta_2}}{S/N_1 + S/N_2} \qquad i = 1, 2, 3, 4, 5$$

From the true and integral values of the reference group follow the *SNR* index and the tolerances as an objective accuracy measure (Table 11.34).

The retransformation of the standardization with average and deviation results in the original dimensions (Table 11.35).

Table 11.30 Application of various kinds of data transformation

Procedure #	Transformation	Normalization
1	Displacement	Subtraction average value of the reference group
2	Displacement	Subtraction of smallest numerical value of the reference group
3	Orthogonalization Data of reference group	Standardization
4	Quotient Reference point or average of reference group	Logarithmic calculation

Table 11.31 Data and its standardization

Individuals	Weight [kg]	Height [m]	Rating number	Parameter 1 standardized	Parameter 2 standardized	*M* standardized
11	73	1.68	25.86	0.0801	−0.7673	0.2685
12	48	1.95	12.62	−1.3495	1.4846	−1.3831
13	98	1.70	33.91	1.5096	−0.6005	1.2720
14	58	1.88	16.41	−0.7777	0.9008	−0.9108
15	81	1.65	29.75	0.5375	−1.0175	0.7534
Average	71.6000	1.7720	23.7120	0	0	0
Standard deviation	17.4883	0.1199	8.0174	1	1	1
16	59	2.00		−0.7205	1.9016	
17	79	1.69		0.4231	−0.6839	
18	46	1.98		−1.4638	1.7348	
19	105	1.75		1.9099	−0.1835	
20	62	1.85		−0.5489	0.6505	

Table 11.32 Orthogonalization of the parameters

Individuals	x_1 standardized (orthogonalized)	x_2 orthogonalized	*M* standardized
11	0.0801	−0.6998	0.2685
12	−1.3495	0.3460	−1.3831
13	1.5096	0.6732	1.2720
14	−0.7777	0.2446	−0.9108
15	0.5375	−0.5640	0.7534
16	−0.7205	1.2937	
17	0.4231	−0.3269	
18	−1.4638	0.4997	
19	1.9099	1.4279	
20	−0.5489	0.1874	

Table 11.33 Characteristic features of the orthogonalized parameters

Orthogonalized parameters	Slope β	Squared standard deviation σ^2	*S/N* index dimension factor $S_2 = 1$
x_1	0.9843	0.0312	31.0189
x_2	−0.0916	0.2797	0.0300

For graphic illustration, the integral values are plotted as a function of the true values including the limit tolerance lines of the reference group (Fig. 11.17).

If all data show the same sign, a data transformation on logarithmic scale is applicable. All data must first be normalized as a quotient without dimensions. The denominator can be the average value of the reference group or a selected reference

Table 11.34 Prognosis accuracy and tolerances of the orthogonalized parameters

Slope β	Squared standard deviation σ^2	Dimension factor S_2	*SNR* index	Tolerance of the slope $\frac{3\sigma}{\sqrt{S_2}}$
$\beta = 1$ $\beta_{min} = 0.48$ $\beta_{max} = 1.52$	0.0303	1	15.1886 [dB]	0.5220

Table 11.35 Prognosticating the key figures for the reference group and unknown object group with tolerances

Individuals	Rating number	Integral values	Lower tolerance limit	Upper tolerance limit
11	25.86	24.4226	24.0517	24.7935
12	12.62	12.7011	18.4491	6.9531
13	33.91	35.9397	29.5565	42.3229
14	16.41	17.36291	20.6773	14.0485
15	29.75	28.1337	25.8255	30.4420
16		17.7396	20.8574	14.6218
17		27.1830	25.3711	28.9950
18		11.7575	17.9981	5.5169
19		39.1332	31.0829	47.1835
20		19.2291	21.5693	16.8888

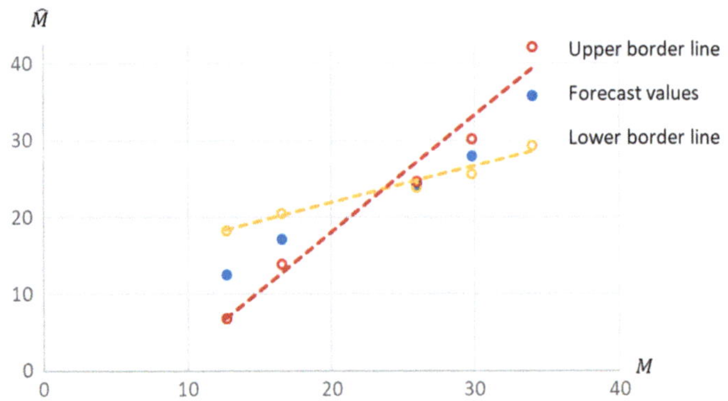

Fig. 11.17 Integral values and tolerances for the objects of the reference group derived from orthogonalized parameters (procedure 3)

point. The subsequent logarithmic transformation generates a zero point for the applicability of a zero-point-proportional equation (Table 11.36).

Weighting factors and parameter-specific key figures are determined from the reference group (Table 11.37).

Table 11.36 Characteristics as quotients and in logarithmic form

Individuals	Weight/Average	Height/Average	Rating/Average	x_1 normalized logarithm	x_2 normalized logarithm	M normalized logarithm
11	1.0196	0.9481	1.0908	0.0084	−0.0232	0.0377
12	0.6704	1.1005	0.5324	−0.1737	0.0416	−0.2738
13	1.3687	0.9594	1.4301	0.1363	−0.0180	0.1554
14	0.8101	1.0609	0.6921	−0.0915	0.0257	−0.1599
15	1.1313	0.9312	1.2547	0.0536	−0.0310	0.0985
Average	1	1	1			
16	0.8240	1.1287		−0.0841	0.0526	
17	1.1034	0.9537		0.0427	−0.0206	
18	0.6425	1.1174		−0.1922	0.0482	
19	1.4665	0.9876		0.1663	−0.0054	
20	0.8659	1.0440		−0.0625	0.0187	

Table 11.37 Characteristics of the logarithmic parameters

Normalized logarithmic parameter	Slope β	Squared standard deviation σ^2	*S/N* index dimension factor $S_2 = 0.02716$
x_1	0.655048	0.0003574	32.6025
x_2	−0.1636	0.0001149	6.3263

Table 11.38 *SNR* index of accuracy of the prognosis and tolerances of the logarithmic parameters

Slope β	Squared standard deviation σ^2	Dimension factor S_2	*SNR* index	Tolerance of the slope $\frac{3\sigma}{\sqrt{S_2}}$
$\beta = 1$	0.0003565	0.02716	18.8180 [dB]	0.3437
$\beta_{min} = 0.66$ $\beta_{max} = 1.34$				

The integral values are calculated accordingly (Sect. 11.2)

$$\widehat{M}_i = \frac{S/N_1 \cdot \frac{x_{1i}}{\beta_1} + S/N_2 \cdot \frac{x_{2i}}{\beta_2}}{S/N_1 + S/N_2} \quad i = 1, 2, 3, 4, 5$$

which results in the *SNR* index for the accuracy of the signals and the associated tolerances (Table 11.38).

The retransformation consists in the formation of the inverse logarithm (power function) and multiplication by the average value (Table 11.39).

The results for the reference group are shown graphically (Fig. 11.18).

Table 11.39 Prognosticated signal values with corresponding tolerances of the reference group and of the unknown object group

Individuals	Rating number	Integral values	Lower tolerance limit	Upper tolerance limit
11	25.86	25.6284	24.9528	26.3223
12	12.62	12.9310	15.9275	10.4982
13	33.91	36.9098	31.7019	42.9733
14	16.41	17.0797	19.1187	15.2581
15	29.75	29.8013	27.5495	32.2371
16		16.4164	18.6281	14.4673
17		28.1851	26.5596	29.9100
18		12.0618	15.2166	9.5611
19		39.1692	32.9625	46.5448
20		18.8995	20.4321	17.4818

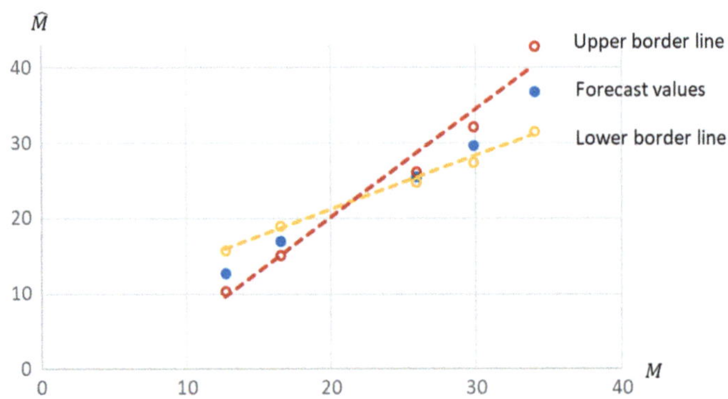

Fig. 11.18 Integral signal values and tolerances for the objects of the reference group calculated from logarithmic parameters (procedure 4)

The results are summarized in the figure. The first bar represents the true value of the object of the reference group (Fig. 11.19).

The relative deviation of the prognosticated results from the true value averaged over the objects of the reference group, is shown in percent below (Fig. 11.20).

$$\text{Relative deviation in percent:} \left| \frac{y_{\text{true}} - y_{\text{prognosis}}}{y_{\text{true}}} \right| * 100\%$$

The smallest possible tolerance interval is decisive for the reliability of a prognosis. The relative tolerance interval is plotted as a percentage of the listed procedures, also averaged over all objects of the reference group (Fig. 11.21).

$$\text{Relative tolerance in percent:}\left|\frac{y_{\text{lower tolerance limit}} - y_{\text{upper tolerance limit}}}{y_{\text{true}}}\right| * 100\%$$

As the comparative figures show, all procedures in this simply chosen example lead to similar results. In principle, however, it should be noted that independently of the numerical examples listed, the procedures must be checked against other data sets. If high accuracy is required, a retrospective analysis with existing data is recommended first. By removing some data points as a test group and predicting their appearance, the most appropriate method for prognosis is to be selected. This

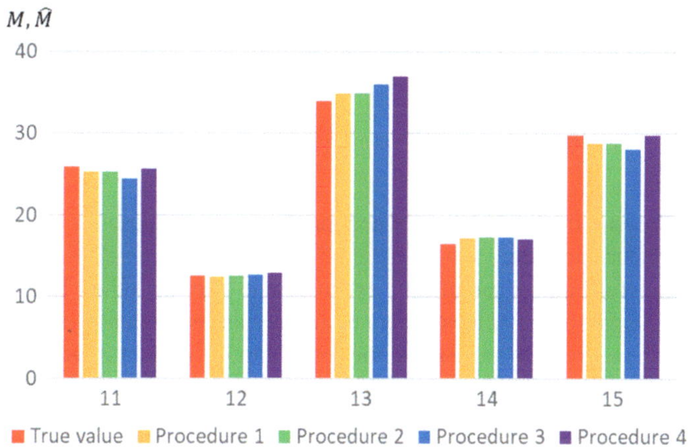

Fig. 11.19 Precision of the integral values in comparison of the different transformation methods

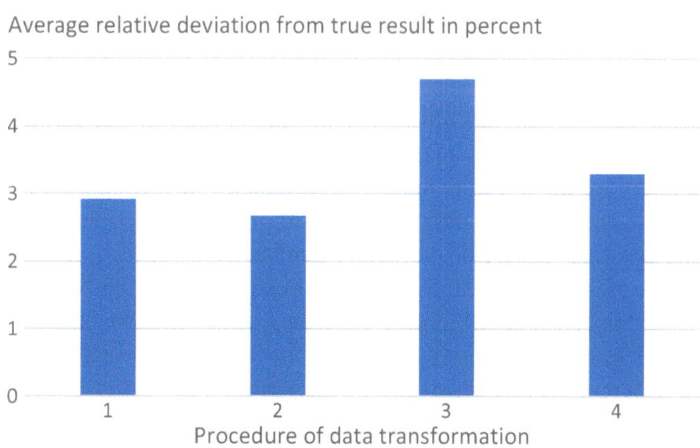

Fig. 11.20 Average values of the relative deviations in percent in comparison of the applied data transformations

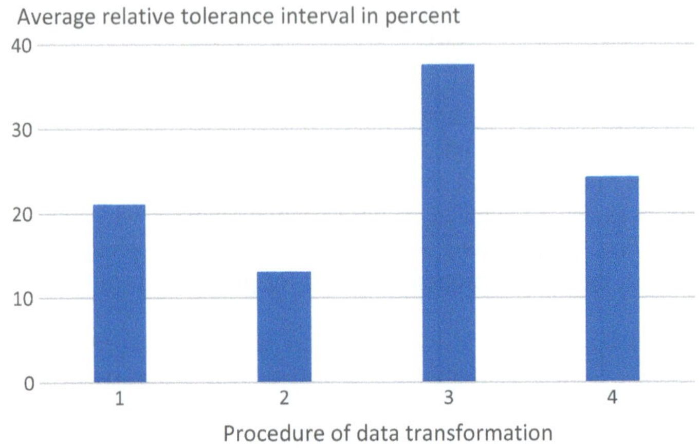

Fig. 11.21 Average relative tolerance interval as a percentage of the different data transformations

guarantees the smallest tolerances when predicting an unknown response from existing characteristic data.

Several transformations can also be executed one after the other. If, for example, the range of the parameter values differs by more than one order of magnitude from the range of the associated outcomes, the logarithm of all data in advance leads to an improved accuracy of prognosis. The reason is that possible relations with a power effect are converted into a linear form. An additional orthogonalization can also lead to a further improvement due to the decoupling of the original parameters.

11.6 Prognosis Algorithm with Squared Mahalanobis Distance

A prerequisite for the applicability of one-sided prognosis procedures is that characteristic features based on parameter data are available as positive numbers. Typical examples are the absolute temperature, which is always positive, as well as the time, if the starting point is defined as the origin from which the course of time can only take positive values, or the defect number of an object, image or sensor, which cannot be negative, to name a few for illustration. Similar objects classified according to certain criteria, which are described with different parameter values, can be regarded as a uniform group. Apart from that, variable values that lead to considerable different results, depart more or less from the previously defined object group. The further the result value deviates, the more different is its appearance. Known objects that do not belong to the same group are combined into a reference group.

Table 11.40 Parameters generating characteristic features and the Mahalanobis metric

Objects	Parameter x_1, x_2, x_3, …, x_k	True value, characteristic feature	Squared Mahalanobis distance
1	Uniform group for defining the zero point; corresponds to the unit group of the Mahalanobis algorithm	M_1	D_1^2
2		M_2	D_2^2
3		M_3	D_3^2
⋮		⋮	⋮
l		M_l	D_l^2
Average		$\overline{M}$	1
$l + 1$	Reference group	M_{l+1}	
$l + 2$		M_{l+2}	
$l + 3$		M_{l+3}	
⋮		⋮	
m		M_m	
$m + 1$	Group of known parameter values with unknown characteristic feature		
$m + 2$			
$m + 3$			
⋮			
n			

From this consideration, it can be expected that the squared Mahalanobis distance calculated from correlated parameter values traces an analog behavior, provided that a relationship between the non-squared Mahalanobis distance and the characteristic feature can be approximated (Table 11.40).

As before with the analysis of dynamic systems, the parameters are reinterpreted into the signal values M. The transformation equations for calculating the squared Mahalanobis distances for all available parameter data are obtained from the unit group. With the squared distance, the direction is only positive on one side, whereby the average value of the squared Mahalanobis distances of the unit group is identical to one. A zero-point-proportional relationship between the characteristic feature and the non-squared Mahalanobis distance arises when the same relation is established with the center value zero. By subtracting the average value of the parameters of the unit group $\overline{M}$ from the parameters of the reference group M_{ref}

$$M_{center} = M_{ref} - \overline{M}$$

and according to the numerical value one of the squared Mahalanobis distances of the reference group, a transformation to the origin is achieved.

$$D^2_{\text{center}} = D^2_{\text{Ref}} - 1$$

Thus, a functional relationship of the parameters and the Mahalanobis metric appears attainable via a calibration factor if the Mahalanobis distance D_{center} is not in squared form.

$$D_{\text{center}} = \sqrt{D^2_{\text{ref}} - 1}$$

If the reference group or group with unknown characteristic features contain objects with a squared Mahalanobis distance smaller than one, then these must be sorted out and assigned to the unit group. This ensures that all squared distances, except objects of the unit group, are greater than zero after the mathematical subtraction. If sorting is not possible, the subtraction of the numerical value one can be omitted as an approximation. Assuming a proportional relation, the approach is chosen

$$M_{\text{center}} \sim D_{\text{center}}$$

$$M_{\text{ref}} - \overline{M} \sim \sqrt{D^2_{\text{ref}} - 1}$$

$$\left(\text{if } D^2_{\text{ref}} > 1, \text{otherwise approximately } M_{\text{center}} \sim \sqrt{D^2}\right)$$

$$\sqrt{D^2_{\text{ref}} - 1} = \beta_0 \left(M_{\text{ref}} - \overline{M}\right)$$

$$\beta_0 = \text{constant of proportionality}$$

The proportionality constant is determined from the objects of the reference group.

$$\beta_0 = \frac{\sqrt{D^2_{\text{ref}} - 1}}{M_{\text{ref}} - \overline{M}}$$

It is used as a calibration factor to predict unknown characteristic features from different parameter values. The calculated value for the signal is $\widehat{M}$:

$$\widehat{M} = \frac{\sqrt{D^2_{\text{ref}} - 1}}{\beta_0}$$

The signal $\widehat{M}$ is transformed back to the original scale with the average value.

To explain the procedure, the two-parameter example, given by weight and height, is used. Table 11.41 shows an extension by a reference group with known parameters and objects with unknown results that are to be predicted. The squared Mahalanobis distances are calculated using the Mahalanobis algorithm (Sect. 7.2).

Table 11.41 Parameter values with the rating number as characteristic feature and Mahalanobis distances of the unit group and reference group; object group with unknown rating numbers

Groups	Individuals	Weight [kg]	Height [m]	Rating number	D^2
1. Group: 1–10					
Setting the zero point	1	55	1.61	21.22	1.6617
	2	60	1.75	19.59	0.3545
	3	50	1.69	17.51	1.8385
	4	65	1.72	21.97	0.2623
	5	68	1.77	21.71	0.0674
	6	62	1.66	22.50	1.1489
	7	75	1.81	22.89	0.6985
	8	76	1.86	21.97	0.8357
	9	61	1.82	18.42	1.8518
	10	79	1.88	22.35	1.2808
Average				21.012	1
2. Group: 11–15					
Known reference group	11	73	1.68	25.8645	4.0748
	12	88	1.95	23.1427	3.3723
	13	98	1.7	33.9100	24.9953
	14	96	1.88	27.1616	8.2639
	15	81	1.65	29.7521	11.7295
3. Group: 16–20					
Group with unknown rating numbers	16	59	1.55		5.4358
	17	79	1.69		7.0165
	18	92	1.98		4.6174
	19	105	1.75		28.4437
	20	62	1.46		15.0237

The characteristic features transformed to the center value zero are interpreted as signals for dynamic behavior (Table 11.42).

The calibration factor results from the slope of the zero-point-proportional equation.

$$\beta_0 = \frac{\sqrt{D_1^2-1}\cdot M_1 + \sqrt{D_2^2-1}\cdot M_2 + \sqrt{D_3^2-1}\cdot M_3 + \ldots + \sqrt{D_i^2-1}\cdot M_i}{M_1^2 + M_2^2 + M_3^2 + \cdots + M_i^2}$$

$$i = 1, 2, 3, \ldots, m \qquad \text{here } m = 5$$

$$\beta_0 = 0.3894$$

The signal values are calculated from the quotient of the non-squared Mahalanobis distance and the slope (Table 11.43).

Table 11.42 Transformed characteristic feature of the reference group and of the unknown object group

Individuals or objects	Transformed characteristic feature, Signal M	D^2	$\sqrt{D^2-1}$
1–10 Average	21.012	1	
11	4.8524	4.0748	1.7535
12	2.1306	3.3723	1.5402
13	12.8979	24.9953	4.8985
14	6.1495	8.2639	2.6952
15	8.7400	11.7295	3.2756
16		5.4358	2.1061
17		7.0165	2.4529
18		4.6174	1.9020
19		28.4437	5.2387
20		15.0237	3.7448

Table 11.43 Calculated signal values less the average value

Individuals, objects	True signal value	Calculated signal value
11	4.8524	4.5036
12	2.1306	3.9558
13	12.8979	12.5810
14	6.1495	6.9221
15	8.7400	8.4128
16		5.4093
17		6.2998
18		4.8849
19		13.4547
20		9.6179

$$\widehat{M}_i = \frac{\sqrt{D_i^2 - 1}}{\beta_0} \qquad i = 1, 2, 3, 4, 5$$

The correspondence between true and calculated signal value is evaluated with the same *SNR* index of dynamic systems as for the two-sided method (Sect. 11.2). The slope is always equal to one since the calculated values are derived from the best possible adjustment of the calibration factor with the reference group. The precision is exclusively described by the squared standard deviation, from which the *SNR* index is determined in dimension-free form (Table 11.44).

Table 11.44 Evaluation of the accuracy of the prognosis with the dynamic *SNR* key figure

Slope β	Squared standard deviation σ^2	Dimension factor S_2	*SNR* index
1	0.8515	61.7291	18.6029 [dB]

Table 11.45 Minimum and maximum slopes for the prognosticated signal values

Slope β	Squared standard deviation σ^2	Dimension factor S_2	Tolerance of the slope
$\beta = 1$	$\sigma^2 = 0.8515$	$S_2 = 61.7291$	$\frac{3\sigma}{\sqrt{S_2}} = 0.3524$
$\beta_{min} = 0.65$ $\beta_{max} = 1.35$			

$$\sigma^2 = \frac{1}{m}\left(\left(\widehat{M}_1 - \beta M_1\right)^2 + \left(\widehat{M}_2 - \beta M_2\right)^2 + \left(\widehat{M}_3 - \beta M_3\right)^2 + \cdots + \left(\widehat{M}_i - \beta M_i\right)^2\right)$$

$$\beta = 1 \quad i = 1, 2, 3, \ldots, m$$

$$\sigma^2 = \frac{1}{m}\left(\left(\widehat{M}_1 - M_1\right)^2 + \left(\widehat{M}_2 - M_2\right)^2 + \left(\widehat{M}_3 - M_3\right)^2 + \cdots + \left(\widehat{M}_i - M_i\right)^2\right)$$

$$S_2 = \frac{1}{m}\left(M_1^2 + M_2^2 + M_3^2 + \cdots + M_i^2\right)$$

$$SNR = 10 \log\left(\frac{\beta^2}{\sigma^2} S_2\right)$$

For the tolerances of the slope, the threefold deviation value is permitted (Table 11.45).

$$\beta_{min/max} = \beta \pm \frac{3\sigma}{\sqrt{S_2}}$$

The original rating numbers (characteristic features) result from the retransformation of the average value of the unity group.

$$\text{Calculated result} = \widehat{M}_i + \text{Average} \quad i = 1, 2, 3, \ldots, m$$

In Table 11.46, the true and calculated characteristic values are supplemented by the upper and lower tolerance limits.

The accuracy of the prognosticated characteristic data is illustrated in the graph (Fig. 11.22).

Table 11.46 Comparison of true and calculated rating numbers with tolerances

Individuals	True value	Calculated value	Lower tolerance limit	Upper tolerance limit
11	25.8645	25.5157	23.9289	27.1026
12	23.1427	24.9679	23.5741	26.3618
13	33.9100	33.5931	29.1602	38.0260
14	27.1616	27.9342	25.4952	30.3732
15	29.7521	29.4249	26.4606	32.3892
16		26.4214	24.5154	28.3273
17		27.3119	25.0921	29.5316
18		25.8969	24.1758	27.6181
19		34.4668	29.7260	39.2075
20		30.6300	27.2412	34.0189

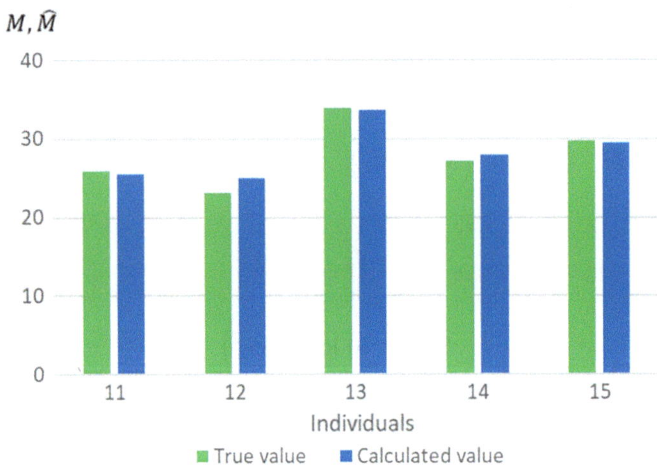

Fig. 11.22 Bar graph of the true and calculated characteristic data of the reference group

The calculated signal values are plotted with the tolerances of the reference objects as a function of the true signal values (Fig. 11.23).

11.7 Size of the Reference Group

For considering correlations between the parameters, the information of the unit group must be included. The reference group is used for scaling. The minimum requirements to be met by both groups must be examined. As shown in Sect. 7.2, an object number greater than the number of parameters is required to calculate the inverse correlation matrix. If the number of objects approaches the number of

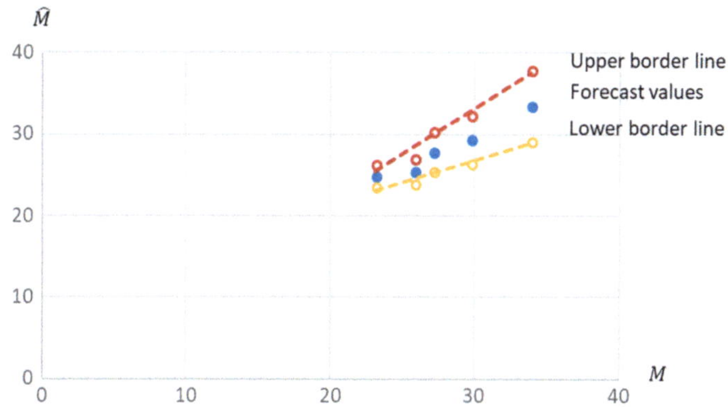

Fig. 11.23 Prognosticated signal values and tolerances for the objects of the reference group calculated according to the Mahalanobis algorithm

Table 11.47 Transformed characteristics of objects 11–15 of the reference group and of objects 16–20 of the test group

Individuals or objects	Transformed characteristic feature, Signal M	D^2	$\sqrt{D^2 - 1}$
1–10 Average	21.012	1	
11	4.8524	4.0748	1.7535
12	2.1306	3.3723	1.5402
13	12.8979	24.9953	4.8985
14	6.1495	8.2639	2.6952
15	8.7400	11.7295	3.2756
16	3.5457	5.4358	2.1061
17	6.6480	7.0165	2.4529
18	2.4549	4.6174	1.9020
19	13.2736	28.4437	5.2387
20	8.0740	15.0237	3.7448

variables, bypassing the inverse correlation matrix with Gram–Schmidt's orthogonalization method provides more precise results. The situation is different for the reference group. Exactly one quantity, the constant of proportionality, is to be determined from it. Thus, one equation is sufficient regardless of the number of parameters. Since only one single value does not necessarily have to be representative of all characteristic data from a reference group, the numerical value of the proportionality constant becomes more reliable if more data is available. The example below lists the results of successive reduction of a reference group. To check the accuracy of the prognosis, another data set is used as a test group with known results. The unit group, reference, and test group are identical to the previously used data from Sect. 11.6, extended by the true characteristic features of the test group. The complete output data are listed (Table 11.47).

Table 11.48 Calibration factor and dynamic *SNR* index derived from the complete test group

Size of the reference group	Calibration factor reference group β_0	Slope test group β	Squared standard deviation test group σ^2	*SNR* index test group $S_2 = 60.8346$
5 Objects: 11–15	0.3894	1.0826	2.3831	14.7595 [dB]
4 Objects: 11–14	0.3942	1.0694	2.2137	14.9734 [dB]
3 Objects: 11–13	0.3856	1.0932	2.5357	14.5748 [dB]
2 Objects: 11–12	0.4198	1.0041	1.6938	15.5881 [dB]
1 Object: 11	0.3614	1.1664	3.9700	13.1909 [dB]

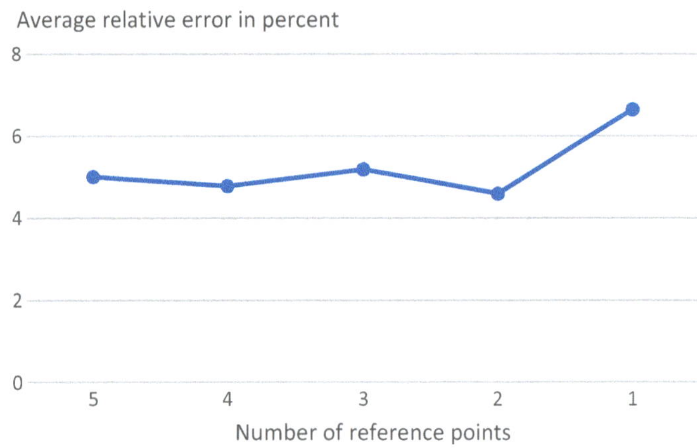

Fig. 11.24 Average relative error in percent as a function of the number of used objects in the reference group

In the introductory numerical example, the entire reference group, consisting of five objects, is used for the determination of the calibration factor β_0. With this constant, the signal values for reference and test group are derived. To carry out an objective comparison, the precision of the prognosis is always evaluated using the test group of the same size. The results of successive reduction of the reference group are plotted (Table 11.48).

The back-transformation with the displacement of the average value of the characteristic feature from the unit group results in the data on the original scale. For a better overview, the average relative error of the calculated characteristic feature in percent here referred to the original scale, is shown graphically (Fig. 11.24).

As it turns out, relatively minor deviations in the predicted results are to be expected as the size of the reference group decreases. If the reference group is reduced to a single object, the predictions to be made are coined by this one object. Depending on its position, the defect size can increase but does not have to, so that a single object can be considered sufficient for calibration purposes.

11.8 Procedure with Orthogonalized Variables According to Gram–Schmidt

If only a few objects with their characteristic data are available for a reference group, i.e., their number falls below four to five times the number of variables, an overestimation in the squared Mahalanobis distances can occur (Sect. 9.7). This is caused by the calculation of the inverse correlation matrix. It can be avoided using Gram–Schmidt's algorithm. To do this, the usual standardization of all variables must be carried out and orthogonalized serially to the Gram–Schmidt procedure based on the first variable that stays unchanged (Sect. 8.1). The squared Mahalanobis distance is derived from the orthogonalized data set of the parameters of the reference group.

$$D_i^2 = \frac{1}{k}\left(\frac{x_{1i}^2}{\sigma_1^2} + \frac{x_{2i}^2}{\sigma_2^2} + \frac{x_{3i}^2}{\sigma_3^2} + \cdots + \frac{x_{ki}^2}{\sigma_k^2}\right)$$

$k =$ Number of variables

$i = 1, 2, 3, \ldots, n$ (Number of objects or individuals)

Thus, the same initial situation is present as with the application of the Mahalanobis algorithm, so that all further steps are identical. If the calculation is to be carried out and the variables are in orthogonalized form, the prognosis algorithm with the linear equation is also recommended for comparison. The larger *SNR* index for the accuracy of the prognosis is decisive for the choice of the method.

11.9 Procedure Applying the Adjoint Correlation Matrix

Apart from the Mahalanobis algorithm, the adjoint matrix can also be used to determine the squared Mahalanobis distance (Sect. 8.5).

$$D_{\text{adj}}^2 = \frac{1}{k} Y R_{\text{adj}} Y^{\text{T}}$$

D_{adj}^2 squared distance with adjoint matrix

k number of parameters

Y standardized parameters (vector)
Y^T transposed standardized parameters (vector)
R_{adj} adjoint of correlation matrix R

Without the determinant of the correlation matrix, the squared distance is not identical but proportional to the Mahalanobis distance with the advantage of bypassing the calculation of the inverse correlation matrix. Consequently, the average value of the squared distances of the objects in the unit group is different from one. After mathematical displacement of the parameters and the squared distances of the objects of the reference group to the origin, the proportionality constant β_0 of the zero-point-proportional relation is determined. From the non-squared distance

$$\beta_0 = \frac{\sqrt{D^2_{\text{adj,ref}} - \overline{D}^2_{\text{adj,objects of the unit group}}}}{M_{\text{ref}} - \overline{M}_{\text{objects of the unit group}}}$$

the signal values are gained with the calibration factor β_0. The accuracy between true and calculated value can be seen from the *SNR* index of dynamic systems. The tolerances are determined equally from three times the standard deviation value with the dimension factor. The signals of the original dimension result with the inverse displacement from the average value of the parameters of the unit group. All the calculation steps lead to an identical result since the method is invariant to constant factors.

11.10 Combined Procedures

Comparing the relatively straightforward procedures for prognosis mentioned above, each one has its own advantages. The linear approach does not require a unit group but ignores possible correlations between the characteristics. The quadratic approach needs a greater effort with the input data but allows the application of an extremely simple relationship to determine the signal values. If one data situation is such that one method leads more frequently to overvaluation and the other to undervaluation, a combination of both methods will prove to be superior, provided the *SNR* indicators are of a similar magnitude. The following numerical example, extended by one more parameter, serves as an illustration. To demonstrate the use of any kind of variable, the added one is a digital character. The example with weight and height, which has already been quoted several times, is complemented by gender. A common coding makes use of the definition (Table 11.49).

Table 11.49 Coding of digital features

Character	Numerical code
Male	1
Female	2

Table 11.50 Data set for prognosis with quadratic equation

Groups	Individuals	Weight [kg]	Height [m]	Male/ female	Rating number, characteristic feature
1. Group: 1–15					
Unit group	1	55	1.61	2	20.69
	2	60	1.75	2	19.10
	3	50	1.69	1	17.94
	4	65	1.72	2	21.42
Determination of the zero point	5	68	1.77	1	22.25
	6	62	1.66	1	23.06
	7	75	1.81	2	22.32
	8	76	1.86	1	22.52
	9	61	1.82	2	17.96
	10	79	1.88	1	22.91
	11	53	1.63	2	19.45
	12	71	1.79	2	21.61
	13	69	1.76	1	22.83
	14	57	1.73	1	19.52
	15	59	1.68	1	21.43
2. Group: 16–20					
Known reference group, training set	16	73	1.68	1	26.51
	17	88	1.95	1	23.72
	18	98	1.7	1	34.76
	19	96	1.88	2	26.48
	20	81	1.65	2	29.01
3. Group: 21–25					
Group with unknown rating numbers, test group	21	59	1.55	1	
	22	79	1.69	2	
	23	92	1.98	1	
	24	105	1.75	2	
	25	62	1.46	1	

Any other numerical values are permissible if they characterize only two different states. However, three parameters require a larger unit group, as shown in Table 11.50.

In the first step, only the quadratic algorithm is used. With the standardized parameters, the squared Mahalanobis distances of all objects are determined. The displacement by the average value of the parameters of the unit group and the displacement of the squared Mahalanobis distances by one and taking the square root is a prerequisite for the applicability of the zero-point-equation. The interim results are given in Table 11.51.

The calibration factor β_0 is obtained from the data of the reference group.

Table 11.51 Squared and non-squared Mahalanobis distances of the characteristic features

Individuals	Characteristic feature, Signal M	Characteristic feature minus average	D^2	$\sqrt{D^2-1}$
1	20.69	−0.3124	1.3723	
2	19.10	−1.8982	0.6781	
3	17.94	−3.0562	1.7682	
4	21.42	0.4218	0.5481	
5	22.25	1.2475	0.3409	
6	23.06	2.0619	1.2102	
7	22.32	1.3205	1.1089	
8	22.52	1.5168	0.9544	
9	17.96	−3.0450	2.1677	
10	22.91	1.9103	1.2867	
11	19.45	−1.5509	0.9787	
12	21.61	0.6049	0.6954	
13	22.83	1.8319	0.4656	
14	19.52	−1.4791	0.8231	
15	21.43	0.4265	0.6017	
Average	21.00	0	1	$\sqrt{D^2-1}$
16	26.51	5.5108	3.7015	1.6436
17	23.72	2.7210	2.8668	1.3663
18	34.76	13.7575	21.2324	4.4980
19	26.48	5.4823	7.3596	2.5218
20	29.01	8.0080	10.2704	3.0447
21			5.4358	2.1061
22			7.0165	2.4529
23			4.6174	1.9020
24			28.4437	5.2387
25			15.0237	3.7448

$$\beta_0 = \frac{\sqrt{D_{16}^2-1}\cdot M_{16}+\sqrt{D_{17}^2-1}\cdot M_{17}+\sqrt{D_{18}^2-1}\cdot M_{18}+\sqrt{D_{19}^2-1}\cdot M_{19}+\sqrt{D_{20}^2-1}\cdot M_{20}}{M_{16}^2+M_{17}^2+M_{18}^2+M_{19}^2+M_{20}^2}$$

$$\beta_0 = 0.3514$$

By inserting the calibration factor for the objects of the reference group and the group with unknown characteristics, the calculated signal values result on the scale shifted by the average value. If the displacement is reversed, the numerical results are in their original form (Table 11.52).

Table 11.52 Prognosticated rating numbers shifted by the average and their true values

Individuals or objects	Signal minus average	Prognosticated signal minus average	True characteristic feature	Calculated characteristic feature
16	5.5108	4.6779	26.51	25.6782
17	2.7210	3.8887	23.72	24.8890
18	13.7575	12.8019	34.76	33.8022
19	5.4823	7.1774	26.48	28.1776
20	8.0080	8.6656	29.01	29.6659
21		5.5568		26.5570
22		6.5446		27.5449
23		4.7666		25.7669
24		13.7494		34.7496
25		9.8841		30.8844

Table 11.53 Precision and tolerance interval on the original scale

Slope β	Squared standard deviation σ^2	Dimension factor S_2	Slope tolerance $3\sqrt{\frac{\sigma^2}{S_2}}$	*S/N* ratio	*SNR* index
1.0091	1.1894	803.2893	0.1154	687.6662	28.3738 [dB]
$\beta_{min} = 0.88$ $\beta_{max} = 1.12$					

The accuracy and the tolerance interval of the predicted values can be determined from the true and calculated rating numbers of the objects in the reference group, assuming an approximate zero-point-proportional (Table 11.53).

Table 11.54 summarizes the results with the corresponding tolerance limits.

In the second step, only the linear algorithm in suitable transformation is to be applied for reasons of comparison. The simplest way is to normalize all quantities without considering a unit group (Table 11.55).

With the data transformation, the zero point is included in all parameter data and in the characteristic feature. From the zero-point-proportional equation, slope, deviation, and the *S/N* indicators result the weighting coefficients (Table 11.56).

The integral signals are calculated with the weighting coefficients

$$M_{i,\text{calculated}} = \frac{S/N_1 \cdot \frac{x_{1i}}{\beta_1} + S/N_2 \cdot \frac{x_{2i}}{\beta_2} + S/N_3 \cdot \frac{x_{3i}}{\beta_3}}{S/N_1 + S/N_2 + S/N_3} \quad i = 16, 17, 18, 19, 20$$

The numeric value for the first object of the reference group is calculated and displayed with all other objects in the Table (Table 11.57)

Table 11.54 True and calculated rating numbers with tolerance limits

Individuals or objects	True characteristic feature	Calculated characteristic feature	Lower tolerance limit	Upper tolerance limit
16	26.5111	25.6782	23.6908	29.8116
17	23.7212	24.8890	21.1977	26.6744
18	34.7578	33.8022	31.0602	39.0849
19	26.4826	28.1776	23.6653	29.7795
20	29.0083	29.6659	25.9223	32.6196
21		26.5570	23.7319	29.8632
22		27.5449	24.6146	30.9740
23		25.7669	23.0257	28.9747
24		34.7496	31.0529	39.0757
25		30.8844	27.5988	34.7293

$$M_{16} = \frac{0.2010 \cdot \left(\frac{-14.2}{1.0235}\right) + 0.6433 \cdot \left(\frac{-0.092}{-0.0201}\right) + 0.005939 \cdot \left(\frac{-0.4}{-0.101}\right)}{0.2010 + 0.6433 + 0.005939} = 0.4553$$

Provided that a zero-point-proportional equation can be applied, the accuracy and tolerances of the calculated characteristics can be derived on the original scale (Table 11.58).

Table 11.59 shows the results for all individuals/objects.

In a graphical comparison of both methods, an under-, respectively, overvaluation of the calculated characteristic features can be stated (Fig. 11.25).

Since for most of the reference objects the true characteristic value is exceeded or not reached, the achievement of a more precise prognosis from a combination of both methods seems promising. What kind of procedure could be applied?

A decisive consideration of the linear algorithm is the use of the *S/N* indicators as weighting coefficients. A weighting according to parameter-individual predictions can be transferred identically to different methods. In most cases, no more than two different procedures must be considered, and this only if there are actual positive/negative differences in the numerical values of the characteristic features with respect to the true value. By adding the corresponding *S/N* indicators as weighting coefficients, the following relationship is established.

$$M_{\text{calculated}} = \frac{S/N_{\text{proced.1}}\left(M_{\text{calculated,proc.1}}\right) + S/N_{\text{proc.2}}\left(M_{\text{calculated,proc.2}}\right)}{S/N_{\text{procedure1}} + S/N_{\text{procedure2}}}$$

Since the individual calculated values already exist for reasons of comparison, they are used in the formula instead of $\left(\frac{x_i}{\beta}\right)$. Another possibility, which can also be applied and compared, is based on the squared calculated values, provided these are

Table 11.55 Original data and normalized data set

Individuals or objects	Weight [kg]	Height [m]	Gender	Rating number, charact. feature	Normalized weight	Normalized height	Normalized gender	Normalized charact. feature
16	73	1.68	1	26.51	−14.2	−0.092	−0.4	−1.5851
17	88	1.95	1	23.72	0.8	0.178	−0.4	−4.3750
18	98	1.7	1	34.76	10.8	−0.072	−0.4	6.6616
19	96	1.88	2	26.48	8.8	0.108	0.6	−1.6136
20	81	1.65	2	29.01	−6.2	−0.122	0.6	0.9121
Average	87.2	1.772	1.4	28.0962	0	0	0	0
21	59	1.55	1		−28.2	−0.222	−0.4	
22	79	1.69	2		−8.2	−0.082	0.6	
23	92	1.98	1		4.8	0.208	−0.4	
24	105	1.75	2		17.8	−0.022	0.6	
25	62	1.46	1		−25.2	−0.312	−0.4	

Table 11.56 Parameters for determining the weighting coefficients

Normalized parameter	Slope β	Squared standard deviation σ^2	*S/N* index $S_2 = 13.8930$
x_1 (Weight)	$\beta_1 = 1.0235$	$\sigma_1^2 = 72.4060$	$S/N_1 = 0.2010$
x_2 (Height)	$\beta_2 = -0.0201$	$\sigma_2^2 = 0.008748$	$S/N_2 = 0.6433$
x_3 (Gender)	$\beta_3 = -0.0101$	$\sigma_3^2 = 0.2386$	$S/N_3 = 0.005939$

Table 11.57 True characteristic features and their associated integral signal values for the reference group and group with unknown values

Objects	Normalized characteristic feature	Integral signal	True characteristic feature	Calculated characteristic feature
16	−1.5851	0.4553	26.5111	28.5515
17	−4.3750	−6.2300	23.7212	21.8662
18	6.6616	5.4779	34.7578	33.5741
19	−1.6136	−2.4424	26.4826	25.6538
20	0.9121	2.7392	29.0083	30.8354
21		2.1086		30.2048
22		0.7736		28.8697
23		−6.4339		21.6623
24		4.5234		32.6196
25		6.1849		34.2811

Table 11.58 *SNR* index standing for the precision with associated tolerances on the original scale

Slope β	Squared standard deviation σ^2	Dimension factor S_2	Slope tolerance $3\sqrt{\frac{\sigma^2}{S_2}}$	*S/N* ratio	*SNR* index
1	2.6061	803.2893	0.1709	308.2307	24.8888 [dB]
$\beta_{min} = 0.83$ $\beta_{max} = 1.17$					

always positive. With the applicability of the quadratic method, this is generally fulfilled.

$$M_{\text{calculated}}^2 = \frac{S/N_{\text{proc.1}}\left(M_{\text{calculated,proc.1}}\right)^2 + S/N_{\text{proc.2}}\left(M_{\text{calculated,proc.2}}\right)^2}{S/N_{\text{procedure 1}} + S/N_{\text{procedure 2}}}$$

$$M_{\text{calculated}} = \sqrt{M_{\text{calculated}}^2}$$

Table 11.59 True and calculated characteristic features and tolerances on the original scale

Individuals, objects	True characteristic feature	Calculated characteristic feature	Lower tolerance limit	Upper tolerance limit
16	26.5111	28.5515	21.9810	31.0413
17	23.7212	21.8662	19.6678	27.7746
18	34.7578	33.5741	28.8185	40.6971
19	26.4826	25.6538	21.9573	31.0078
20	29.0083	30.8354	24.0514	33.9651
21		30.2048	25.0435	35.3661
22		28.8697	23.9366	33.8029
23		21.6623	17.9607	25.3639
24		32.6196	27.0457	38.1936
25		34.2811	28.4232	40.1389

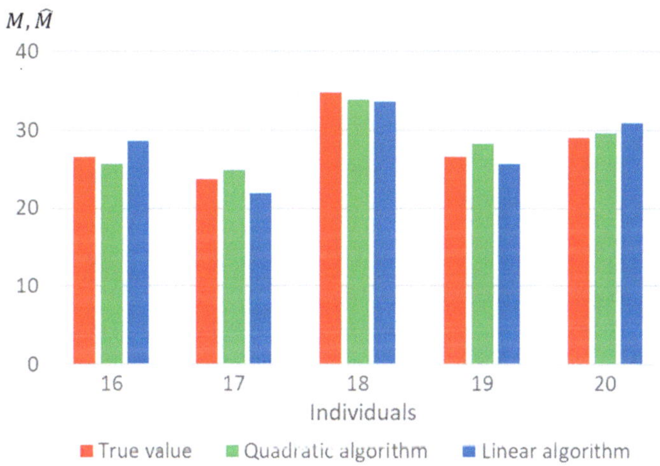

Fig. 11.25 Comparison of the quadratic and linear prognosis algorithms

Both combination procedures are carried out, whereby the *S/N* indicators, derived from the individual procedures on the original scale, are used as weighting coefficients. The numerical values can be taken from the previous tables. The following applies to the linear combination

$$M_{\text{calculated}} = \frac{687.6662 \cdot \left(M_{\text{calc.,squared algorithm}}\right) + 308.2307 \cdot \left(M_{\text{calc.,linear algorithm}}\right)}{687.666 + 308.2307}$$

and correspondingly for the squared combination

Table 11.60 Results from the linear and the quadratic combination of prognosis algorithms

Individuals, objects	True characteristic feature	Squared combination	Linear combination
16	26.5111	26.6007	26.5675
17	23.7212	23.9942	23.9534
18	34.7578	33.7317	33.7316
19	26.4826	27.4213	27.3965
20	29.0083	30.0327	30.0279
21		27.7373	27.6860
22		27.9616	27.9549
23		24.5699	24.4965
24		34.1046	34.0904
25		31.9742	31.9356

Table 11.61 Accuracy of prognosis and slope tolerances of the combined methods

Slope β	Squared standard deviation σ^2	Dimension factor S_2	Slope tolerance $3\sqrt{\frac{\sigma^2}{S_2}}$	*S/N* ratio	*SNR* index
Squared combination of the prognosis algorithms					
1.0069	0.5749	803.2893	0.08025	1416.8133	31.5131 [dB]
$\beta_{min} = 0.92$ $\beta_{max} = 1.08$					
Linear combination of the prognosis algorithms					
1.0063	0.5656	803.2893	0.07961	1438.0410	31.5777 [dB]
$\beta_{min} = 0.92$ $\beta_{max} = 1.08$					

$$M^2_{\text{calc.}} = \frac{687.6662 \cdot \left(M_{\text{calc.,squared algorithm}}\right)^2 + 308.2307 \cdot \left(M_{\text{calc.,linear algorithm}}\right)^2}{687.666 + 308.2307}$$

The improved prognosticated values are shown in Table 11.60.

Precision and tolerances of the prognosis are derived from the results of both the linear and the quadratic combination (Table 11.61).

The complete values of the reference group and the group with unknown characteristic features are listed (Table 11.62).

A comparison is easier to judge in the graphic illustration, whereby the results of the linear prognosis algorithm and the linear combination with corresponding tolerances are displayed (Fig. 11.26 and 11.27).

Table 11.62 Calculated characteristic features and tolerances of the quadratic and linear combination of the prognosis algorithms

Individuals, objects	True characteristic feature	Calculated characteristic feature	Lower tolerance limit	Upper tolerance limit
Quadratic combination of the prognosis algorithms				
16	26.5111	26.6007	24.6498	28.9193
17	23.7212	23.9942	22.2344	26.0856
18	34.7578	33.7317	31.2579	36.6720
19	26.4826	27.4213	25.4103	29.8115
20	29.0083	30.0327	27.8302	32.6505
21		27.7373	25.7031	30.1551
22		27.9616	25.9109	30.3989
23		24.5699	22.7679	26.7115
24		34.1046	31.6034	37.0773
25		31.9742	29.6292	34.7613
Linear combination of the prognosis algorithms				
16	26.5111	26.5675	24.6187	28.8485
17	23.7212	23.9534	22.1964	26.0100
18	34.7578	33.7316	31.2573	36.6277
19	26.4826	27.3965	25.3869	29.7487
20	29.0083	30.0279	27.8253	32.6060
21		27.6860	25.6552	30.0631
22		27.9549	25.9044	30.3551
23		24.4965	22.6996	26.5997
24		34.0904	31.5898	37.0173
25		31.9356	29.5931	34.6776

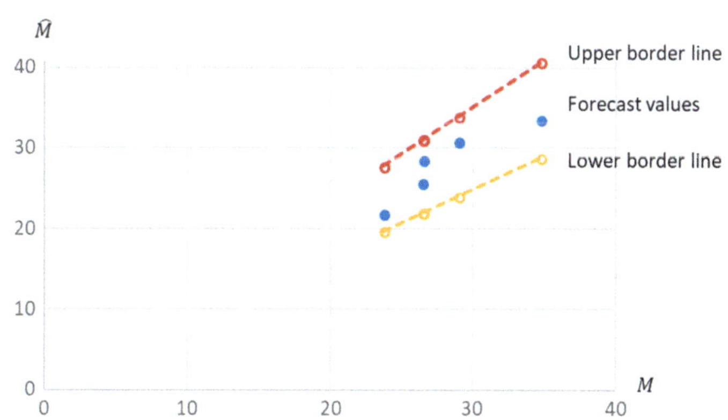

Fig. 11.26 Values of the characteristic feature and tolerances of the reference group with linear prognosis algorithm

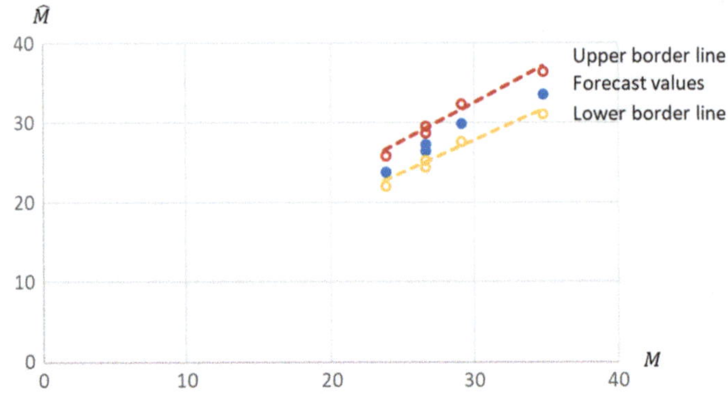

Fig. 11.27 Values of the characteristic feature and tolerances of the reference group with a linear combination of prognosis algorithms

It should be noted that no claim to universal validity can be made from this numerical example. Based on experience, every other data set will behave differently. Nevertheless, it can always be a worthwhile undertaking to apply both procedures.

11.11 Variables Without Deviation

In clinical and pharmaceutical studies, it is often the case that one or more variables are constants and consequently, without dispersion. This occurs if, for example, only male individuals are represented or if the persons involved take a certain drug or not. These characteristics are digital in nature and have zero deviation, provided that only one such group is considered. For prediction purposes, all prognosis methods that require standardization of the data are useless. If only one variable, or at least a few with sufficiently large amounts of data, should have a digital character, a separation into two or according to the number of digital characteristics, several data sets is advantageous. The results are more accurate, as digital behavior corresponds to step-functions that cause larger tolerances. If separation into several datasets is not possible, the only way out is to start with a heterogeneously composed reference group to obtain the weight coefficients. The object group with unknown characteristic values can also be processed with uniform digital variables, i.e., with zero deviation and the corresponding characteristic features can be calculated. The same applies to the quadratic method, which also requires a heterogeneous unit group, together with a heterogeneous reference group for calibrating the method.

Important items to memorize A prognosis based on a linear equation can be carried out without a unit group. However, the presence of a unit group offers the advantage of being able to apply a squared equation method as well and thus choosing the more suitable procedure.

Relatively large value ranges of the variables in comparison of the numerical spreading of the output characteristic features can be used as an indicator that a logarithmic data transformation can be used to improve the accuracy of the prognosis.

Chapter 12
Expanded Options

Contents

Abstract Frequently, a prognosis is applied for sets of data composed of many variables. Screening of variables is useful to get rid of those without meaning. This reduces the calculation effort and sometimes improves the accuracy of the prognosis. For some applications, the characteristic outcomes do not exist. For this reason, standardized *SNR* indices are calculated to take over a similar function. When applying the Mahalanobis algorithm, the squared distance stands for the non-existing numerical outcome and pattern recognition is practicable. Another valuable property of the prognosis is—presupposing that the accuracy of the results is high enough proven by the corresponding *SNR* index—to analyze existing relationships between variables and the characteristic outcome. According to the remaining parameters (after screening), the adequate size of an orthogonal array is applied. With the assignment of the variables to the orthogonal array and choosing appropriate value settings, all characteristic outcomes are prognosticated and analyzed with the *SNR* indices. This displays the kind and strength of the contribution of each individual variable to the characteristic outcome without the need of performing any experiment.

© Springer Nature Switzerland AG 2019

H. Ruefer, *Living Without Mathematical Statistics*,
https://doi.org/10.1007/978-3-319-99632-5_12

12.1 Parameter Screening

As with a data-analytical approach, the most complete possible set of parameters facilitates a reliable prognosis statement. Since direct cause-and-effect relationships are usually not apparent for predictions, very extensive data sets are often put together. This may also include variables that are not in relation to the result or even disturb it, i.e., change in opposite directions and give rise to misinterpretations. Before a data set is processed in its entirety with a prognosis algorithm, a screening process that has been carried out first can have an advantageous effect. Unsuitable variables are sorted out and at the same time, the effort is reduced, which often improves the precision of the prognosis. The basic procedure has already been introduced in Sect. 7.4 for improving discrimination power.

The starting point is a known data set with measured true results. The parameters are assigned to the columns of a sufficiently large orthogonal array, since parallel testing is more efficient and reliable than the serial individual test, i.e., the omission of one parameter at a time. The conformity with the true result is evaluated using the *SNR* index **(3-3*)**. The parameter configuration to be used is based on the same definition that takes setting value 1 to accept the parameter and reject it for setting value 2. For the evaluation of the orthogonal array, the parameters whose setting value 1 assumes a higher numerical value than setting value 2 must be retained, otherwise in the opposite case, they must be sorted out. In the case of a reduced parameter set, the result, i.e., the accuracy of prognosis, must be confirmed.

12.2 Screening with Linear Equation

Based on a multi-stage technical process, the procedure of a parameter screening is explained. The target value is the production yield, which is not constant despite predefined machine settings and deviates around an average of 85%. Possible reasons are assumed, among others, in machine vibrations. Over a certain production period, corresponding additional measurements are recorded for data analysis. Since a relation does not necessarily have to exist, the parameters are screened initially. In this case, 6 parameters are available in 30 data sets (Table 12.1).

Six parameters are assigned to an orthogonal array of minimum size L8 **(3-4*)** (Tables 12.2 and 12.3).

According to the constraints of the orthogonal array, eight configurations of parameters are to be used in the prognosis procedure in question.

With the origin as the average value from the respective columns and representation of the yield as a decimal number, the first test calculation for all parameters is to be carried out (Table 12.4).

The weighting coefficients are calculated in the first step (Table 12.5).

The integral signal values for the batches $i = 1\ldots 30$ are gained from

Table 12.1 Data set for screening

Data set	*A*	*B*	*C*	*D*	*E*	*F*	Yield %
1	11.29	10.26	60.67	56.57	14.42	19.31	80.6
2	21.72	2.95	35.69	44.07	45.26	6.1	81.1
3	19.38	21.49	37.74	29.38	1.81	23.1	92.7
4	7.33	28.32	11.37	40.81	43.41	14.48	89.6
5	15.28	42.94	80.48	25.62	4.15	27.63	88.1
6	3.69	22.2	58	11.97	49.93	18.68	84.4
7	5.85	38.2	67.7	24.97	14.23	4.08	88.8
8	48.25	10.92	46.39	26.72	46.62	15.9	83.1
9	51.69	39.54	45.73	56.06	23.31	17.54	86.5
10	58.61	20.83	45.57	24.67	64.1	28.7	81.9
11	39.79	39.02	22.46	54.17	15.37	16.43	91.7
12	41.29	23.65	39.77	43.54	13.06	30.12	88.7
13	11.06	17.65	44.15	56.1	58.67	25.13	77.5
14	51.8	30.06	54	25.36	24.63	12.45	88.4
15	29.87	41.13	23.6	12.51	52.79	18.56	92.6
16	43.51	22.82	82.88	11.08	40.88	44.32	80.8
17	61.82	37.87	16.36	50.7	47.15	39.44	88
18	54.4	17.63	67.99	17.22	60.24	19.91	79.1
19	9.11	29.77	70.87	28.42	20.72	38.48	85.3
20	40.87	29.93	65.19	18.65	59.09	38.61	81.6
21	28.19	37.97	57.37	54.81	34.46	44.23	82.7
22	26.93	21.8	58.98	12.01	15.07	39.9	89.8
23	14.91	9.98	83.27	41.2	45.56	30.07	73.3
24	25.76	18.03	57.4	19.46	27.36	32.08	86.5
25	38.78	38.02	4.98	61.76	13.05	17.36	93.2
26	20.03	25.88	63.52	45.31	6.78	27.14	85.5
27	9.45	39.55	72.49	56.84	34.9	5.01	79.5
28	41.64	19.52	74.2	31.04	52.6	36.05	77.4
29	18.9	11.46	13.65	37.76	17.49	23.95	91.2
30	25.41	34.36	52.65	27.25	63.85	18.8	82.4
Average	29.22	26.12	50.5	34.87	33.7	24.45	85.06

$$\widehat{M}_i = \frac{S/N_A \cdot \frac{x_{Ai}}{\beta_A} + S/N_B \cdot \frac{x_{Bi}}{\beta_B} + S/N_C \cdot \frac{x_{Ci}}{\beta_C} + S/N_D \cdot \frac{x_{Di}}{\beta_D} + S/N_E \cdot \frac{x_{Ei}}{\beta_E} + S/N_F \cdot \frac{x_{Fi}}{\beta_F}}{S/N_A + S/N_B + S/N_C + S/N_D + S/N_E + S/N_F}$$

The *SNR* index represents the prognosis accuracy (Tables 12.6 and 12.7).

The further test calculations of the given parameter configurations must be carried out in the same way. Since the intermediate results of all parameters are already complete with the first test calculation, the integral signal values 1–30 can be calculated from these for the tests #2–#8. The precision of prognosis is derived

Table 12.2 Orthogonal array L8

Test #	*A*	*B*	*C*	*D*	*E*	*F*	Error column
1	1	1	1	1	1	1	1
2	1	1	1	2	2	2	2
3	1	2	2	1	1	2	2
4	1	2	2	2	2	1	1
5	2	1	2	1	2	1	2
6	2	1	2	2	1	2	1
7	2	2	1	1	2	2	1
8	2	2	1	2	1	1	2

Table 12.3 Selected parameters for the screening process

Test #	Parameter
1	*A*, *B*, *C*, *D*, *E*, *F*
2	*A*, *B*, *C*
3	*A*, *D*, *E*
4	*A*, *F*
5	*B*, *D*, *F*
6	*B*, *E*
7	*C*, *D*
8	*C*, *E*, *F*

from the true and calculated results as numerical value of the *SNR* index (Table 12.8).

The interpretation of the orthogonal array is done numerically, and the results transferred into the graphic (Fig. 12.1).

With the definition of setting values, variables *A* and *F* are omitted for prognostic purposes. If not all columns of the orthogonal array are assigned due to the available number of parameters, an error estimation is possible which can be used to further reduce the effort. For confirmation, a test calculation is to be carried out using only the reduced data set, in this case, variables *B*, *C*, *D*, and *E* (Table 12.9).

The reduced number of parameters leads to an equivalent or improved prognosis precision, so that the elimination of parameters *A* and *F* appears to be justified. The remaining parameter set consists of *B*, *C*, *D*, and *E*. As an illustration, all numerical values can be transformed back to the original scale and displayed graphically (Fig. 12.2).

From the standard deviation of the prognosticated accuracy on the true scale result the tolerances as limit lines for further application considerations (Fig. 12.3).

Table 12.4 Normalized set of data

Data set	*A*	*B*	*C*	*D*	*E*	*F*	Yield
1	−17.9262	−15.8679	10.1693	21.7038	−19.2803	−5.1463	−0.0449
2	−7.4972	−23.171	−14.8094	9.2062	11.5574	−18.3479	−0.0399
3	−9.8438	−4.6391	−12.7636	−5.4865	−31.8893	−1.3546	0.0764
4	−21.8873	2.1922	−39.1327	5.9413	9.7099	−9.9735	0.0449
5	−13.9453	16.8153	29.9729	−9.2443	−29.5462	3.1811	0.0306
6	−25.53	−3.9213	7.4931	−22.8973	16.2288	−5.7741	−0.0063
7	−23.372	12.0726	17.1992	−9.9005	−19.4672	−20.3688	0.0373
8	19.0311	−15.2058	−4.1126	−8.148	12.9167	−8.5498	−0.0199
9	22.4708	13.417	−4.7774	21.1875	−10.3869	−6.908	0.0146
10	29.3934	−5.295	−4.9311	−10.2006	30.3986	4.2486	−0.0319
11	10.5722	12.8955	−28.0483	19.3012	−18.3292	−8.0187	0.0665
12	12.0693	−2.472	−10.7313	8.6693	−20.6339	5.6682	0.0366
13	−18.1565	−8.4785	−6.3543	21.2326	24.9752	0.6803	−0.076
14	22.5771	3.9312	3.4925	−9.5055	−9.0732	−12.0033	0.0338
15	0.6455	15.0037	−26.9022	−22.3576	19.0905	−5.8964	0.075
16	14.2886	−3.308	32.3783	−23.7856	7.1769	19.8641	−0.0425
17	32.6028	11.7482	−34.1447	15.8342	13.4549	14.9851	0.0296
18	25.1777	−8.4992	17.4843	−17.648	26.5462	−4.5437	−0.0591
19	−20.108	3.6486	20.3681	−6.4492	−12.9833	14.0299	0.002
20	11.6469	3.803	14.6849	−16.2157	25.3883	14.1609	−0.0345
21	−1.0359	11.8472	6.8657	19.9377	0.7618	19.7747	−0.0236
22	−2.2906	−4.3207	8.4719	−22.8589	−18.6253	15.4511	0.047
23	−14.3079	−16.1421	32.7696	6.3346	11.8635	5.616	−0.1179
24	−3.4619	−8.0947	6.895	−15.404	−6.3405	7.626	0.0139
25	9.5585	11.8953	−45.5232	26.8924	−20.6481	−7.0869	0.0809
26	−9.1907	−0.2447	13.0143	10.4376	−26.9181	2.6868	0.0047
27	−19.7691	13.422	21.9828	21.9737	1.2021	−19.4452	−0.0554
28	12.4203	−6.6012	23.697	−3.8278	18.9	11.5939	−0.077
29	−10.3245	−14.6683	−36.8493	2.8944	−16.2055	−0.499	0.0616
30	−3.8073	8.2374	2.1413	−7.6172	30.1561	−5.6507	−0.0265

12.3 Screening with Quadratic Equation

To carry out parameter screening using the Mahalanobis algorithm, the previously used object group must be extended by a unit group. If the composition of the objects is appropriate uniform, a number of 30 appear to be sufficient (Sect. 9.7). For example, manufacturing data with a lower yield but with smaller relative deviation, is suitable for this purpose. The average value and deviation of the objects in the unit group are the basis for the standardization of all data (Table 12.10).

Table 12.5 Calculation of weighting coefficients for the first test

Calculation test #1	Parameter	Slope β	Squared standard deviation σ^2	*S*/*N* ratio $S_2 = 0.002601$
1	*A*	11.4559	289.0046	0.001181
	B	95.8941	94.9248	0.2520
	C	−269.0927	279.9369	0.6729
	D	−12.4651	248.3632	0.001627
	E	−221.9866	242.5821	0.5285
	F	−34.0065	120.1970	0.02503

Table 12.6 Comparison of the true and predicted results for the first test calculation

Data set	True results	Calculated results
1	−0.04491	−0.01494
2	−0.03986	−0.0269
3	0.07636	0.06504
4	0.04493	0.05726
5	0.03055	0.02498
6	−0.00626	−0.04258
7	0.03733	0.03304
8	−0.01989	−0.0345
9	0.01456	0.05169
10	−0.03188	−0.04909
11	0.06651	0.10271
12	0.03661	0.04415
13	−0.076	−0.04793
14	0.03382	0.02403
15	0.07498	0.0463
16	−0.04247	−0.07885
17	0.02957	0.05029
18	−0.05914	−0.0817
19	0.00204	−0.01485
20	−0.03451	−0.06364
21	−0.02364	−0.00345
22	0.04697	0.00214
23	−0.1179	−0.10737
24	0.01391	−0.01848
25	0.08091	0.13296
26	0.00469	0.01796
27	−0.05543	−0.00888
28	−0.07699	−0.08665
29	0.06159	0.06151
30	−0.02647	−0.03425

Table 12.7 Prognosis accuracy of test calculation #1

Calculation test #	Slope β out of true and calculated value	Squared standard deviation σ^2	*SNR* index $S_2 = 0.002601$
1	1	0.000662	5.9447 [dB]

Table 12.8 Complete evaluations of the orthogonal array L8

Calculations test #	*SNR*
1	5.9447 [dB]
2	0.7412 [dB]
3	−2.4317 [dB]
4	−15.3334 [dB]
5	−5.2882 [dB]
6	−0.6298 [dB]
7	−1.4923 [dB]
8	3.5002 [dB]

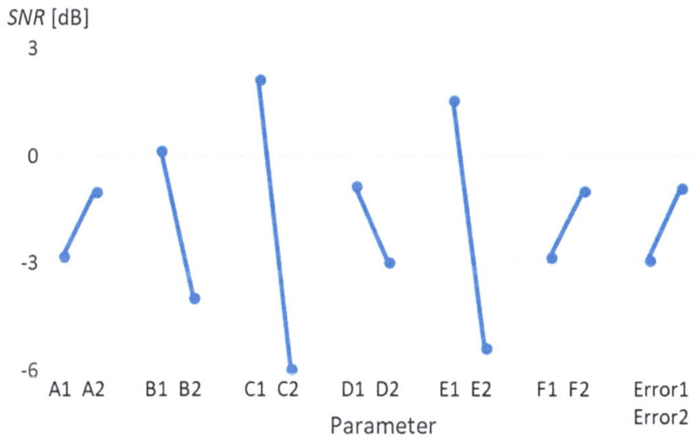

Fig. 12.1 Graphical representation of the parameter screening

Table 12.9 Calculation for confirmation the screening results

Parameter	Slope β	Squared standard deviation σ^2	*S/N* ratio $S_2 = 0.002601$	*SNR* index $S_2 = 0.002601$
B	95.8941	94.9248	0.25201	6.2831 [dB]
C	−269.0927	279.9369	0.67291	
D	−12.4651	248.3632	0.001627	
E	−221.9866	242.5821	0.52846	

Fig. 12.2 Comparative representation of true and prognosticated production yields

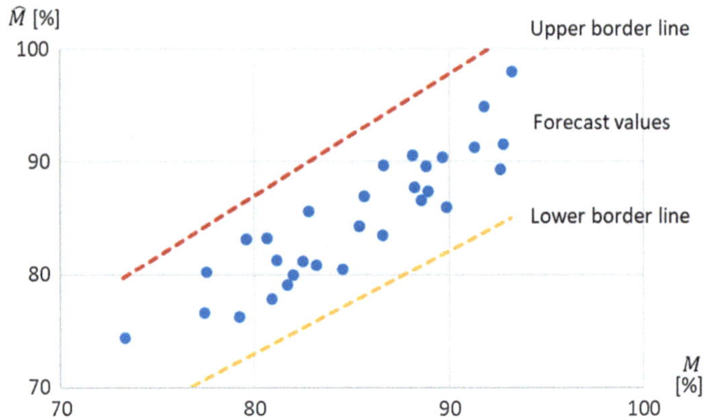

Fig. 12.3 Calculated values as a function of the true values with upper and lower limit lines

When using the orthogonal array L8, all six parameters must be included for the first test calculation. Since a further column is available without parameter assignment, it can be evaluated for error estimation in comparison of the effect amplitudes. The result of the squared Mahalanobis distances is listed (Table 12.11).

For the application of a zero-point-proportional equation, the Mahalanobis distance must be in non-squared form with zero-point reference. For reasons of comparison, the reference group uses the decimal value with the corresponding average value instead of the percentage figure. Assuming proportional behavior to the normalized output characteristic, the yield, the proportionality constant β_0 can be determined (Sect. 11.6; Table 12.12).

With the proportionality constant β_0, the signal variables are calculated from the non-squared Mahalanobis distances. The *SNR* index of dynamic systems describes

Table 12.10 Unit group supplement to the previous used data set (Sect. 12.2)

Data set unit group	*A*	*B*	*C*	*D*	*E*	*F*	Yield %
1	32.51	11.97	60.05	77.44	44.61	13.45	71.9
2	16.16	10.18	69.13	71.01	53.49	8.35	69.8
3	33.66	11.48	57.59	68.09	68.22	11.67	70
4	42.24	1.42	52.45	38.9	78.1	15.68	72.5
5	20.8	21.17	74.78	64	62.65	10.46	70
6	22.77	0.1	63.75	22.66	78.46	6.27	72.7
7	59.68	7.01	67.52	48.32	63.73	13.36	72
8	44.21	30.24	66.46	73.08	66.89	7.84	70.4
9	58.62	3.55	69.85	77.66	24.71	22.84	71.8
10	51.3	33.46	57.45	66.22	76.01	4.91	72.1
11	21.89	9.28	35.61	71.93	66.39	21.37	73
12	27.53	25.35	67.16	67.88	65.25	17.42	70.9
13	12.57	3.3	61.27	51.32	62.96	14.74	72.3
14	35.64	3.79	75.7	45.16	56.54	10.64	71.6
15	43.3	20.46	56.36	75.16	58.97	27.43	71.9
16	33.98	22.43	59.16	77.2	65.76	21.47	69.9
17	38.42	30.66	66.96	74.65	59.14	8.12	71.5
18	45.99	7.9	61.33	78.85	39.32	16.56	71.6
19	10.76	5.83	49.32	60.22	79.98	26.3	69.5
20	32.58	7.3	76.34	25.06	74.23	13.02	71.7
21	47.98	13.31	69.63	75.9	36.79	14.9	71.9
22	17.85	21.47	62.63	75.99	50.92	12.94	72.1
23	29.41	25.32	27.94	76.4	79.57	13.83	72.7
24	43.86	1.74	77.09	35.65	60.92	26.47	71.8
25	21.67	2.77	47.57	48.15	76.45	14.8	72.4
26	40.62	6.26	76.08	73.92	34.14	23.5	70.3
27	48.5	5.45	77.94	45.07	56.76	19.3	71.3
28	40.11	9.51	42.74	78.93	59.32	12.23	71.6
29	30.47	3.24	76.24	60.71	36.25	16.86	72.1
30	54.77	25.77	71.03	58.27	67.28	26.13	71.7
Average	35.8	12.72	62.57	62.13	60.13	17.76	71.48
Standard deviation	13.1765	9.9274	12.3277	16.3032	14.5035	6.2219	0.946

Table 12.11 Squared Mahalanobis distances of the unit and reference group for the first test calculation

Calculation test #1	Unit group D^2	Reference group D^2
1	0.3997	23.5827
2	1.2492	15.8201
3	1.1192	110.8486
4	0.9204	72.6444
5	0.8729	100.331
6	1.5386	45.7668
7	1.0977	87.5215
8	0.9271	23.717
9	1.4123	49.9027
10	1.444	21.8651
11	1.1828	91.8136
12	0.5986	68.3331
13	0.6446	8.9502
14	0.4708	63.1271
15	0.8894	101.1545
16	0.7239	37.9086
17	0.7758	59.4018
18	0.5936	13.6078
19	2.0374	66.4453
20	1.0772	31.9464
21	0.5696	40.6654
22	1.1239	93.2418
23	1.8314	6.028
24	1.1106	57.7507
25	0.6192	106.8924
26	0.7798	57.5678
27	0.5253	23.7997
28	1.0823	13.1672
29	0.8695	86.6068
30	1.5131	29.6844
Average	1	

the precision of the correspondence between the true value of the characteristic output related to the zero point and the calculated signal values (Table 12.13).

Furthermore, test calculations #2–#8 with the differently configured unit groups according to Table 12.3 must be worked out in the same way. The intermediate results and associated precision of prognosis are put together (Table 12.14).

The evaluation of the orthogonal array L8 is shown graphically (Fig. 12.4).

From a simple viewpoint, it seems justified because of

Table 12.12 Normalized characteristic output (yield) and non-squared Mahalanobis distance of the reference group for determining the proportionality constant β_0

Calculation test #1	Normalized output of reference group $y - \bar{y}$	Non-squared Mahalanobis distance of reference group $\sqrt{D^2 - 1}$
1	0.0906	4.7521
2	0.0957	3.8497
3	0.2119	10.4809
4	0.1805	8.4643
5	0.1661	9.9665
6	0.1293	6.6908
7	0.1729	9.3017
8	0.1156	4.7662
9	0.1501	6.9931
10	0.1037	4.5678
11	0.202	9.5296
12	0.1721	8.2057
13	0.0595	2.8196
14	0.1694	7.8821
15	0.2105	10.0077
16	0.0931	6.0752
17	0.1651	7.6421
18	0.0764	3.5507
19	0.1376	8.0898
20	0.101	5.563
21	0.1119	6.2981
22	0.1825	9.6043
23	0.0176	2.2423
24	0.1494	7.5333
25	0.2164	10.2904
26	0.1402	7.5212
27	0.0801	4.7749
28	0.0585	3.4882
29	0.1971	9.2524
30	0.1091	5.3558
	$\beta_0 = 49.8622$	

$$SNR_{\text{level 1}} > SNR_{\text{level 2}}$$

to retain all parameters for the prognosis algorithm. However, a comparison of the parameter effects shows that the error amplitude is twice as large as the effect amplitudes of parameters *A* and *F*. Their contribution thus appears to be irrelevant for the precision of prognosis. However, the latter must be confirmed with a test calculation of the remaining parameters *B*, *C*, *D*, and *E*. If the *SNR* index for

Table 12.13 Calculated signal values to derive the prognosis accuracy

Calculation test #1	Normalized output of reference group	Calculated signal values
1	0.0906	0.0953
2	0.0957	0.0772
3	0.2119	0.2102
4	0.1805	0.1698
5	0.1661	0.1999
6	0.1293	0.1342
7	0.1729	0.1865
8	0.1156	0.0956
9	0.1501	0.1402
10	0.1037	0.0916
11	0.202	0.1911
12	0.1721	0.1646
13	0.0595	0.0565
14	0.1694	0.1581
15	0.2105	0.2007
16	0.0931	0.1218
17	0.1651	0.1533
18	0.0764	0.0712
19	0.1376	0.1622
20	0.101	0.1116
21	0.1119	0.1263
22	0.1825	0.1926
23	0.0176	0.045
24	0.1494	0.1511
25	0.2164	0.2064
26	0.1402	0.1508
27	0.0801	0.0958
28	0.0585	0.07
29	0.1971	0.1856
30	0.1091	0.1074
Slope β	Squared standard deviation σ^2	*SNR* index $S_2 = 0.02097$
1	0.000213	19.9343 [dB]

prognosis accuracy decreases, the parameters *A* and *F* must be retained, otherwise with an unchanged or improved index, *A* and *F* are to be removed (Table 12.15).

For comparison, the true and calculated production yields are back-transformed to the original scale (Fig. 12.5).

From the remaining residual deviation, the tolerances can be supplemented as minimum and maximum straight lines (Fig. 12.6).

Table 12.14 Intermediate results and precision of prognosis of the test calculations #1–#8

Calculation test #	Proportionality constant β_0	Squared standard deviation σ^2	*SNR* index [dB] $S_2 = 0.02097$
1	49.8622	0.000213	19.9343
2	9.3599	0.00489	6.323
3	14.6521	0.003065	8.3514
4	8.673	0.01639	1.0716
5	15.3239	0.00651	5.0803
6	12.9745	0.002831	8.697
7	12.5056	0.003009	8.4327
8	14.9685	0.002161	9.8693

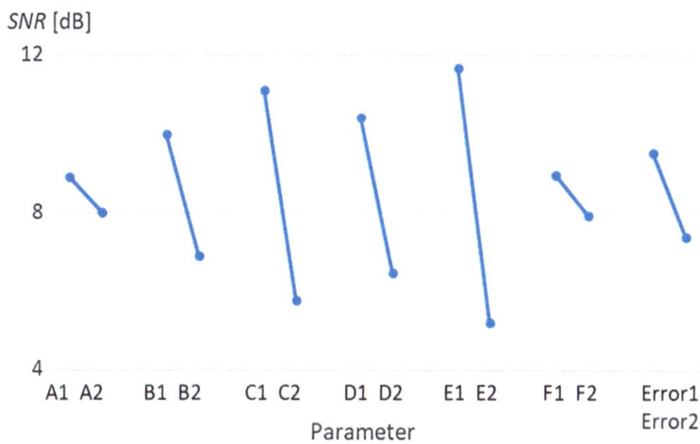

Fig. 12.4 Evaluation of the parameter screening according to the quadratic procedure

Table 12.15 Confirmation result with reduced parameter sets *B*, *C*, *D*, and *E*

Confirmation for parameters	Proportionality constant β_0	Squared standard deviation σ^2	*SNR* index $S_2 = 0.02097$
B, *C*, *D*, *E*	53.9267	6.63E−05	24.9989 [dB]

12.4 Screening with Quadratic Equation Without Correlation Matrix

The remarks in the second part are dedicated to data sets with noticeable correlation patterns. The focus is on data discrimination for diagnostic purposes, with the Mahalanobis algorithm proving to be a powerful approach. If the coupling quantities between the variables are small or nonexistent, a similar index can be created as the squared Mahalanobis distance in mathematically simplified form **(3-2)**. The

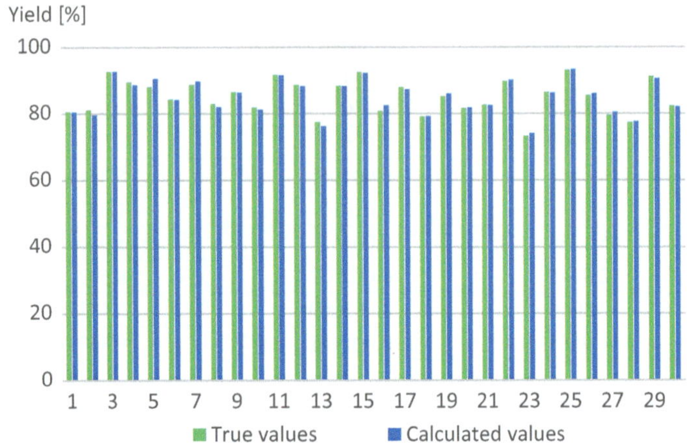

Fig. 12.5 Comparison of true and calculated production yield

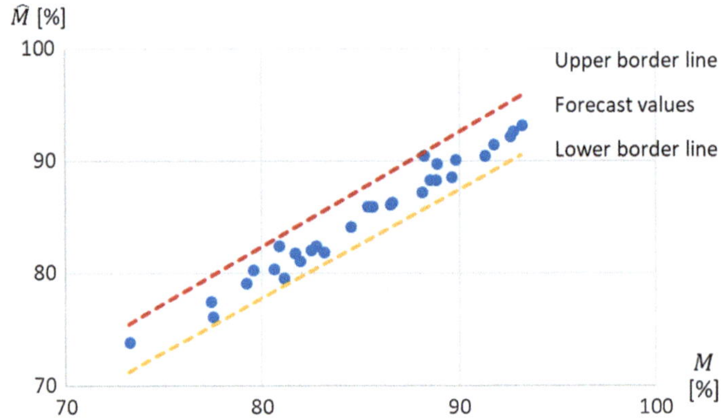

Fig. 12.6 Calculated production yield as a function of the true yield with corresponding tolerance range

starting point is the equally standardized data set without deriving the inverse correlation matrix or using indirect methods to circumvent it. The sum of the squares of the standardized parameters of the unit group divided by the number of variables is referred to below as the squared pseudo-Mahalanobis distance D_p^2.

$$D_p^2 = \frac{1}{k}\left(\left(\frac{x_1 - \bar{x}_1}{\sigma_1}\right)^2 + \left(\frac{x_2 - \bar{x}_2}{\sigma_2}\right)^2 + \left(\frac{x_3 - \bar{x}_3}{\sigma_3}\right)^2 + \ldots + \left(\frac{x_k - \bar{x}_k}{\sigma_k}\right)^2\right)$$

k = number of variables

Table 12.16 Comparison of the evaluations in non-squared form of the Mahalanobis distances of the original algorithm and the pseudo-Mahalanobis distances

Patient	D	D_p	Patient	D	D_p
1	0.9855	0.9116	21	1.3557	1.4264
2	1.0133	0.7052	22	1.5001	1.3802
3	0.9501	0.8319	23	1.058	0.992
4	1.0844	1.0121	24	1.4115	1.875
5	1.0014	1.1156	25	0.8593	0.8916
6	0.831	0.8694	26	0.6654	0.5643
7	1.1543	0.8415	27	0.6744	0.7156
8	0.9693	1.0077	28	1.1509	0.9912
9	1.1218	0.9311	29	1.0269	0.8071
10	0.8824	0.9592	30	0.7692	0.8856
11	0.9363	0.7032	31	0.6852	0.7705
12	1.3148	1.6668	32	0.9896	0.9428
13	0.9105	0.9553	33	1.1728	0.8525
14	0.9583	0.9973	34	0.655	0.495
15	0.8628	0.8327	35	3.2162	2.6946
16	0.7268	0.8083	36	5.0607	3.8182
17	1.0967	1.1182	37	2.0562	1.6186
18	0.7935	1.1458	38	2.7587	2.0155
19	0.748	0.882	39	2.4375	2.8568
20	0.9188	0.7993			

The average value $\overline{D_p^2}$ of objects of the unit group is close to one or equal to one, like the original algorithm. The case study consisting of 39 individuals, known from Sect. 7.3, is used to compare both definitions (Table 12.16; Fig. 12.7).

The comparison shows that although the result is not identical, it can be regarded as relatively similar. In individual cases, specific data sets must be used to check which approach is more suitable, i.e., comes closest to the true facts. Due to the simplicity of the procedure, it is worth to test the ability of parameter screening with the same data set.

The unit group (Sect. 12.3) provides average value and standard deviation for the standardization of all data.

$$X = \frac{x - \bar{x}}{\sigma}$$

The unit group is only retained for comparison purposes, since otherwise two objects are already sufficient for the standardization of the reference group. The same orthogonal array L8 with the derived parameter configurations is used for parameter screening. For the first test calculation, all existing parameters must be considered.

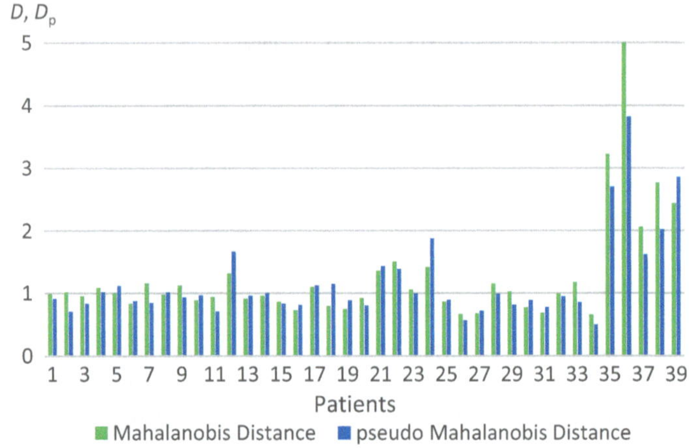

Fig. 12.7 Graphical representation of the numerical results of the Mahalanobis distances D and D_{p} of the clinical case study

$$D_{\mathrm{p}}^2 = \frac{1}{6}\left(\left(\frac{x_1 - \bar{x}_1}{\sigma_1}\right)^2 + \left(\frac{x_2 - \bar{x}_2}{\sigma_2}\right)^2 + \left(\frac{x_3 - \bar{x}_3}{\sigma_3}\right)^2 + \left(\frac{x_4 - \bar{x}_4}{\sigma_4}\right)^2 + \left(\frac{x_5 - \bar{x}_5}{\sigma_5}\right)^2 + \left(\frac{x_6 - \bar{x}_6}{\sigma_6}\right)^2\right)$$

For comparison, in addition to the squared pseudo-Mahalanobis distances of the unit and reference group, the values of the original algorithm are also listed (Table 12.17).

The proportionality constant β_0 results from the relation of the standardized parameters and the pseudo-Mahalanobis distances reduced by 1 (rounded) (Table 12.18).

Derivation of the signal values is performed just as for the Mahalanobis algorithm.

$$\widehat{M} = \frac{\sqrt{D_{\mathrm{p}}^2 - 1}}{\beta_0}$$

Similarly, further test calculations must be carried out with the configurations specified by the orthogonal array L8 (Tables 12.19 and 12.20).

The parameter effects are displayed (Fig. 12.8).

The parameters A and F can be omitted because

$$SNR_{\text{level 1}} < SNR_{\text{level 2}}$$

In a test calculation with parameters B, C, D, and E the retention or improvement of the precision of prognosis must be confirmed (Table 12.21).

Table 12.17 Squared Mahalanobis distances D^2 and D_p^2 for the first test calculation

Objects	Unit group		Reference group	
	D^2	D_p^2	D^2	D_p^2
1	0.3997	0.4363	23.5827	2.2756
2	1.2492	0.8939	15.8201	2.1085
3	1.1192	0.2682	110.8486	4.5544
4	0.9204	0.9811	72.6444	4.6166
5	0.8729	0.7371	100.331	6.0374
6	1.5386	2.2456	45.7668	2.828
7	1.0977	0.8427	87.5215	5.3277
8	0.9271	1.1382	23.717	1.3871
9	1.4123	1.9567	49.9027	2.867
10	1.444	1.9078	21.8651	2.3351
11	1.1828	1.1498	91.8136	4.584
12	0.5986	0.4005	68.3331	3.4299
13	0.6446	0.7888	8.9502	1.2587
14	0.4708	0.733	63.1271	2.8021
15	0.8894	0.7075	101.1545	4.6526
16	0.7239	0.4023	37.9086	5.6448
17	0.7758	1.0713	59.4018	6.2994
18	0.5936	0.6653	13.6078	1.6893
19	2.0374	1.5033	66.4453	5.0421
20	1.0772	1.3832	31.9464	3.5903
21	0.5696	0.7835	40.6654	4.7341
22	1.1239	0.7264	93.2418	5.5228
23	1.8314	2.1168	6.028	1.9961
24	1.1106	1.2643	57.7507	3.0481
25	0.6192	0.9772	106.8924	6.4846
26	0.7798	1.057	57.5678	3.3431
27	0.5253	0.7053	23.7997	3.2131
28	1.0823	0.7761	13.1672	2.3493
29	0.8695	0.8407	86.6068	4.8784
30	1.5131	1.063	29.6844	1.7817
Average	1	1.0174		

The true parameters and the calculated values are compared on the original scale for comparison (Fig. 12.9).

The tolerances for the prognosis can be defined as straight lines for the lower and upper limit (Fig. 12.10).

In principle, a parameter screening according to the same sequence can also be carried out with other algorithms, whereby the respective screening result does not have to be identical when using different procedures. This depends on the coupling strength of possible interactions between the parameters, considering their correlation patterns and, if applicable, the composition of a unit group.

Table 12.18 Outcome values normalized to the average value of the unit group and pseudo-Mahalanobis distances of the reference group for determination of the proportionality constant β_0

Calculation test #1	Normalized response of the reference group $x - \bar{x}$	Pseudo-Mahalanobis distance of the reference group $\sqrt{D_p^2 - 1}$
1	0.0906	1.1294
2	0.0957	1.0529
3	0.2119	1.8853
4	0.1805	1.9017
5	0.1661	2.2444
6	0.1293	1.352
7	0.1729	2.0803
8	0.1156	0.6222
9	0.1501	1.3664
10	0.1037	1.1555
11	0.202	1.8931
12	0.1721	1.5588
13	0.0595	0.5086
14	0.1694	1.3424
15	0.2105	1.9112
16	0.0931	2.1552
17	0.1651	2.302
18	0.0764	0.8302
19	0.1376	2.0105
20	0.101	1.6095
21	0.1119	1.9324
22	0.1825	2.1267
23	0.0176	0.998
24	0.1494	1.4311
25	0.2164	2.3419
26	0.1402	1.5307
27	0.0801	1.4877
28	0.0585	1.1616
29	0.1971	1.9694
30	0.1091	0.8841
	$\beta_0 = 10.9245$	

12.5 Procedure Selection

As the numerical example shows, the prognosis algorithms presented provide identical conclusions but with different degrees of accuracy. It appears advisable to test the methods and to select and apply the one with the largest *SNR* index as an

Table 12.19 Determination of the *SNR* key figure from the standardized parameters and calculated signal values for the prognosis accuracy

Calculation test #1	Normalized response of the reference group	Calculated signal values
1	0.0906	0.1034
2	0.0957	0.0964
3	0.2119	0.1726
4	0.1805	0.1741
5	0.1661	0.2054
6	0.1293	0.1238
7	0.1729	0.1904
8	0.1156	0.057
9	0.1501	0.1251
10	0.1037	0.1058
11	0.202	0.1733
12	0.1721	0.1427
13	0.0595	0.0466
14	0.1694	0.1229
15	0.2105	0.1749
16	0.0931	0.1973
17	0.1651	0.2107
18	0.0764	0.076
19	0.1376	0.184
20	0.101	0.1473
21	0.1119	0.1769
22	0.1825	0.1947
23	0.0176	0.0914
24	0.1494	0.131
25	0.2164	0.2144
26	0.1402	0.1401
27	0.0801	0.1362
28	0.0585	0.1063
29	0.1971	0.1803
30	0.1091	0.0809
Slope β	Squared standard deviation σ^2	*SNR* index $S_2 = 0.02097$
1	0.001563	11.2760 [dB]

objective criterion. For each data set, it is always necessary to check again which approach comes closest to the true facts. In addition, it should be noted that the nomenclature introduced here is based on mathematical principles and not on historical developments **(3-5*)**.

The linear approach can also be considered as if the unit space represents the origin in the middle, i.e., it is placed on the average value of the characteristic

Table 12.20 Proportionality constant, squared standard deviation, and prognosis accuracy of test calculations 1–8

Calculation test #	Proportionality constant β_0	Squared standard deviation σ^2	*SNR* [dB] $S_2 = 0.02097$
1	10.9245	0.001563	11.276
2	9.2489	0.004292	6.8896
3	10.4376	0.002681	8.9339
4	7.7396	0.017878	0.6931
5	9.7454	0.008152	4.1035
6	11.7052	0.003033	8.3983
7	11.0826	0.003608	7.6442
8	12.3815	0.002644	8.993

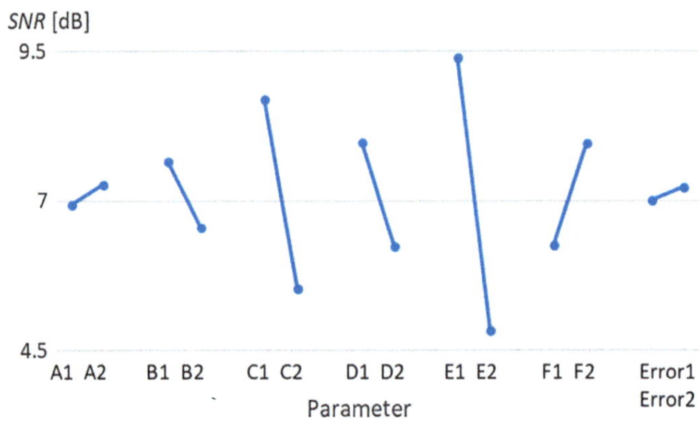

Fig. 12.8 Result of the parameter screening according to the method of the squared pseudo-Mahalanobis distances

Table 12.21 Prognosis precision with the set of parameters *B*, *C*, *D*, and *E*

Confirmation for parameters	Proportionality constant β_0	Squared standard deviation σ^2	*SNR* index $S_2 = 0.02097$
B, *C*, *D*, *E*	11.8301	0.000492	16.2943 [dB]

feature and the parameter values, while retaining the directional information. A reference point can also be selected instead of the average value. Since no further conditions are coupled to a data set, it can already be evaluated in minimum form, even if the number of variables exceeds that of the objects. Data transformations, such as the logarithm including the outcome—assuming positive data—and/or orthogonalization without the outcome can improve the prognosis precision. This can always be achieved by a mathematical displacement of the largest negative

Fig. 12.9 Comparative representation of the true parameters and the prognosticated values

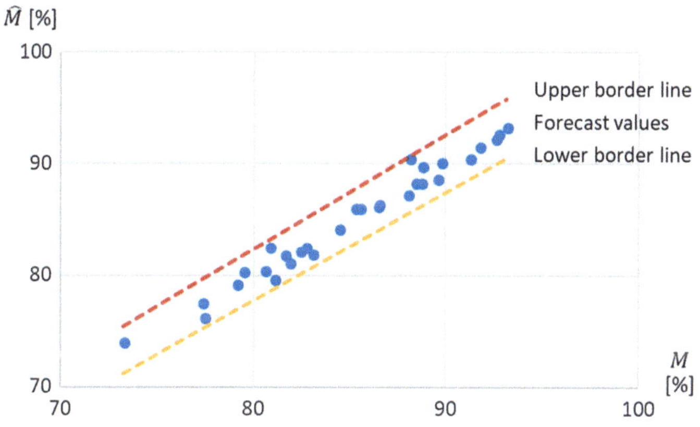

Fig. 12.10 Calculated values as a function of the true response with the tolerance lines

number. By using a two-level orthogonal array of appropriate size, a variable-specific assignment of a mathematical transformation (such as the logarithm or similar operations) can be determined for the best possible approximation of the prognosticated value to the true response (Table 12.22).

If, for a large range of parameter settings, the precision of prognosis even on a logarithmic scale lacks on precision, a sufficiently large data set can also be divided into several smaller groups. If this improves the accuracy that means the *SNR* index, weighting coefficients specific to the value range can be used. An additional aspect is the implementation of parameter screening. Here is a chance to get rid of variables with random character, which only increase the noise and worsen the *SNR* index.

Table 12.22 Definition of the setting values of a two-level orthogonal array for the derivation of variable-specific data transformations

Level	Mathematical transformation
1	Applied
2	Not applied

The quadratic procedure is sensitive to the concept of the unit space, which is usually not given. Regarding an existing data set, such a record is to be defined based on specific expertise or extracted as a subset from the data set itself. Provided that a data set with respect to the number of objects is of sufficient size, it is often possible to identify a set of output features with associated parameter values, which are, for example, on the side of minor values or relate to smaller relative dispersion. A preliminary application of the Mahalanobis algorithm for data discrimination shows whether the selected choice made already leads to a unique identification of an object with respect to belonging or not belonging to the unit group. The unit group may have to be supplemented or reduced by one or the other object until proper discrimination is achieved. In other words, the unit space itself has the function of a control parameter, which can make a decisive contribution to improving the prognosis precision. In the case of a relatively small unit group, which must always consist of more objects than variables, the avoidance of the inverse correlation matrix with Gram–Schmidt's orthogonalization method can be advantageous (Sect. 8.1). For mathematical transformations of the variables and the outcome figure, the same facts apply as already described under the linear method. The accuracy of a prediction depends decisively on the specific data set itself, but there is often an improvement in precision with logarithmic data. When performing a parameter screening, the situation will arise for some test configurations with a reduced number of variables that the squared Mahalanobis distance is less than one. Instead of using the non-squared form, the square root must be applied directly to the squared Mahalanobis distance.

The quadratic procedure based on the pseudo-Mahalanobis distance appears similar the accuracy of the results obtained by the mathematically more sophisticated Mahalanobis algorithm. In addition, there are no restrictions that must otherwise be observed when setting up the inverse correlation matrix. In this way, the unit group can be reduced to two objects that are sufficient for standardizing the data. For the selection of objects in the unit group, parameters at the lower end of the value range also appear practicable here. The comparison of the squared pseudo-Mahalanobis distances of the unit and reference groups displays whether the determination made causes a sufficient spreading or whether it must be improved by exchanging objects. Mathematical transformations and parameter screening with the resulting effects on the *SNR* index must be carried out in the same way as with other methods. All in all, the method is characterized by a safe and, above all, simple mathematical applicability when correlations between the characteristics are of minor importance.

Apart from the prognosis algorithms listed here in the context of the method according to G. Taguchi, there are numerous approaches of differently conceived mathematical procedures. The spectrum ranges from multivariate regression analyses to artificial neural networks, depending on the application (data discrimination, pattern recognition, and forecasts). The suitability or performance must be measured and compared with the identical criterion, the *SNR* index. It is not possible to see from the outset of an unknown data set which structures, relations, and correlations are hidden. Therefore, it is not possible to make a statement of the type:

Mathematical procedure *A* is superior to mathematical procedure *B*

Even with the use of software **(3–6*),** the path can be arduous and leaves much to be desired. This reflects the real strength of the approach according to G. Taguchi. With relatively small (computer) effort, it is possible to achieve results very quickly, which are at least equal to more complicated algorithms. Particularly noteworthy are problem solutions for which no other mathematical approaches are known. These include situations that require dealing with

- the multicollinearity of variables (Sect. 8.5)
- the number of variables exceeding the number of equations (= objects),(Sect. 11.2)
- reduction of large data sets to exactly two variables

The topic of the following chapter deals with the latter point.

12.6 Prognosis Algorithm Without Numerical Key Figures

Both the linear and the quadratic method require a characteristic feature that can be approximated as closely as possible by using a prognosis algorithm. This can be a technical production yield, a crime rate, or the water level of a river, to name a few examples. For other objects such as letters and numbers or faces of persons, the abovementioned approaches are not applicable due to a nonexistent numerical characteristic feature. On the contrary, at most one object can be identified by the fact that others are discarded. From a methodological point of view, the intention is to classify the object in question in the appropriate category with the greatest reliability. Whether alphanumeric characters are present or the recognition of any object or even a person, the object itself is coded, whereby the data set consists of a generally very large number of variables (Sect. 9.3). Even a large number of features with repetitions of the object under different angles of view or handwritten signs, executed by different people etc. can be mastered in principle with elaborate hardware and software use. Nevertheless, methodical approaches with low and manageable effort have their advantages, especially when simple adjustments are required due to unusual characteristics or a very small number of objects. Thus, the questions must be answered:

- Is there a minimum number of variables that can substitute a much larger ensemble and allowing still prognostic applications?
- Does a prognosis algorithm work with the smallest possible number of variables in a data set without numerical parameters?

12.7 Procedure with One-Step Reduction of Variables

In all previous considerations, the *S/N* key figure and the corresponding slope in dynamic behavior have always been the focus for evaluations and decisions to be made. From this point of view, it seems logical to compress a multiparameter data set to the two values mentioned above and to process them further. How to obtain the *S/N* ratio and slope? The path is already lined out with the derivation of the standard *S/N* key figure (Sect. 5.1). For each variable listed, the average value is calculated using the number of objects and defined as a signal value. This enables the determination of both slope and *S/N* ratio for each of the objects contained in the unit space for all existing variables.

$$\beta = \frac{x_1\bar{x}_1 + x_2\bar{x}_2 + x_3\bar{x}_3 + \cdots + x_k\bar{x}_k}{\bar{x}_1^2 + \bar{x}_2^2 + \bar{x}_3^2 + \cdots + \bar{x}_k^2}$$

$$\sigma^2 = \frac{1}{k}\left((x_1 - \beta\bar{x}_1)^2 + (x_2 - \beta\bar{x}_2)^2 + (x_3 - \beta\bar{x}_3)^2 + \cdots + (x_k - \beta\bar{x}_k)^2\right)$$

$$S_2 = \frac{1}{k}\left(\bar{x}_1^2 + \bar{x}_2^2 + \bar{x}_3^2 + \cdots + \bar{x}_k^2\right)$$

k = Number of variables

$$S/N = \frac{\beta^2}{\sigma^2} S_2$$

The *S/N* ratio already contains the slope in squared form. To provide equivalent variables for an evaluation algorithm, a direct reference to the deviation range of the objects appears obvious in addition to the slope as such. This results in a new definition of exactly two variables (Tables 12.23 and 12.24).

For discriminating against objects of any kind, the Mahalanobis algorithm is recommended, whereby only the variables Y_1 and Y_2 are used to calculate the squared distance (Table 12.25).

If characteristic sets of data of unknown objects exist that are to be identified, that is, if their belonging or not belonging to the unit space of a corresponding category is to be determined, they are dealt with in the same sequence. In the first step, the slope and the *S/N* ratio with the signal values of the objects of the previously defined unit space are to be derived. In the second step, the conversion to the new variables Y_1 and Y_2 is performed. Finally, in the third step, the same

Table 12.23 *S/N* key figures for a set of data without numerical characteristic features

Object #	Parameter x_1 x_2 x_3 ... x_k	Slope β	Squared standard deviation σ^2	*S/N* ratio $S/N = \frac{\beta^2}{\sigma^2} S_2$
1	x_{11} x_{12} x_{13} ... x_{1k}	β_1	σ_1^2	S/N_1
2	x_{21} x_{22} x_{23} ... x_{2k}	β_2	σ_2^2	S/N_2
3	x_{31} x_{32} x_{33} ... x_{3k}	β_3	σ_3^2	S/N_3
.	.	.	.	.
.	.	.	.	.
.	.	.	.	.
n	x_{n1} x_{n2} x_{n3} ... x_{nk}	β_n	σ_n^2	S/N_n
Average	$\bar{x}_1$ $\bar{x}_2$ $\bar{x}_3$...$\bar{x}_k$	1		

Table 12.24 Definition of a pair of variables for an evaluation algorithm

Output variable	Redefinition of variables
β	$Y_1 = \beta$
S/N	$Y_2 = \frac{1}{\sqrt{S/N}} = \sigma \frac{1}{\beta\sqrt{S_2}}$

Table 12.25 Variable pairs Y_1 and Y_2 of the unit group objects with their corresponding squared Mahalanobis distance

Object #	Variable Y_1 Y_2	Squared Mahalanobis distance D^2
1	Y_{11} Y_{12}	D_1^2
2	Y_{21} Y_{22}	D_2^2
3	Y_{31} Y_{32}	D_3^2
.	.	.
.	.	.
.	.	.
n	Y_{n1} Y_{n2}	D_n^2
Average	1	$\bar{D}^2 = 1$

Table 12.26 Distinguishing criterion for object identification

Threshold $D_T^2 = 3$	Classification of objects
$D^2 < 3$	Object identified; object belongs to the unit group
$D^2 \geq 3$	Object does not belong to the unit group category

Mahalanobis algorithm must now be applied to the reference group. The threshold value D_T^2 that has already been introduced serves as a criterion for belonging or not belonging to the unit group, Table 12.26 (Sect. 7.2).

Typical applications can be found, as briefly mentioned, in speech recognition or the identification of handwritten characters or in image processing. For handwritten

Table 12.27 Variants of numerical characters

Standard	1	2	3	4	5	6	7	8	9	0
Person 1										
Person 2										
Person 3										
…										
Person *n*										

characters, a unit space must be defined for each element (numbers, letters, and symbols) (Table 12.27).

The character itself is rasterized, i.e., converted into a code. Depending on the resolution and type of coding, approximately 10–100 parameters per character are generated in analog or digital form. Repetitions in the most diverse forms serve to ensure the greatest possible degree of unambiguous assignment. The parameters must always have the same dimension or be dimensionless. As signal values, the average is calculated from the parameter values of the writing samples to derive the slope and the standardized *S/N* ratio. The total number of parameters is thus already reduced to 2, which are transferred to the new variables Y_1 and Y_2. The Mahalanobis algorithm is only applied to this and the character-specific unit space is generated. After performing the same formalism for each character, these can be read from an unknown record. For this purpose, the unknown data set for each character must be prepared in such a way that the slope and the *S/N* index are set up based on the signal values in question. With the same transformation, Y_1 and Y_2 are used as reference group parameters for the calculation of the squared Mahalanobis distances. Ideally, all identical characters are recognized at a distance of less than 3. The distance is larger for all deviating characters.

The procedure as such can be applied to any parameter sets such as screening, creditworthiness, clinical appearance, product characteristics or similar, provided that the recognition characteristic feature is essentially based on different relative scattering values. The advantage of this approach, however, only comes into effect when the parameter numbers reach thousands or even far higher. Only then does the compression to exactly two variables result in a decisive gain in processing time while retaining similarly valuable information.

Since a larger numeric example with all individual steps appears too extensive for demonstration, the starting situation is limited to four parameters for exercise purposes only.

During the synthesis process of industrial diamonds for core bits, the majority of these are produced with satisfactory properties in terms of hardness and further mechanical workability. With a few batches, however, the products or the required dimensions cannot be obtained by grinding. The process parameters have already been examined without success. Although the concentrations of trace elements look similar on average, more complex raw material analyses are carried out related to the product properties for a thorough investigation over a limited time. For this purpose, 3 groups with the described properties are formed and 12 samples are

Table 12.28 Three unit groups of 10 samples each characterized by 4 parameters

	Products alright				Products split				Products too hard			
Sample	Parameter				Parameter				Parameter			
	A	*B*	*C*	*D*	*A*	*B*	*C*	*D*	*A*	*B*	*C*	*D*
1	5.34	7.22	6.88	7.7	2.08	8.85	2.04	9.6	7.86	1.14	1.04	9.33
2	2.97	7.61	3.98	9.46	5.67	2.96	9.9	7.8	3.62	10.79	4.57	1.63
3	2.74	8.61	7.32	9.02	4.49	5.02	5.47	12.24	1	11.04	3.78	2.18
4	4.46	8.67	5.2	9.87	5.18	3.25	8.92	9.43	0.74	0.5	1.74	11.78
5	6.59	6.07	4.24	6.04	5.89	8.35	2.47	12.14	1.91	4.6	11.7	2.99
6	4.43	4.58	5.72	5.81	5.9	10.74	1.64	8.53	11.6	2.18	3.49	3.6
7	4.72	8.45	3.44	8.96	4.2	3.47	10.49	7.88	3.43	1.31	11.18	2.8
8	3.03	6.67	7.68	5.94	7.21	5.09	10.66	6.24	6.6	8.55	2.2	1.53
9	3.05	4.07	3.79	6.31	3.31	11.84	1.41	9.55	7.9	6.05	1.89	3.86
10	6.95	8.3	7.76	9.4	8.34	9.57	10.57	5.07	0.89	10.34	0.07	6.62
Average	4.43	7.03	5.6	7.85								
	Retained samples to confirm predictions											
11	4.63	5.8	6.77	7.72	2.54	8.82	2.11	10.96	3.12	1.3	0.29	15.16
12	4.79	4.34	3.14	6.24	5.1	6.59	12.99	5.44	3.08	5.53	8.49	2.51

collected. Two samples are held back to check the predictions so that each unit group consists of 10 samples. If the group belonging is correctly identified, it can be assumed that the properties of the synthesized solids are determined not only by the purity of the raw materials but also by the interaction of the trace elements (Table 12.28).

The average values of the samples, which are defined as a unit group, must be used as signal values for all available test pieces in order to derive the slope and the standardized *S*/*N* ratio (Table 12.29).

For the new variables to be evaluated, the slope β for Y_1 and the square root of the reciprocal *S*/*N* ratio for Y_2 are taken. The Mahalanobis algorithm is to be applied to these with the result of the squared distance D^2 (Table 12.30).

Applying the new variables in a scatter diagram already shows clear differences between the three different product groups (Fig. 12.11).

The differences are extracted using a key figure, the squared Mahalanobis distance. As confirmation, the retained samples are listed together (Fig. 12.12).

12.8 Procedure with Multiple Reduction of Variables

The handling of larger volumes of data is considerably facilitated if it is broken down into smaller subunits, which are preprocessed and reassembled (Sect. 9.3). The procedure is explained using an example.

Table 12.29 Characteristic terms for the derivation of the standardized *S/N* ratio

Sample	Products alright				Products split				Products too hard			
	β	σ^2	S_2	*S/N*	β	σ^2	S_2	*S/N*	β	σ^2	S_2	*S/N*
1	1.07	0.43	40.49	107.7	0.98	6.1	40.49	6.32	0.75	14.88	40.49	1.54
2	1.01	1.92	40.49	21.42	1	9.15	40.49	4.46	0.8	12.08	40.49	2.17
3	1.14	1.64	40.49	31.94	1.12	5.22	40.49	9.79	0.74	13.16	40.49	1.7
4	1.16	0.75	40.49	72.07	1.05	6.99	40.49	6.36	0.67	17.29	40.49	1.06
5	0.88	2.13	40.49	14.85	1.2	6.47	40.49	8.97	0.8	16.67	40.49	1.56
6	0.8	0.91	40.49	28.46	1.1	7.63	40.49	6.39	0.71	20.87	40.49	0.97
7	1.05	1.9	40.49	23.41	1.01	9.13	40.49	4.52	0.67	18.24	40.49	1.01
8	0.93	2.29	40.49	15.17	1.09	9.58	40.49	5.01	0.7	11.03	40.49	1.81
9	0.7	0.35	40.49	56.04	1.12	10.67	40.49	4.72	0.73	7.75	40.49	2.79
10	1.27	0.72	40.49	91.08	1.25	10.94	40.49	5.83	0.8	12.22	40.49	2.1
Average	1											
	Retained samples to confirm predictions											
11	0.99	0.73	40.49	54.31	1.06	7.02	40.49	6.43	0.89	28.51	40.49	1.12
12	0.73	1.05	40.49	20.6	1.14	14.5	40.49	3.62	0.74	7.49	40.49	2.96

Table 12.30 New variables Y_1 and Y_2 with their corresponding squared Mahalanobis distance

Sample	Products alright			Products split			Products too hard		
	Y_1	Y_2	D^2	Y_1	Y_2	D^2	Y_1	Y_2	D^2
1	1.07	0.1	0.97	0.98	0.4	8.66	0.75	0.81	65.33
2	1.01	0.22	0.31	1	0.47	16.09	0.8	0.68	41.81
3	1.14	0.18	0.42	1.12	0.32	4.93	0.74	0.77	57.29
4	1.16	0.12	0.68	1.05	0.4	9.39	0.67	0.97	103.8
5	0.88	0.26	1.08	1.2	0.33	6.87	0.8	0.8	65.35
6	0.8	0.19	0.77	1.1	0.4	9.99	0.71	1.02	117.3
7	1.05	0.21	0.3	1.01	0.47	15.81	0.67	1	111
8	0.93	0.26	1.01	1.09	0.45	14.62	0.7	0.74	51.88
9	0.7	0.13	2.9	1.12	0.46	16.5	0.73	0.6	28.27
10	1.27	0.1	1.56	1.25	0.41	14.62	0.8	0.69	43.54
Average	1		1						
	Retained samples to confirm predictions								
11	0.99	0.14	0.32	1.06	0.39	9.31	0.89	0.95	103.2
12	0.73	0.22	1.32	1.14	0.53	24.9	0.74	0.58	26.03

With advances in digital imaging on the sensor side, supported by high readout speed, image dimensions of 4000 × 6000 pixels are the rule rather than the exception. If the intensity of each pixel is read out for the automatic image analysis of thousands of images, a stream of data is created that is difficult to handle using standard equipment. For most applications, the number of pixels is reduced to about

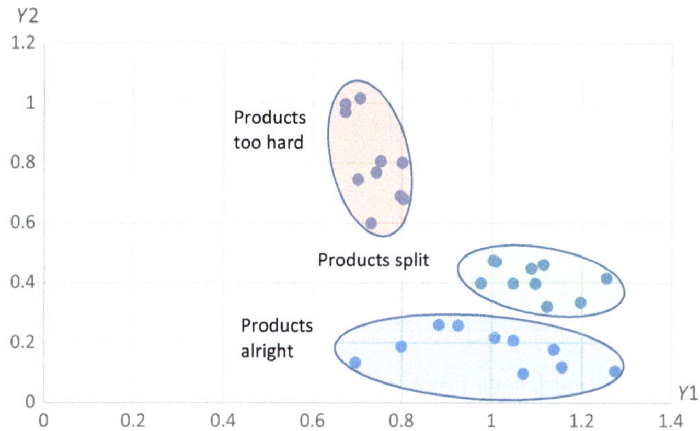

Fig. 12.11 Illustration of the new variables in mutual dependency

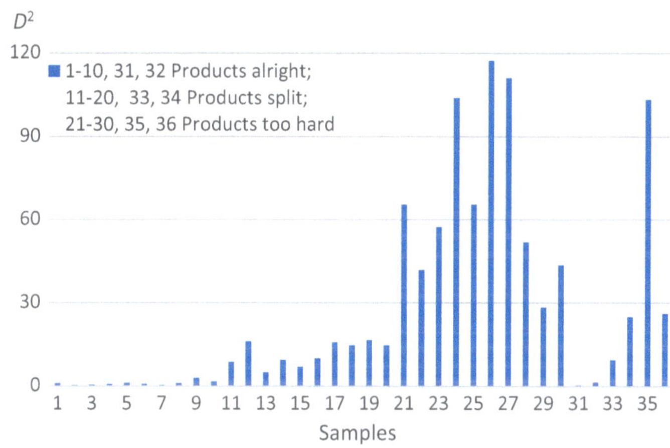

Fig. 12.12 Squared Mahalanobis distances of the product groups, each with two samples for confirmation

1/10 by compression, e.g., in the JPG format, with almost negligible loss of information. With the idea that a certain person or object can be identified with two variables based on individual dispersion differences, the amount of data becomes manageable. If the remaining pixel number as such is 1200×1800 pixels, this still exceeds the processing capabilities of common standard software. The best way out is to split the image into partial images, whereby the size of the partial image must be selected in such a way that it can be processed by software. If the pre-compressed image is split into 20×20 sub-images, each of them has a size of 60×90 pixels (Table 12.31).

Table 12.31 Splitting into partial pictures for multiple variable reduction

1. Partial picture 60×90 Pixel	2. Partial picture 60×90 Pixel	… …	20. Partial picture 60×90 Pixel
21. Partial picture 60×90 Pixel	22. Partial picture 60×90 Pixel	… …	40. Partial picture 60×90 Pixel
… …	… …	… …	… …
381. Partial picture 60×90 Pixel	382. Partial picture 60×90 Pixel	… …	400. Partial picture 60×90 Pixel

Table 12.32 Variable reduction for first partial picture

	1. Partial picture: variable = pixel				New variable
Picture sequence	Pixel 1	Pixel 2	…	Pixel 5400	$Y_{1,1}$ $Y_{2,1}$
1	$\text{Intensity}_{1,1}$	$\text{Intensity}_{1,2}$	…	$\text{Intensity}_{1,\ 5400}$	$Y_{1,1,1}$ $Y_{2,1,1}$
2	$\text{Intensity}_{2,1}$	$\text{Intensity}_{2,2}$	…	$\text{Intensity}_{2,\ 5400}$	$Y_{1,1,2}$ $Y_{2,1,2}$
…	…	…	…	…	…
n	$\text{Intensity}_{n,1}$	$\text{Intensity}_{n,2}$	…	$\text{Intensity}_{n,\ 5400}$	$Y_{1,1,n}$ $Y_{2,1,n}$
Average	$\overline{\text{Intensity}_1}$	$\overline{\text{Intensity}_2}$	…	$\overline{\text{Intensity}_{5400}}$	1

Table 12.33 Pairs of variables for each partial picture

Picture sequence	1. Partial picture	2. Partial picture	…	400. Partial picture	Y_1 Y_2
1	$Y_{1,1,1}$ $Y_{2,1,1}$	$Y_{1,2,1}$ $Y_{2,2,1}$	…	$Y_{1,400,1}$ $Y_{2,400,1}$	$Y_{1,1}$ $Y_{2,1}$
2	$Y_{1,1,2}$ $Y_{2,1,2}$	$Y_{1,2,2}$ $Y_{2,2,2}$	…	$Y_{1,400,2}$ $Y_{2,400,2}$	$Y_{1,2}$ $Y_{2,2}$
…	…	…	…	…	
n	$Y_{1,1,n}$ $Y_{2,1,n}$	$Y_{1,2,n}$ $Y_{2,2,n}$	…	$Y_{1,400,n}$ $Y_{2,400,n}$	$Y_{1,n}$ $Y_{2,n}$
Average	$\overline{Y_{1,1}}$ $\overline{Y_{2,1}}$	$\overline{Y_{1,2}}$ $\overline{Y_{2,2}}$	…	$\overline{Y_{1,400}}$ $\overline{Y_{2,400}}$	1

If the intensity of each pixel is read as a numerical variable, these are arranged in the form of 5400 columns in a sequence of n image repetitions (Table 12.32).

The average values are used as signal to determine the slope and the standardized *S/N* key figure for each pixel 1–5400. The procedure is carried out for all repetitions from 1 to n for the first partial image. Both variables are converted to the new variable pair $Y_{1,1}$ and $Y_{2,1}$. The process must be repeated for each partial image until all variable pairs $Y_{1,400}$ and $Y_{2,400}$ of the 400th partial picture are complete (Table 12.33).

Table 12.34 Squared Mahalanobis distances for each picture repetition 1–n

Picture sequence	New variable	Squared Mahalanobis distance D^2
1	$Y_{1,1}$ $Y_{2,1}$	D_1^2
2	$Y_{1,2}$ $Y_{2,2}$	D_2^2
…	…	…
n	$Y_{1,n}$ $Y_{2,n}$	D_n^2
Average	1	1

The average values resulting from the new pairs of variables determine the slope and the standardized *S/N* ratio. With the new variable pair Y_1 and Y_2, they are ready for the application of the Mahalanobis algorithm (Table 12.34).

The result is a single parameter, the squared Mahalanobis distance. The image sequence with n repetitions of the same object or person under different viewing angles defines the unit space. Other images are treated in the same way with the average values of the previously determined unit group and are included as an unknown group in the same Mahalanobis algorithm. The resulting size of the squared Mahalanobis distance, compared with the threshold value, determines the identification of the person, i.e., to the belonging or not belonging to the object category of the unit group.

12.9 Prediction and Modeling

Apart from the possibility of generating prognosticated values from a data set, the question: to which extent the observed parameters are causally responsible for the result can be investigated. If specific changes to the parameter values are permitted, this could also be used to steer a prognosis in a certain direction. A parameter screening gives an idea of the effect contribution without being able to say anything about the directional tendency of the prognosis result. If the *SNR* index of the prognosis is sufficiently large, it is suggesting itself to carry out a simulation calculation of a parameter analysis with an orthogonal array based on the parameter settings. As an example, the result adjusted by screening, consisting of four parameters, is used (Sect. 12.1). One orthogonal array of size L9 is sufficient for four parameters with three setting values each. With generally negligible expenditure, often larger orthogonal arrays such as the L18 are used to keep possible interactions to a minimum by their uniform redistribution to all columns. Conversely, when using an L81, all possible interactions of all paired parameter combinations can be read out. If the available data set is of sufficient size, it can also be halved to create two separate prognosis models. This gives the chance to include deviation as additional information for the reliability of a parameter analysis. For reasons of clarity, the orthogonal array L9 is used once, whereby it should be noted that the prognosticated result is obtained as a fraction or percentage. The chosen

Table 12.35 Selection of the level settings

	Levels		
Parameter	1	2	3
B	10	25	40
C	10	45	80
D	15	35	55
E	10	35	60

prognosis algorithm is the one with the best *SNR* index. For each parameter, three setting values from the lower edge position, the middle and the upper edge position of the existing value range are to be determined. In the case of linear behavior and satisfactory model confirmation, the value range can also be exceeded, and the target value extrapolated (Table 12.35).

The application of the quadratic prognosis algorithm to the orthogonal array L9 is tabulated. The slope β_0 of the reference group is used for the result to be calculated, and the average value of the production yield of the unit group is used for the back transformation to the original scale (Table 12.36).

The effect table of the calculated production yield is displayed numerically and graphically (Table 12.37; Fig. 12.13).

Within the value range of the parameters, the effects can be interpreted in such a way that for the combination of the level settings

$$B3 + C1 + D1 + E1$$

the maximum production yield is achieved. If the setting values can be controlled, the maximum value can be derived from the model. To avoid over or underestimation, the calculation must be carried out in the logarithmic scale [Omega transformed target value: $10\log\left(\frac{Y}{1-Y}\right)$] and the result must be transformed back again (Table 12.38).

Transformed back, this result in a maximum yield of 97%, which must be compared with the previous average yield of the reference group of 85% with the rarely occurring peak value of 93%.

What is the value of posterior modelling? There are many situations that do not allow experiments for different reasons, e.g., historical data, budget situations, availability of resources, etc. Instead, however, large data collections are often found. If these appear to be trustworthy, a reference group can be put together from them and the *SNR* index of various prognosis algorithms can be determined. Assuming sufficient precision, a parameter analysis is feasible, which reveals causal dependencies or relations. Such a robust model is also available for interpolation and extrapolation, so that valuable knowledge can be gained in condensed form at virtually no cost without a single experiment.

Table 12.36 Analysis of the orthogonal array L9

Simulation calculation #	B	C	D	E	Squared Mahalanobis distance D^2	$\sqrt{D^2 - 1}$	Calculated value	Calculated yield	Omega transformed yield
1	10	10	15	10	175.68	13.22	0.25	0.96	13.82
2	10	45	35	35	37.23	6.02	0.11	0.83	6.79
3	10	80	55	60	1.46	0.68	0.01	0.73	4.27
4	25	10	35	60	65.83	8.05	0.15	0.86	8.04
5	25	45	55	10	65.73	8.05	0.15	0.86	8.04
6	25	80	15	35	44.07	6.56	0.12	0.84	7.1
7	40	10	55	35	101.36	10.02	0.19	0.9	9.58
8	40	45	15	60	72.15	8.44	0.16	0.87	8.31
9	40	80	35	10	74.89	8.6	0.16	0.87	8.43

Table 12.37 Parameter effects of the calculated yields

Levels	*B*	*C*	*D*	*E*
1	0.84	0.91	0.89	0.9
2	0.85	0.85	0.85	0.86
3	0.88	0.81	0.83	0.82

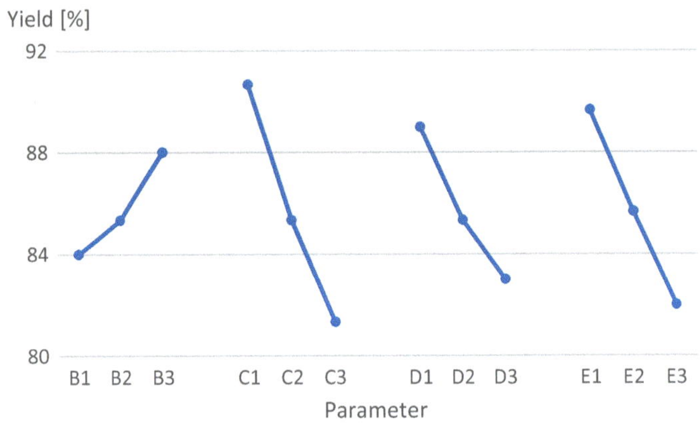

Fig. 12.13 Graphical presentation of the parameter effects in percent yield

Table 12.38 Parameter effects in the logarithmic scale

Levels	*B*	*C*	*D*	*E*
1	8.29	10.48	9.74	10.1
2	7.73	7.71	7.75	7.82
3	8.77	6.6	7.3	6.87
Yield maximum	$8.77 + 10.48 + 9.74 + 10.10 - 3\times 8.26 = 14.31$ [dB]			

Table 12.39 Properties of the algorithms for prognosis

Algorithm	Linear equation	Quadratic equation	Quadratic equation without correlation matrix	Equation for reduction of variables
Mathematical function	*S/N* characteristic figure, sensitivity	Inverse correlation matrix	Standardization	Reduction to 2 variables
Unit group	None	Required	In a minimal form	Required
Correlation pattern	Not considered	Considered	Not considered	Considered to a reduced extent
Signal	Required	None	None	Generated as mean
Multicollinearity	Tolerated	Modified algorithm	Tolerated	Tolerated
Direction	Both-sided	One-sided	One-sided	One-sided
n objects, k parameter	n can be smaller than k	$n \gg k$	$n > k$	n can be smaller than k
Sensitive to	Deviations related to the linearity	Deviations related to the correlation pattern	Deviations related to average and standard deviation	Deviations related to the squared standard deviation

12.10 Properties of the Algorithms for Prognosis

Since the question and scope of multiparametric data sets are not subject to any standardization, it is not possible to provide information about the most suitable prognosis algorithm. However, if there are special boundary conditions such as the knowledge of a unit group or more parameters than data records, etc., the table makes the selection of a method to be preferred easier (Table 12.39).

Important items to memorize For a data set used for prognosis, parameter screening is essential if the number of variables comes close to the number of objects. In case the number of parameters can be reduced, the *SNR* index of the prognosis improves with reduced effort.

Assuming a sufficiently large *SNR* index of the prognosticated characteristic features, a subsequent data analysis using an orthogonal array preferably at three levels leads to deeper insights into the cause-effect relationships.

Epilogue

Dealing with multiparameter data as developed in the chapters is characterized by simple mathematical approaches. Contrary to today's widespread belief that only large amount of data is accepted as proof of or against a hypothesis to achieve statistical significance, Genichi Taguchi succeeds in obtaining decisive information even from a small amount of data using objective criteria. He follows a path that now and then required corrections but was ultimately able to tackle and resolve persistent problems with unusual solutions. With increasing demands on technical precision, yield, speed, costs, and reliability, methodical aspects have been continuously developed further. This process has by no means been completed so far, but also encourages users to experiment themselves, to open up new territory with specific *SNR* indices suitable for this purpose, to penetrate deeper into medical diagnostics or, finally, to prognosticate natural events (trajectories of cyclones, the triggering of tsunamis, and much more) for better damage prevention more reliably. With such an intended continuous improvement with simple means and increasing dissemination of the comprehensively methodical approaches, Taguchi's life's dream would come true.

© Springer Nature Switzerland AG 2019

H. Ruefer, *Living Without Mathematical Statistics*,
https://doi.org/10.1007/978-3-319-99632-5

Appendix A

Further notes and mathematical supplements
Part 1

1-1*)

The magnification function of almost any image software allows to see and evaluate the pixels of digital images. The so-called point-shaped imaging is simple and instructive. When shooting a section of the night sky (point a digital camera on a tripod at some stars, open aperture, setting sensitivity at about 800 ASA, exposure time approx. 5 s, and infinite distance adjusting), stars should be imaged as points, i.e., ideally a star with only a few pixels should be represented. With a sufficiently high magnification of the digital image, it can be seen that blurred light spots of irregular shape, consisting of several 100 pixels, appear especially at the edge of the image. These distortions are usually not noticeable when taking pictures of the daily environment. For this reason, exact object dimensions must be used in the form of test images for the objective evaluation of an imaging system.

1-2*)

Carl Friedrich Gauss (1777–1855) examined four different average values with regard to their appropriateness for measurement technology. The best known is the arithmetic average, i.e., the sum of the individual values divided by their number. In addition, there are the quadratic average value, the harmonic average value, the geometric average value, and the concept of the median. With the quadratic average, the squares are added, and the square root is taken from the result. The harmonic average results from the number of measured values divided by the sum of the reciprocal individual values. In geometric averaging, the nth-root is taken from the product of n-individual values. The median is the central value of a data series if existing values are arranged according to its size. The latter can sometimes be advantageous if a measurement series does not converge.

© Springer Nature Switzerland AG 2019

H. Ruefer, *Living Without Mathematical Statistics*,
https://doi.org/10.1007/978-3-319-99632-5

Definitions of various averages

Arithmetic average
$\bar{y} = \frac{y_1 + y_2 + \cdots + y_n}{n}$
Quadratic average
$s = \sqrt[2]{\frac{1}{n}\left(\sum_i^n y_i^2\right)}$
Harmonic average
$h = \frac{n}{\sum_i^n \frac{1}{y_i}}$
Geometric average
$g = \sqrt[n]{\prod_i^n y_i}$
n = Number of data y_i

It applies $h \leq g \leq \bar{y} \leq s$.

The average value from the squares is also called the second moment; an example of this is the calculation of the squared standard deviation. The squared standard deviation is the average value of the squared deviations from the average value.

$$\sigma^2 = \frac{(y_1 - \bar{y})^2 + (y_2 - \bar{y})^2 + \cdots + (y_n - \bar{y})^2}{n}$$

The third moment is the averaging in the third power, etc., as introduced in Sect. 5.1 dealing with nonlinear system behavior.

The first moment is the arithmetic average. In technical mechanics, the meaning of the first moment lies in the center of gravity of a considered mass distribution, the second moment in the moment of inertia with rotation of the mass distribution around an axis perpendicular to it.

The squared standard deviation is also called variance. All versions are based on the definition given above, even if no total population of data is available. The acquisition of meaningful measures for the characterization of data is the concern of the descriptive statistics used here.

A reduced data set or a sample is considered in the stricter statistical view by granting a larger numerical value; mathematically, the denominator is reduced by one. Therefore, the symbolic abbreviations σ_{n-1} or s can be found in the corresponding literature.

1-3*)

If measured values are to be found less frequently with increasing distance from the average value, an approximately symmetrical or asymmetrical bell-shaped distribution of the data is present. The distribution function describes the frequency of expected events in a certain area. This is not to be equated with the well-known Gaussian normal distribution, which leads to a rapid decrease of events the further away they are from the average value. The probability density of the normal distribution is described by the function

$$y = \frac{1}{\sqrt{2\pi}} e^{-\frac{1}{2}x^2}$$

Normal form with average $\bar{y} = 0$ and
standard deviation $\sigma = 1$.

For mathematical reasons, the pre-factor has been selected in such a way that the total area beneath the curve of the graphical representation of the function is exactly one. The practical consequence is that with a sufficiently large distance from the average value—often greater than six times the standard deviation (6σ) is quoted—events simply can not occur. A typical example is the distribution of the body height of the adult male or female population. There are no adults smaller than 50 cm or taller than 3 m. Of course, this does not mean that all distributions have these properties.

As practice has shown, most distributions follow a bell curve, but by no means a Gaussian curve. It must be said that it is difficult even for specialists to distinguish graphical representations of bell curves from the Gaussian curve. One reason for this may be a certain suggestion since schoolbooks and textbooks of statistics deal with nothing else than the Gauss curve. In reality, other power coefficients exist. In general, applies

$$y = \frac{1}{\gamma} e^{-\frac{1}{2}x^\alpha}$$

$$\text{with} \quad 0 < \alpha \leq 2 \quad \text{and} \quad \gamma = \text{pre - factor}$$

For clarification, some bell curves with different exponents are plotted.

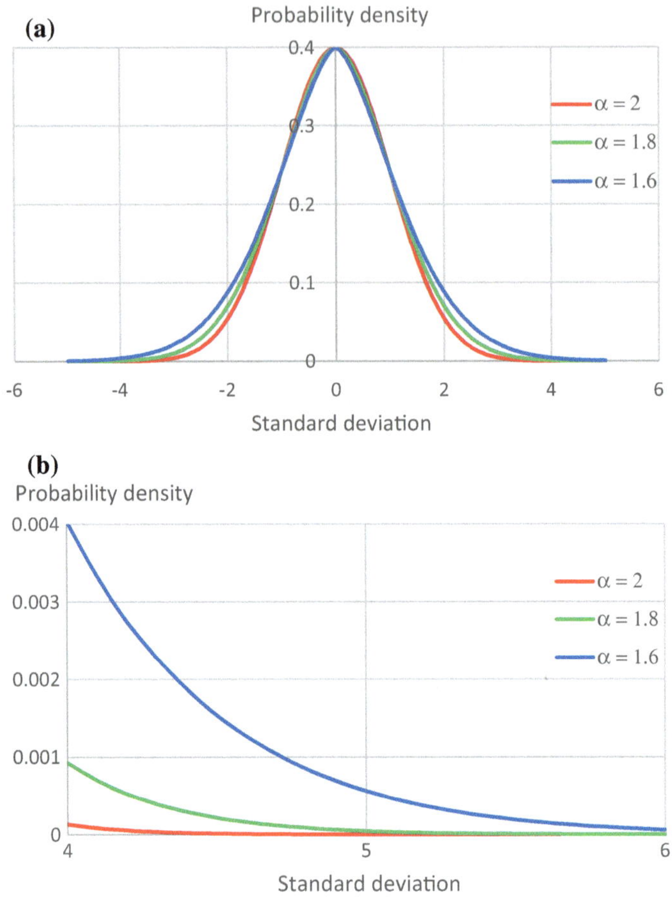

Probability density for the exponents ∝ = 2.0, 1.8, 1.6

(a) for standard deviations ± 6 σ

(b) resolved for the range 4-6 σ

For coefficients $\alpha < 2$, the curve tails—also known as fat tails—are above the Gaussian normal distribution. As a result, relatively rare events certainly can be observed although they cannot occur following the Gauss curve. For this reason, it is of almost extraordinary importance not to discuss existing, possibly unusual measurement results as outliers away just because they assume higher values than the unproven assumption of a Gaussian normal distribution far from the average value corresponds. The problem of outlier tests with subsequent elimination of data is a pseudo-problem that does not arise. In fact, exactly the opposite view is appropriate: events that are far away and correspond to the true facts must be considered with great care for an evaluation.

All calculations carried out under the implied condition of the validity of the Gaussian distribution, for example, the analysis of variance, are therefore to be regarded as questionable. In principle, this also includes the calculation of the standard deviation. However, in a broader sense, it can be used for a family of distribution functions of bell-like form, since the intersections of Gaussian-like curves approximately coincide with the turning points that occupy the distance of one standard deviation from the average value. Here, the practical aspect of applying a known and easily accessible characteristic quantity prevails. In connection with a small number of measured variables, typically 2 per test, profound statistical interpretations are also not appropriate. For this reason, the average value and squared standard deviation can be calculated without concern whether the prerequisite of a Gaussian normal distribution is fulfilled or not.

Another relatively common distribution function is the Lorentz function, widely used in electrical engineering and optics (spectroscopy). This describes resonance curves which, in electrotechnical jargon, are characterized with the ±3[dB] points (=curve width at half peak height). In physics, the equivalent is called "full width at half maximum" (FWHM), and the dimension-free ratio is used as an evaluation characteristic.

$$\frac{\lambda}{\Delta\lambda} \quad \lambda = \text{Wavelength}$$

This also shows that, depending on the task, an area-specific characterization of the spread of measured values is preferable to considerations of average and standard deviation.

1-4*)

The non-squared reciprocal value σ/y is also referred to as coefficient of variation; this stands as a dimension-free measure of relative dispersion.

1-5*)

For a better understanding of the key figure *S/N*, the numerator and denominator can first be considered separately. *S* in the numerator stands for signal. Therefore, the first question is: what is a signal? A signal has to do with energy. Signals are sent or received, e.g., during an electrical discharge (radio, television, mobile phone, GPS), location or speed measurements by radar, measurement of the depths of the sea by means of sonar or, more generally, structured energy changes which vary in space and/or time. A signal is more than a form of energy; it stands for everything that transmits information.

The term *N* in the denominator stands for noise. Then, the second question is: what is noise? Noise is also a signal, but in the usual sense an unloved one. The acronym *S/N* is consistently used in technical and scientific terminology whereby *N* stands for glitches, noise, loss. Noise becomes visible as flickering when one switches to a television channel without a transmitter. Noise is a signal without information. If noise does nothing else but destroying the transmission of information, why are not noise-free systems designed and built? Indeed, one is anxious to push back the noise component, as far as technically possible, expressed by the S/N ratio as the decisive factor. A signal can be extremely weak, some nanowatts or less (10^{-9} W), e.g., for the transmission of planetary images such as Jupiter and Saturn through the space probes to Earth; the only important thing that matters is the signal-to-noise ratio, i.e., S/N that must be as much as possible greater than one. Basically, noise-free systems are not allowed by nature due to the always existing thermal energy (kT). Pioneering research results on this topic date back to the pioneer of information theory Claude Elwood Shannon, "A Mathematical Theory of Communication", The Bell System Technical Journal **27**, 379–423, 623–656 (July, October 1948); for interested parties: Claude Ellwood Shannon—Collected Papers, Edited by N. J. A. Sloane, Aaron D. Wyner, John Wiley & Sons, Inc. (1993).

Noise in the context of the experimental methodology presented here includes everything that leads to undesirable deviations of an ideal system function. This can be caused by material inhomogeneities, mechanical tolerances, temperature fluctuations, changes in humidity, changes in concentration, loss of catalyst activity, time dependencies, side reactions, vibrations, side effects, etc.

In addition, it should be noted that noise can also have a good property for a few special applications. Thus, sensor systems with a threshold characteristic can be hypersensitized by a slight superposition of the input signal by noise. The effect is referred to in technical literature as stochastic resonance. This amazing effect is described with numerous other references in: Bart Kosko and Sanya Mitaim, "Stochastic resonance in noisy threshold neurons", Neural Networks **16**, 755–761 (2003).

1-6*)

Straight line through the origin	$y = M$
Signal values	M_i
Response values	y_i; $i = 1, 2, 3, \ldots, n$
Slope	β

Least squares method: derivative of the slope of the compensation line

$$(y_1 - \beta M_1)^2 + (y_2 - \beta M_2)^2 + \cdots + (y_n - \beta M_n)^2 = \text{Minimum}$$

$$\frac{dy}{d\beta}\left[(y_1 - \beta M_1)^2 + (y_2 - \beta M_2)^2 + \cdots + (y_n - \beta M_n)^2\right] = 0$$

$$2(y_1 - \beta M_1)(-M_1) + 2(y_2 - \beta M_2)(-M_2) + \cdots + 2(y_n - \beta M_n)(-M_n) = 0$$

$$\beta\left(M_1^2 + M_2^2 + \cdots + M_n^2\right) = y_1 M_1 + y_2 M_2 + \cdots + y_n M_n$$

$$\beta = \frac{y_1 M_1 + y_2 M_2 + \cdots + y_n M_n}{M_1^2 + M_2^2 + \cdots + M_n^2}$$

Squared standard deviation for dynamic system behavior

$$\sigma^2 = \frac{1}{n}\sum_i^n (y_i - \beta M_i)^2$$

The symbol σ for the standard deviation is used without additional subindex since the calculation is obvious in the context. For static systems, the squared standard deviation is related to the average value, in the dynamic case to the least squares compensation line defined by the slope β.

1-7*)

Conversion of the dynamic into the static *S/N* key figure

$$S/N_{\text{dyn}} = \frac{\beta^2}{\sigma^2} S_2 = \frac{\frac{(y_1 M_1 + y_2 M_2 + \cdots + y_n M_n)^2}{\left(M_1^2 + M_2^2 + \cdots + M_n^2\right)^2}}{\frac{(y_1 - \beta M_1)^2 + (y_2 - \beta M_2)^2 + \cdots + (y_n - \beta M_n)^2}{n}} \frac{\left(M_1^2 + M_2^2 + \cdots + M_n^2\right)}{n}$$

$$M_i \to M$$

$$\to \frac{\frac{(y_1 + y_2 + \cdots + y_n)^2 M^2}{(nM^2)^2}}{\frac{(y_1 - \bar{y})^2 + (y_2 - \bar{y})^2 + \cdots + (y_n - \bar{y})^2}{n}} \frac{(nM^2)}{n} = \frac{\bar{y}^2}{\sigma^2} = S/N_{\text{stat}}$$

Conversely, if static analysis is developed around a signal value, i.e., repeated static analyses are performed for different states of the same system, the dynamic behavior is mapped. The dependence of the output variable on the selected input variable can show a linear or any nonlinear relation. Such point-by-point scanning is rare in practice due to the relatively high experimental effort involved. A more general approach for systems with nonlinear behavior is derived in Sect. 5.1.

1-8*)

Besides to its ability to provide a dimension-free, dynamic *S/N* ratio, the dimension factor has an additional compensation function. Thus, an identical test as repetition but with different signal values must lead to identical results regarding the *S/N* index. The dimension factor considers the runs of an orthogonal array when they are carried out with individual signal values. This situation, the same test conditions with changes in both the number and the signal values themselves, is proven by simulation calculations. Due to the numerical compensation by the dimension factor, any signal settings can be selected for each test.

Only for the simulation calculations, the following functional system response is assumed for the noise conditions N_1, N_2 (presupposed is the knowledge of Sect. 2.8, evaluation of orthogonal arrays).

$$N_1: \quad y = 0.9M$$
$$N_2: \quad y = 1.1M$$

1. Case: Four signals, each with two noise conditions N_1 and N_2

	$M1 = 1$	$M2 = 2$	$M3 = 3$	$M4 = 4$
N_1	0.9	1.8	2.7	3.6
N_2	1.1	2.2	3.3	4.4

$$S_2 = \frac{1}{8}\left(1^2 + 2^2 + 3^2 + 4^2 + 1^2 + 2^2 + 3^2 + 4^2\right) = 7.5$$

$$\beta = \frac{0.9 * 1 + 1.8 * 2 + 2.7 * 3 + 3.6 * 4 + 1.1 * 1 + 2.2 * 2 + 3.3 * 3 + 4.4 * 4}{1^2 + 2^2 + 3^2 + 4^2 + 1^2 + 2^2 + 3^2 + 4^2} = \frac{60}{60} = 1$$

$$\sigma^2 = \frac{1}{8}\Big((0.9 - 1 * 1)^2 + (1.8 - 1 * 2)^2 + (2.7 - 1 * 3)^2 + (3.6 - 1 * 4)^2 + (1.1 - 1 * 1)^2 + (2.2 - 1 * 2)^2 + (3.3 - 1 * 3)^2 + (4.4 - 1 * 4)^2\Big) = 0.075$$

$$\frac{S}{N} = \frac{\beta^2}{\sigma^2} S_2 = \frac{1^2}{0.075} 7.5 = 100$$

2. Case: Two signals, each with two noise conditions N_1 and N_2

	$M1 = 1$	$M2 = 2$
N_1	0.9	1.8
N_2	1.1	2.2

$$S_2 = \frac{1}{4}\left(1^2 + 2^2 + 1^2 + 2^2\right) = 2.5$$

$$\beta = \frac{0.9 * 1 + 1.8 * 2 + 1.1 * 1 + 2.2 * 2}{1^2 + 2^2 + 1^2 + 2^2} = \frac{10}{10} = 1$$

$$\sigma^2 = \frac{1}{4}\Big((0.9 - 1 * 1)^2 + (1.8 - 1 * 2)^2 + (1.1 - 1 * 1)^2 + (2.2 - 1 * 2)^2\Big) = 0.025$$

$$\frac{S}{N} = \frac{\beta^2}{\sigma^2} S_2 = \frac{1^2}{0.025} 2.5 = 100$$

3. Case: Three different signals, each with two noise conditions N_1 and N_2

	$M1$ = 0.5	$M2$ = 1.5	$M3$ = 2.5
N_1	0.45	1.35	2.25
N_2	0.55	1.65	2.75

$$S_2 = \frac{1}{6}2\left(0.5^2 + 1.5^2 + 2.5^2\right) = \frac{8.75}{3} = 2.9167$$

$$\beta = \frac{0.45 * 0.5 + 1.35 * 1.5 + 2.25 * 2.5 + 0.55 * 0.5 + 1.65 * 1.5 + 2.75 * 2.5}{2(0.5^2 + 1.5^2 + 2.5^2)} = \frac{17.5}{17.5} = 1$$

$$\sigma^2 = \frac{1}{6}2\left((0.45 - 1 * 0.5)^2 + (1.35 - 1 * 1.5)^2 + (2.25 - 1 * 2.5)^2\right) = 0.029167$$

$$\frac{S}{N} = \frac{\beta^2}{\sigma^2}S_2 = \frac{1^2}{0.029167}2.9167 = 100$$

4. Case: Four signals, each with one noise condition

	$M1$ = 1	$M2$ = 2	$M3$ = 3	$M4$ = 4
N_1	0.9		2.7	
N_2		2.2		4.4

Although this situation will not occur in experimental practice, it can happen in process simulations with functional determination of the noise conditions. If test repetitions are calculated for different signal values, the slope β_1 must be determined from the smaller response values under noise condition N_1 and β_2 from the larger response values for N_2. The following applies for the actual slope:

$$\beta = \frac{\beta_1 \mid \beta_2}{2}$$

$$S_2 = \frac{1}{4}\left(1^2 + 2^2 + 3^2 + 4^2\right) = 7.5$$

$$\beta_1 = \frac{0.9 * 1 + 2.7 * 3}{1^2 + 3^2} = \frac{90}{10} = 0.9$$

$$\beta_2 = \frac{2.2 * 2 + 4.4 * 4}{2^2 + 4^2} = \frac{22}{20} = 1.1$$

$$\beta = \frac{0.9 + 1.1}{2} = 1$$

$$\sigma^2 = \frac{1}{4}\left((0.9 - 1 * 1)^2 + (2.2 - 1 * 2)^2 + (2.7 - 1 * 3)^2 + (4.4 - 1 * 4)^2\right) = 0.075$$

$$\frac{S}{N} = \frac{\beta^2}{\sigma^2}S_2 = \frac{1^2}{0.075}7.5 = 100$$

Interestingly, the kind of calculation of the *S/N* ratio or *SNR* index used in Japan in its modern view leads to identical numerical results; see Makoto Maeda: "Proposal of *S/N* Ratio for Zero-point Proportional Equation", Journal of Quality Engineering Society (QES) **16**(4), 62–69 (2008).

1-9*)

As already mentioned in Appendix A **(1-5*)**, fundamental information theory is attributed to Claude E. Shannon. Using the information unit Bit, he recognized the relationship between the following quantities:

C Channel capacity (Bit/s),
B Bandwidth (Hz),
S Signal power, and
N Noise power

with the maximum transmittable information

$$C = B \log_2\left(\frac{S+N}{N}\right)$$

The logarithm to base 2 is obvious for digital reason, but the logarithm to base 10 is more common. The conversion is done with the module and results in

$$\log_2(x) = \frac{\log_{10}(x)}{\log_{10}(2)} = 10\frac{\log_{10}(x)}{3}$$

substituted above

$$C = \frac{B}{3} 10 \log_{10}\left(\frac{S}{N} + 1\right)$$

In common practice, the index for identifying the base 10 is omitted. Furthermore, it can be assumed that the signal power of almost all systems under consideration exceeds by far the noise power, i.e.,

$$S \gg N$$

$$C \approx \frac{B}{3} 10 \log\left(\frac{S}{N}\right)$$

The logarithmic form of the *S/N* key figure is defined as

$$SNR = 10 \log\left(\frac{S}{N}\right)$$

If the amplitude is considered instead of the power or energy, the quadratic relationship must be taken. The average useful effect, equivalent to the signal to be transmitted, was described at the beginning with $\bar{y}$. The influence of noise was introduced as standard deviation σ. If the signal and noise power are substituted accordingly, the logarithmic form of the static signal-to-noise ratio results.

$$SNR = 10 \log\left(\frac{\bar{y}^2}{\sigma^2}\right)$$

An introduction to modern communication technology is provided, for example, by Allan R. Hambley; "Electrical Engineering", Prentice Hall (ISBN: 978-0-273-79325-0, 6th Edition 2013).

1-10*)

Historical starting points for the evaluation of signal strength are sensory sensations such as hearing. Weber and Fechner investigated the relationship between sensory stimulation with physical signal amplitude and the perceived intensity. They discovered the relation that a tenfold increase in the pressure amplitude causes a similarly felt increase in volume, also called Weber–Fechner's law in honor of them. The following applies to the volume:

$$L = 10 \log\left(\frac{p^2}{p^2_{\text{hearing threshold}}}\right)$$

Pressure p in Pascal [Pa]

Volume L in decibels [dB]

The hearing threshold, just above the thermal noise of the air molecules, lies at the maximum sensitivity of the auditory organ at a frequency of 4 kHz in the pressure range of 20 μPa. A volume change of 1 dB [decibel] is recognized by healthy people as barely distinguishable. The unit decibel goes back to Alexander Graham Bell (1847–1922). The original definition was the Bel without the pre-factor 10; these numbers were too large for practical use, so that the tenth part was quickly taken, from which the naming decibel emerged. Instead of the pressure ratio, the volume can be described identically by the sound intensity. This is measured as power per area with the unit W/m^2. A barely perceptible noise must have a minimum sound intensity of 10^{-12} W/m^2; the pain threshold is reached at about 1 W/m^2.

$$L = 10 \log\left(\frac{\text{Sound intensity}}{\text{Hearing threshold}}\right)$$

The table compares the logarithmic decibel scale with the actual, linear sound pressure and sound intensity.

Volume dB (decibel)	Sound source at a distance of	Relative change of sound pressure with respect to the hearing threshold	Sound intensity W/m^2
0	Hearing threshold close to the ear	–	10^{-12}
30	Whisper 1 m from the ear	33	10^{-9}
60	Normal conversation 1 m from the ear	1000	10^{-6}
90	High volume machines 1 m from the ear	33,000	10^{-3}
120	Threshold of pain close to the ear	1,000,000	10^{0}

As can be seen, a very large numerical range is transformed into a clear manageable range by the logarithm. More exact relations between the stimulus intensity and the sensation through the sensory organs can be found in: Stanley S. Stevens, Psychological Review **64**, 153 (1957).

1-11*)

Relationship of the second moment with the squared average and the squared standard deviation.

$$\frac{1}{n}\left(y_1^2+y_2^2+\cdots+y_n^2\right)=\frac{1}{n}\sum_{i=1}^{n}y_i^2=\frac{1}{n}\sum_{i=1}^{n}((y_i-\bar{y})+\bar{y})^2$$
$$=\frac{1}{n}\sum_{i=1}^{n}\left((y_i-\bar{y})^2+2(y_i-\bar{y})\bar{y}+\bar{y}^2\right)=\sigma^2+\bar{y}^2$$

The first term in the resulting sum is the definition of the squared standard deviation, the second term is omitted, since the sum of all positive and negative deviations from the average value adds up to zero, the third term is the square of the arithmetic average value.

The following relationship applies to the reciprocal values:

$$\frac{1}{n}\left(\frac{1}{y_1^2}+\frac{1}{y_2^2}+\cdots+\frac{1}{y_n^2}\right)=\frac{1}{n}\sum_{i=1}^{n}\frac{1}{y_i^2}=\frac{1}{n}\sum_{i=1}^{n}\left(\frac{1}{((y_i-\bar{y})+\bar{y})^2}\frac{\bar{y}^2}{\bar{y}^2}\right)$$
$$=\frac{1}{n}\sum_{i=1}^{n}\frac{1}{\bar{y}^2}\left(\frac{((y_i-\bar{y})+\bar{y})}{\bar{y}}\right)^{-2}$$
$$=\text{Series expansion }(1+x)^{-2}=1-2x+3x^2-4x^3+5x^4-\cdots$$
$$=\frac{1}{n}\sum_{i=1}^{n}\frac{1}{\bar{y}^2}\left(1-2\left(\frac{y_i-\bar{y}}{\bar{y}}\right)+3\left(\frac{y_i-\bar{y}}{\bar{y}}\right)^2-\cdots\right)\approx\frac{1}{\bar{y}^2}\left(1+\frac{3\sigma^2}{\bar{y}^2}\right)$$

For a nominal value m, this results in:

$$\frac{1}{n}\left((y_1-m)^2+(y_2-m)^2+\cdots+(y_n-m)^2\right)=\frac{1}{n}\sum_{i=1}^{n}(y_i-m)^2$$
$$=\frac{1}{n}\sum_{i=1}^{n}((y_i-\bar{y})+(\bar{y}-m))^2$$
$$=\frac{1}{n}\sum_{i=1}^{n}\left((y_i-\bar{y})^2+2(y_i-\bar{y})(\bar{y}-m)+(\bar{y}-m)^2\right)$$
$$=\sigma^2+(\bar{y}-m)^2$$

1-12*)

By increasing the *SNR* index, it is possible to reduce the dispersion of processes or product properties, which increases the reliability of the intended function. An increase in reliability is reflected in a longer life expectancy of products. The relative gain in reliability is characterized as follows:

$$\text{Actual state}: \quad SNR_{\text{initial}}=10\log\left(\frac{y_0^2}{\sigma_{\text{initial}}^2}\right)$$
$$\text{Improved state}: \quad SNR_{\text{new}}=10\log\left(\frac{y_0^2}{\sigma_{\text{new}}^2}\right)$$
$$\text{Increment}: \quad \Delta SNR=10\log\left(\frac{\sigma_{\text{initial}}^2}{\sigma_{\text{new}}^2}\right)$$

$$\text{Log base 2}: \quad \Delta SNR=10\cdot 0.3\log_2\left(\frac{\sigma_{\text{initial}}^2}{\sigma_{\text{new}}^2}\right)$$

$$\frac{\sigma_{\text{new}}^2}{\sigma_{\text{initial}}^2}=\left(\frac{1}{2}\right)^{\frac{\Delta SNR}{3[\text{dB}]}}$$

$$\sigma_{\text{new}}^2=\left(\frac{1}{2}\right)^{\frac{\Delta SNR}{3[\text{dB}]}}\sigma_{\text{initial}}^2$$

$$\sigma_{\text{new}}=\left(\frac{1}{2}\right)^{\frac{\Delta SNR}{6[\text{dB}]}}\sigma_{\text{initial}}$$

Relative reliability factor r: $\frac{\sigma_{\text{new}}}{\sigma_{\text{initial}}}=\frac{1}{r}$.

Relative reliability gain: $\frac{\sigma_{\text{initial}}-\sigma_{\text{new}}}{\sigma_{\text{initial}}}=1-\left(\frac{1}{2}\right)^{\frac{\Delta SNR}{6[\text{dB}]}}$.

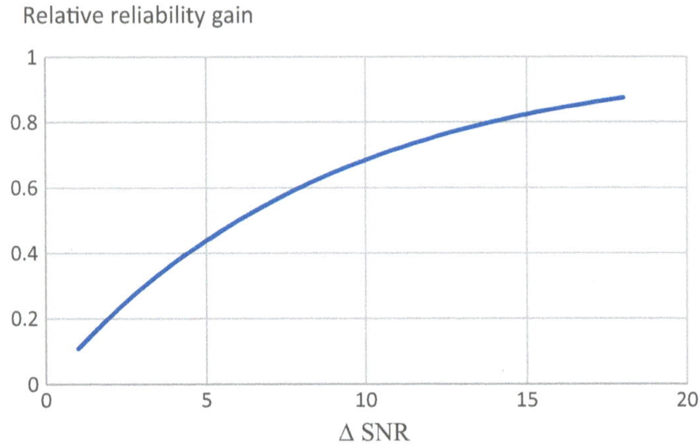

Relative gain in reliability as a function of the increase in the SNR key figure $= 10\log\left(\frac{y_0^2}{\sigma^2}\right)$

1-13*)

Further approaches, known in statistics as design of experiments (DoE), are described in detail, e.g., in George E. P. Box, J. Stuart Hunter, and William G. Hunter: "Statistics for Experimenters: Design, Innovation, and Discovery", Wiley & Sons (2005).

1-14*)

Motosaburo Masuyama has set up the first orthogonal arrays. Spatially, the test points are arranged in such a way that each two-dimensional cross section of the orthogonal array represented as a cube (or hypercube) always contains the same number of test points. Masuyama's starting point were the so-called Latin squares. The name is attributed to Leonhard Euler, who introduced Latin letters as symbols for permutations arranged in an equal number of rows and columns. According to their origin, the orthogonal arrays carry the label L with an appended test number, e.g., L8. L8 stands for the assignment of up to seven parameters with two levels each in eight experiments. An algorithm for deriving orthogonal arrays can be found at Motosaburo Masuyama: "On different sets for constructing orthogonal arrays of index two and of strength two". Rep. Statist. Appl. Res. Un. Jap. Sci. Eng. **5**, 27–34 (1957).

Latin squares recently celebrated their resurrection in the form of the number puzzles called Sudoku.

1-15*)

If a variable with more than four levels is to be resolved, the next larger L16 is to be used. Due to its size and symmetry, this allows a factor with up to eight levels to be accommodated. The procedure is completely analog, whereby instead of 2 in this case 3 independent, i.e., not-coupled columns must be selected because of the required combination number of eight possibilities (→levels). To make it easier to understand, columns 1, 2, and 4 are selected for substitution. The number of variants stands for the number of levels of a new variable. The assignment, here from 1 to 8, can be any but one unique setting value/variant.

Modification of an orthogonal array L16 with one variable to eight levels

A 1	B 2	D 4	→	New variable *N*
1	1	1	→	1
1	1	2	→	2
1	2	1	→	3
1	2	2	→	4
2	1	1	→	5
2	1	2	→	6
2	2	1	→	7
2	2	2	→	8

The key point is to remove both the selected independent columns and their interactions with any further couplings from the original orthogonal array. This guarantees that the number of unknowns does not exceed the number of equations (=number of tests) and the modified orthogonal array remains evaluable. The interaction table indicates which other columns are to be eliminated, since they are no longer available for independent parameters.

Substitution of columns of an orthogonal array L16 for one new variable

Column	Column	→	Interaction on column
1	2	→	3
1	4	→	5
2	4	→	6
3	4	→	7

Thus, columns 1–7 of the original orthogonal array are omitted and are replaced by one new variable *N* with eight levels, whereby the corresponding setting value is to be substituted with columns 1, 2, and 4 according to the definition. The modified orthogonal array L16 thus offers space for one variable with eight levels and the remaining eight variables with two levels each. If more than eight levels are necessary, the orthogonal array L32 or L64 must be used.

The same rules apply to orthogonal arrays with three or more levels. Since the total number of possible combinations increases with three instead of two levels, two columns each for the resolution of the interaction or the modification to more levels must be considered in the accompanying interaction table. Thus, with the orthogonal array L27, an existing coupling of two variables can be described by a loss of four columns. In other words, the modification consists of resolving one variable with nine levels. Here, too, a column pair must be selected, the corresponding interaction columns identified and substituted. For simplicity's sake, one can also start with columns 1 and 2; from the interaction table, the corresponding coupled columns are 3 and 4. Columns 1–4 must therefore be replaced by a new parameter whose levels can be defined as follows.

Modification of an orthogonal array L27 with one variable to nine levels

Column 1	Column 2	→	New variable N
1	1	→	1
1	2	→	2
1	3	→	3
2	1	→	4
2	2	→	5
2	3	→	6
3	1	→	7
3	2	→	8
3	3	→	9

After deleting columns 1–4 and assigning the levels according to the table above, the modified orthogonal array L27 contains one variable of nine levels with the remaining nine columns of three levels each.

Since the "loss" of columns in pure orthogonal arrays increases with the level number in the question of possible interactions, systems with two levels are preferably used. This is shown in the corresponding interaction table by the number of interaction columns:

One interaction column for orthogonal arrays with two levels each,
Two interaction columns for orthogonal arrays with three levels each,
Three interaction columns for orthogonal arrays with four levels each, and
Four interaction columns for orthogonal arrays with five levels each, etc.

If several parameters need higher resolution, further columns with their corresponding interaction columns can be modified in the same way. It should be noted that this cannot be extended at will, since, in elementary terms, the number of unknowns (parameters and their levels) cannot exceed the number of equations (number of tests). The test number n of a selected orthogonal array limits the number of parameters with their corresponding levels k, whereby the condition must be fulfilled:

$$n \geq (k_1 - 1)_{\text{parameter 1}} + (k_2 - 1)_{\text{parameter 2}} + (k_3 - 1)_{\text{parameter 3}} + \cdots + 1$$

Starting from the basic form of orthogonal arrays, the table gives an overview of existing modification options. The nomenclature commonly found in literature is used as a short form. The letter L stands in memory of the Latin squares as precursor of orthogonal arrays. The following number indicates the test number, i.e., the size of the orthogonal array. The numbers in parentheses are the setting values (=levels) with the number of parameters as a superscript. The numerical product represents the total number of all existing parameter combinations (with the example 2^{63} as mentioned in the introduction).

Selected examples of orthogonal arrays with modified level number

Basic orthogonal array	Modified orthogonal arrays with larger number of levels			
L8 (2^7)	L8 ($2^4 \times 4^1$)			
L16 (2^{15})	L16 ($2^{12} \times 4^1$)	L16 ($2^9 \times 4^2$)	L16 ($2^6 \times 4^3$)	L16 ($2^3 \times 4^4$)
	L16 (4^5)	L16 ($2^8 \times 8^1$)		
L32 (2^{31})	L32 ($2^{28} \times 4^1$)	L32 ($2^{19} \times 4^4$)	L32 ($2^7 \times 4^8$)	L32 ($2^1 \times 4^9$)
	L32 ($2^{24} \times 8^1$)	L32 ($2^{12} \times 4^4 \times 8^1$)	L32 ($4^8 \times 8^1$)	
L64 (2^{63})	L64 ($2^{60} \times 4^1$)	L64 ($2^{27} \times 4^{12}$)	L64 (4^{21})	L64 ($2^7 \times 8^8$)
	L64 ($2^{20} \times 4^5 \times 8^4$)	L64 (8^9)		
L27 (3^{13})	L27 ($3^9 \times 9^1$)			
L81 (3^{40})	L81 ($3^{32} \times 9^1$)	L81 ($3^{16} \times 9^6$)	L81 ($3^{27} \times 27^1$)	L81 (9^{10})
L18 ($2^1 \times 3^7$)	L18 ($3^6 \times 6^1$)			
L36 ($2^{11} \times 3^{12}$)	L36 ($2^3 \times 3^{13}$)	L36 ($2^1 \times 3^{12} \times 6^1$)	L36 ($3^{12} \times 12^1$)	
L50 ($2^1 \times 5^{11}$)	L50 ($5^{10} \times 10^1$)			
L54 ($2^1 \times 3^{25}$)	L54 ($3^{24} \times 6^1$)	L54 ($3^{18} \times 18^1$)		

For example, the modification of the orthogonal array L81 (9^{10}) requires 10 quadruple columns to provide space for 10 parameters with nine levels each.

One possibility of combining the columns of the L81 (3^{40}) to form the orthogonal array L81 (9^{10})

Combined columns L81 (3^{40})	1 8	2 14	3 23	4 32	5 24	6 18	7 22	11 19	12 16	13 21
	9 10	17 20	27 31	37 39	15 33	28 35	30 38	25 40	29 36	26 34
New parameters nine levels each L81 (9^{10})	A (1)	B (2)	C (3)	D (4)	E (5)	F (6)	G (7)	H (8)	I (9)	J (10)

The nine setting combinations of the first two columns determine the nine levels of the new variable. All four columns are to be removed from the orthogonal array L81 (3^{40}), whereby the nine combinations of the levels 1, 2, and 3 of the first two columns are to be replaced by the nine levels. The result is the orthogonal array L81 (9^{10}) with 10 columns of nine levels each with 81 tests to be performed (Appendix B). It should be noted that interactions between the parameters should be negligible or at least very small, as they can affect all levels.

If from the mathematical side are more levels than can be used practically, then the surplus state is assigned more often however clearly with one of the existing levels. The orthogonal array is no longer completely balanced but can still be evaluated. This state is also called a pseudo-level. If there are more columns in an orthogonal array than can be assigned to parameters, they are simply not assigned, but can also be evaluated. Effects on unassigned columns (=pseudo-parameters) thus form a valuable error estimation in comparison with the real parameter effects.

Other techniques allow one parameter with three levels to be split into two parameters with two levels each (combination design) or the reverse process of converting several two-level parameters into three-level parameters with reduced column loss (idle-column technique). Both methods provide evaluable arrays, but with loss of pure orthogonality. Since these situations occur very rarely and are basically "saving" at the wrong corner, reference is made to accompanying literature such as William Y. Fowlkes and Clyde M. Creveling: "Engineering Methods for Robust Product Design", Addison-Wesley Publishing Company (1995).

The above explanations are intended to deepen the understanding, offer simple and quick modification possibilities to extend the number of levels and facilitate the "correct" selection of orthogonal arrays according to the problem. In mathematical terms, the layouts of the main effects with strength 2 are important in practice. These are orthogonal arrays in which all pairs of levels are equally frequent for any pair of columns. A more detailed mathematical treatment can be found in: Hedayat, A., Sloane, N., and Stufken, J.: Orthogonal Arrays: Theory and Applications, Springer, New York (1999), and: Dey, A., and Mukerjee, R.: Fractional Factorial Plans, Wiley, New York (1999). Apart from the mathematical–theoretical aspects, Prof. N. Sloane provides a library of both existing orthogonal arrays and software for modification on his web page. Warren F. Kuhfeld has put together a largely complete collection consisting of more than 117,000 orthogonal arrays containing practically all variants. The information is provided by SAS Institute Inc. The web page is: Orthogonal Array —SAS with the link: support.sas.com/techsup/technote/ts723.html.

1-16*)

The mathematical model of orthogonal arrays is additive, i.e., the superposition of the apparently favorable effects is only additive if there are no interactions. Logarithmic data (*SNR*) generally assume a multiplicative effect of the parameters, as is often the case. If parameters are already additively related, the model prediction may result in an overestimation due to the superposition. In the case of deviating confirmation experiments, the test-specific direct *S*/*N* index can therefore also be interpreted tentatively as a "second-best" result.

1-17*)

The principle of parameter decoupling is explained using the example of an optical image to be improved on a photographic film. The exposure time and intensity, set by the aperture of the lens of a camera, are the variables. Since the light-sensitive film adds up the light energy (photons) for the duration of the exposure time, the formula below is valid over a relatively large range:

$$\text{Energy (of light)} = \text{Intensity} \cdot \text{Time}$$

Exposure time and aperture (corresponding to the used lens diameter) are therefore coupled via the total energy. If a test series were performed without decoupling, some of the tests could not be evaluated due to overexposure and underexposure that means lack of measurement data. The decisive variable to be varied is the light energy, which is tested for illustration with three setting values, whereby the parameters are assigned to the orthogonal array L9. The decoupling table shows standardized aperture values and exposure times. The parameters A and B are equivalent and can be interchanged with their corresponding settings in the table.

Decoupling scheme for two variables

Aperture	Energy of light		
	1	2	3
4	1/500	1/250	1/100
5.6	1/250	1/100	1/50
8	1/100	1/50	1/25
Parameter A	Parameter B: Exposure time [s]		

Experimental design of an orthogonal array L9 with and without parameter decoupling

Without decoupling of parameters					With decoupling of parameters				
Test #	A	B	C	D	Test #	A	B	C	D
1	4	1/250			1	4	1/500		
2	4	1/100			2	4	1/250		
3	4	1/50			3	4	1/100		
4	5.6	1/250			4	5.6	1/250		
5	5.6	1/100			5	5.6	1/100		
6	5.6	1/50			6	5.6	1/50		
7	8	1/250			7	8	1/100		
8	8	1/100			8	8	1/50		
9	8	1/50			9	8	1/25		

Without decoupling, parameters A and B each have three identical settings, which means that test 3 does not allow evaluation due to overexposure and test 7 due to underexposure. With decoupling, one of the two parameters can assume a different setting value for each test. Decoupling eliminates the interaction between A and B and the evaluation of the orthogonal array provides reliable parameter effects.

1-18*)

If the measured response already exists as energy or power instead of a signal amplitude, additional squaring is not necessary.

1-19*)

The calculation can be performed with the corresponding formulas either from the sum of the squared response values or the squared arithmetic average with the corresponding squared standard deviation **(1-11*)**. There is a slight difference between the two since the latter results in an approximation of an infinite series. In practice, both formulas are completely equivalent.

1-20*)

Binary or digital variables stand for two states. Mathematically symbolized by 0 and 1. The digital world of communication technology uses the same abbreviations, although these states can be measured, transmitted, and received less abstractly. Due to the properties of electronic components (diodes, transistors), the state 0 is defined as a lower range, when the electrical voltage of a signal does not exceed 0.8 V (0–0.8 V) and the state 1 as a higher range, if the signal voltage does not fall below 2.4 V (2.4–5 V). Even if strongly disturbed signals keep to these ranges, the signal is regenerated error-free. The intermediate range does not "exist" that means it is undefined. A signal value in this voltage interval leads to transmission errors (which can be detected and frequently eliminated using special test algorithms). Fact is that every physical signal is analog, and it is only a matter of agreement to interpret it analog or digital.

1-21*)

Claude E. Shannon defined a certain information composed of two states using the term entropy. The information content is zero, if only zeros or ones are transmitted, and, however, increases with the degree of mixing. Easier to handle than entropy is the squared standard deviation defined by the frequency p of occurrence of an observable state.

$$\sigma^2 = p(1 - p)$$

This formula behaves in the same way as the actual entropy, which is revealed in a more difficult derivation. The maximum value is set for $p = 1/2$ and is zero for $p = 0$ and $p = 1$. The conversion of digital states into continuous frequencies transforms digital into analog behavior with completely similar treatment as analog system.

1-22*)

The pressure Π to be applied for separation substances at low concentrations is

$$\Pi = R \cdot T \cdot c \cdot i$$

R	Ideal gas constant [=0.08478 atm l/K mol]
T	Absolute temperature [K]
c	Concentration of substance [mol/l]
i	Number of ions (particles) dissociated in water per molecule

With complete dissociation of the salt (NaCl) in seawater, the osmotic concentration is 1.2 mol/l due to i = 2, from which a pressure of 30 atm (approximately 30 bar) is derived for a functioning reverse osmosis. This sets the operating conditions. Experimental investigations focus on the membrane to achieve the greatest selectivity and chemical and mechanical stability. See, e.g., the textbook: Gavin Towler and Ray Sinnott; "Chemical engineering design: principles, practice, and economics of plant and process design", Elsevier, USA (ISBN 978-0-7506-8423-1, 2008).

1-23*)

Solution of a quadratic equation:

$$ax^2 + bx + c = 0$$

$$x_{1/2} = \frac{-b \pm \sqrt{b^2 - 4ac}}{2a}$$

The sign must be chosen in a way that the result is a positive, real number.

1-24*)

The determination of a certain ratio of the types of errors is based on the raw material- and processing costs. If the raw material is not a decisive cost factor, the wrongly classified fraction is just waste. If the element to be extracted is rare, as can be seen, for example, from the name of the rare earths group, the second type of error is kept very small in the separation method to be used. This ensures that the loss of valuable ore due to misclassification is as low as possible, although a relatively large proportion of worthless rock must be accepted due to the first type of error.

If, in addition to a valuable raw material, the separation power is very low, several separation stages are connected in series, typically for gaining enriched or pure isotopes, to increase the concentration of the valuable component in the output and to deplete it further in the waste. The output of one stage with the enriched product is fed to the next higher separation stage. The waste stream is recirculated and supplied with the input of the lower stage. The serial connection of several stages is also known as a cascade.

1-25*)

With the advances in medical technology in the field of cell differentiation from stem cells and their conversion into pluripotent cells, new ways of developing therapeutic procedures are emerging. More detailed descriptions can be found in: Stadtfeld, M. and Hochedlinger, K. "Review: Induced pluripotency: History, Mechanisms, and Applications" in Genes & Development 24, 2239–2263 (2010). John Gurdon and Shinya Yamanaka were awarded the Nobel Prize in Medicine in 2012 for demonstrating that body cells can be reset into induced pluripotent stem cells. When this processing is applied to human cells, appropriately designed cell cultures could serve as test models, especially for diseases that have not yet been treated successfully. The cells would come from the patient himself and would first be converted into pluripotent stem cells in the laboratory. Subsequently, differentiated cell growth is stimulated to develop tailored and, if necessary, individualized active pharmaceutical drugs for a robust therapy.

1-26*)

The final velocity of a rocket v_t when using a chemical fuel (e.g., liquid hydrogen or ethyl alcohol as fuel together with liquid oxygen) depends on the exit velocity v_g (=jet velocity) at which the recoil gases leave the rocket nozzle and the mass ratio given from the starting mass m_0 and the remaining mass m_t of the rocket at combustion end.

$$v_t = v_g \ln\left(\frac{m_0}{m_t}\right)$$

The jet velocity itself is proportional to the square root of the thermal power/mass of the chemical reaction. By the way: the mass ratio of the historically well-known V2 rocket was m_0:m_t = 3.23:1. It can be shown that the highest efficiency based on chemical reactions is at a velocity ratio v_t:v_g = 1.6 : 1 and a mass ratio close to 5:1. First time, the so-called rocket equation was derived by Konstantin Ziolkowsky in 1903; additional information can be found in Daniel Kleppner and Robert Kolenkow: "An Introduction to Mechanics", McGraw Hill, ISBN 0-07-035048-85 (1973).

1-27*)

The temperature dependence of the rate constant k is described by the Arrhenius equation:

$$k = k_{\infty} e^{-\frac{E_A}{RT}}$$

E_A Activation energy
R Ideal gas constant
T Absolute temperature [K].

1-28*)

Various chemical components intervene in the activation energy and change the rate constant k. This could be compensated or increased accordingly by adjustment of the reaction temperature, however, kinetic control is limited for higher reaction temperatures by diffusion processes. For even higher temperatures chemical reactions are boundary layer controlled. Consequently, the relative increase in the rate constant k with temperature becomes lower and lower.

1-29*)

In the case of several simultaneous reactions from the same reactants, one or another reaction path can also be of a higher order. Macroscopically, this results in a mixed exponent that can be determined most reliably from experimental data at high conversion. The exponents of the reaction order are treated at first as free parameters and the corresponding set of differential equations are solved by iterative integration methods (e.g., Runge–Kutta method). The percentage of parallel reactions (if present) and simultaneous adjustment of the exponents of the reaction order are changed until the numerical values of all concentrations of the components reflect the experimental data as good as possible. The rate constants determined from this allow a quantitative description of the reaction process with optimization option for a given target product. Software packages such as MATLAB or Mathematica from Wolfram Research facilitate the execution of the calculation steps but require some effort for training and postprocessing of the results.

1-30*)

Since each periodic process can be represented by superpositions of sine and cosine functions (=Fourier synthesis), it is sufficient to consider a sinusoidally alternating electric voltage or current. The substitution of trigonometric functions by exponential functions is based on the application of Euler's formulas.

$$e^{i\varphi} = \cos(\varphi) + i\sin(\varphi)$$

$$e^{-i\varphi} = \cos(\varphi) - i\sin(\varphi)$$

Complex coordinate system $y = Ae^{i\varphi} = A(\cos(\varphi) + i\sin(\varphi)) = a + ib$.

$$\text{with} \quad i^2 = -1 \quad \varphi = \text{phase angle} \quad A = \text{amplitude}$$

For the transition from DC to AC, all electrical laws apply if electrical current, voltage, and resistivity are present in complex form.

$$J_{AC} = Ie^{i\varphi} = I_{effective} + iI_{reactive}$$

$$E_{AC} = Ue^{i\varphi} = U_{effective} + iU_{reactive}$$

$$Z_{complex\ resistivity} = Re^{i\varphi} = R_{effective} + iR_{reactive}$$

The advantageous use of a complex notation is not limited to the field of alternating current but is generally valid for electromagnetic waves. If, for example, a light wave enters the interface against an absorbing medium (e.g., metal), the surface of constant phase differs from the surface of constant amplitude, since the amplitude is damped exponentially with the penetration depth. The optical laws remain applicable if instead of the simple refractive index.

$$n = \frac{\text{Speed of light in vacuum}}{\text{Propagation velocity in a medium}}$$

a complex refractive index is introduced

$$n_{complex} = n - ik \quad k = \text{extinction constant}$$

From reflection measurements under oblique light incidence, n and k can be determined experimentally. If the complex dielectric constant is known, n and k can be calculated from it.

1-31*)

The complex slope is calculated from the same Gaussian formalism as described in Sect. 2.1. The signal variable M can be real or complex. The complex conjugated figure is marked with X^*.

$$\beta = \frac{y_1 M_1 + y_2 M_2 + \cdots + y_n M_n}{M_1 M_1^* + M_2 M_2^* + \cdots + M_n M_n^*}$$

$$\beta^* = \frac{y_1^* M_1^* + y_2^* M_2^* + \cdots + y_n^* M_n^*}{M_1 M_1^* + M_2 M_2^* + \cdots + M_n M_n^*}$$

$$\sigma^2 = \frac{(y_1 - \beta M_1)^2 + (y_2 - \beta M_2)^2 + \cdots + (y_n - \beta M_n)^2}{n}$$

$$\sigma^{2*} = \frac{(y_1^* - \beta^* M_1^*)^2 + (y_2^* - \beta^* M_2^*)^2 + \cdots + (y_n^* - \beta^* M_n^*)^2}{n}$$

1-32*)

The number of applied signals can also be greater than 2, although it is rather rare due to the increasing test effort. More generally, the system response can be composed of terms of different sensitivity coefficients.

$$y = \beta_1 M_1 + \beta_2 M_1 M_2 + \beta_3 \frac{M_1}{M_3} + (\beta_4 M_1 \ldots)$$

Proportional behavior for single signal:	$\beta_2 = 0$	$\beta_3 = 0$
Proportional behavior for double signal:	$\beta_1 = 0$	$\beta_3 = 0$
Indirectly proportional behavior for one of two signals:	$\beta_1 = 0$	$\beta_2 = 0$
Additive behavior for:	$\beta_3 = 0$	

1-33*)

The ideal distance between different noise levels is zero. From an information theoretical viewpoint, this is equivalent to no information about the source of noise. Robustness is characterized by the nonexistence of additional information.

1-34*)

The ideal value m is the origin that the MacLaurin form of the Taylor series uses as development point. The distance from the target value is $(y - m)$, and the first and higher derivatives are designated by $L'(0)$, $L''(0)$, and so on.

$$L(y) = L(0) + \frac{L'(0)}{1!}(y - m) + \frac{L''(0)}{2!}(y - m)^2 + \frac{L'''(0)}{3!}(y - m)^3 + \frac{L''''(0)}{4!}(y - m)^4 + \cdots$$

1. Term: The loss on the target value m is zero.
2. Term: The slope at the target value m is zero for symmetry reasons.
3. Term: The second derivative stands for a curvature; the coefficient is referred to as k.
4. Term: The slope is zero for symmetry reasons.
5. Term: The contribution of the fourth power and all higher orders is negligibly small.

Thus remains

$$L(y) = k(y - m)^2$$

1-35*)

The loss of a product due to a deviation of the ideal numerical value is

$$L(y) = k(y - m)^2$$

$$L(y) = k((y - \bar{y}) + (\bar{y} - m))^2$$

Averaging for several pieces n by adding up and dividing by n

$$\frac{1}{n}\sum_{i=1}^{n} L_i(y) = k\frac{1}{n}\sum_{i=1}^{n}((y_i - \bar{y}) + (\bar{y} - m))^2$$

Average loss L

$$L = k\frac{1}{n}\sum_{i=1}^{n}\left((y_i - \bar{y})^2 + 2(y_i - \bar{y})(\bar{y} - m) + (\bar{y} - m)^2\right)$$

$$L = k\left(\frac{1}{n}\sum_{i=1}^{n}(y_i - \bar{y})^2 + \frac{2}{n}\sum_{i=1}^{n}(y_i - \bar{y})(\bar{y} - m) + \frac{1}{n}\sum_{i=1}^{n}(\bar{y} - m)^2\right)$$

The first term is the squared standard deviation σ^2, the second term is omitted, since all values $(y_i - \bar{y})$ summed up are zero, the third term is added n times and divided by n.

$$L = k\left(\sigma^2 + (\bar{y} - m)^2\right)$$

1-36*)

The loss analysis refers to a situation to be characterized by the *SNR* index of the type nominal-the-best (*NTB I*). In addition, system behavior may also require a smaller-the-better (*STB*) or larger-the-better (*LTB*) description. The loss function for dynamic behavior can be defined accordingly with the slope.

Loss functions for static and dynamic *SNR* key figures

SNR key figures for system evaluation	Loss function	Average loss for ensemble
Nominal-the-best (*NTB* I)	$L(y) = \frac{A}{\Delta^2}(y - m)^2$	$L(y) = \frac{A}{\Delta^2}\left(\sigma^2 + (\bar{y} - m)^2\right)$
Smaller-the-better (*STB*)	$L(y) = \frac{A}{\Delta^2}y^2$	$L(y) = \frac{A}{\Delta^2}(\sigma^2 + \bar{y})^2$
Larger-the-better (*LTB*)	$L(y) = A\Delta^2\frac{1}{y^2}$	$L(y) \approx A\Delta^2\left(\frac{1+\frac{3\sigma^2}{\bar{y}^2}}{\bar{y}^2}\right)$
Dynamic *SNR* index	$L(y) = \frac{A}{\Delta^2}(\beta - \beta_m)^2$	$L(y) = \frac{A}{\Delta^2}\left(\sigma^2 + (\bar{\beta} - \beta_m)^2\right)$

For the standardized *SNR* index, the loss function must be composed proportionally of the deviation of slope β_1 from one and the deviation of slope β_2 from zero.

1-37*)

The spread of the settings around the nominal value at level 2 for a tolerance analysis follows from the definition of the standard deviation.

$$x_i = \bar{x} \pm \Delta \quad \bar{x}: \text{central setting value;} \quad \Delta: \text{distance}$$

$$\sigma^2 = \frac{1}{3}\sum_{i=1}^{3}(x_i - \bar{x})^2$$

$$\sigma^2 = \frac{1}{3}\left(((\bar{x} - \Delta) - \bar{x})^2 + (\bar{x} - \bar{x})^2 + ((\bar{x} + \Delta) - \bar{x})^2\right)$$

$$\sigma^2 = \frac{(-\Delta)^2 + 0 + (+\Delta)^2}{3} = \frac{2}{3}\Delta^2$$

$$\Delta = \pm\sqrt{\frac{3}{2}}\sigma$$

A separation of the linear and quadratic component of the effect on the response can only be calculated for parameters with three levels. If a tolerance analysis is for screening purpose only, two levels are sufficient. In this case, two setting values are chosen.

$$\sigma^2 = \frac{(-\Delta)^2 + (+\Delta)^2}{2} = \Delta^2$$

$$\Delta = \pm\sigma$$

1-38*)

If it is foreseeable that functional loss exceeds the observation time frame, artificial stress conditions can be applied to shorten the life expectancy of the product function also referred to as accelerated life test. Under such unrealistic test conditions, however, caution is required. The test group recognized as more advantageous, i.e., with a longer lifetime, has its properties based on certain predefined values of material parameters. Normal operating and environmental conditions mean nothing more than less drastic levels of stress in the sense of noise, i.e., an existing interaction between material parameters and noise factors can affect the result.

To keep the effort to a minimum, the (material) parameters and setting values are to be defined and the objects to be inspected are to be provided according to the specifications of the orthogonal array used. Transferability of the results from a high stress level to a low stress level can only be seen if a sufficient number of samples is used for each test configuration, considering graduated stress conditions, eventually with more stress parameters, then assigned to an outer orthogonal array. During parameter analysis, broken down by stress parameters and their setting levels, the coupling strength becomes apparent. This is negligible if no change in the effect contributions is apparent and transferability is permitted or if the parameter–noise interaction must be considered in the case of opposing effects. The number of samples and the logistical effort can be reduced if the stress parameters are included in the same pure orthogonal array together with the material parameters. The assignment of the parameters to the columns is to be observed, so that associated interactions considered important can be read out without confounding; see, e.g., Phadke, M.S., Swann, D.W., and Hill, H.A.: "Design and Analysis of an Accelerated Life Test Using Orthogonal Arrays", Annual Proceedings of the Meeting of the American Statistical Association, Toronto, Canada (1983).

1-39*)

The success of a therapy is measured by the patient's condition. An overall appearance (called outcome parameter in the medical field) should be used for the evaluation, e.g., length of stay in a hospital, length of stay in the intensive care unit, survival time, quality of life in the sense of the degree of restriction in habitual activities or similar. If there is a sufficiently large difference between a test group and a control group, the therapy or surgery is assessed as positive. As remarkably simple as the evaluation in the form of the difference in the *SNR* indices appears, the execution of a clinical study is just as complicated.

It is acknowledged and accepted that expectations of the responsible scientist cannot be excluded, especially in clinical research. To avoid impairment of perception and data interpretation, studies are only conducted under so-called blind conditions. This term refers to the secrecy of the individual assignment of the actual real drug and a pharmacologically inert substance for the duration of the test, both on the patient's side and on the medical responsibility side. Such a double-blind placebo-controlled study looks like an objective impression when a treatment effect (therapy) can be clearly distinguished from a placebo effect. Experience shows that a placebo treatment is accompanied by identical changes in physiological parameters and typical side effects, in rare cases even only the side effects occur. This means that medically relevant statements are only permissible in comparison to a test and control group or to various therapies for the same disease pattern in general. Consequently, there are always at least two data sets that must be processed identically—from a mathematical viewpoint—and displayed as a comparison or ratio.

An additional aspect is biodiversity, which is addressed with a larger group of patients. Groups consisting of more than 10 patients up to several hundred in a hospital are common. As a decision for a therapeutic effect on the outcome parameter, a difference of the *SNR* indicators of the test and control group greater than 3 [dB] can also be assumed here. However, this is not the case for detailed medical proof of changes in individual physiological parameter values that contribute to the appearance. By recording the physiological data before a therapy starts or before the surgery trauma, the initial values can be determined individually for each patient and leveled for objective comparability. However, the clinical relevance is decisive, which must be assessed from a professional viewpoint.

Part 2

2-1*)

The purpose of the introduction of a distance measure is to separate (=discriminate against) given, slightly different groups of objects or elements as good as possible using measurable and observable features. For known setting values of the variables, elements that have not yet been classified must be determined with a high degree of accuracy regarding their affiliation. Numerous methods have been developed for this, known as proximity measures (see Michel Marie Deza and Elena Deza: "Encyclopedia of Distances", Springer 2009). Well known is the Euclidean distance measure as distance between two points in the Cartesian coordinate system. The distance concept is the assignment of an element to the group that is closest to it. Usually, squared distances are used. In addition to distances in a Cartesian system, distances can also be calculated from characteristic variables. However, different units of measurement and their dispersion values as well as possible correlations between the variables must be considered. A generalized distance measure suitable for any variable (continuous or discrete) is the Mahalanobis distance. The Mahalanobis distance of an element increases with increasing dissimilarity to a defined group if the correlation decreases or a deviating correlation pattern is present. To ensure comparability of the variables, standardization is to be carried out. The average value and the standard deviation are calculated from the elements of a defined group. With the transformation

$$y = \frac{x - \bar{x}}{\sigma}$$

is achieved that each variable has an average value of zero and a standard deviation of 1. For further details, textbooks of multivariate data analysis can be used, e.g., T. W. Anderson: "An Introduction to Multivariate Analysis", Wiley, New York (3rd Edition ISBN 0471360910, 2003).

2-2*)

All intermediate values of the correlation coefficient

$$-1 \leq r \leq +1$$

mean different compression factors, which convert the original dispersion circle into an ellipse. The correlation coefficient can also be read graphically using the ellipse rule.

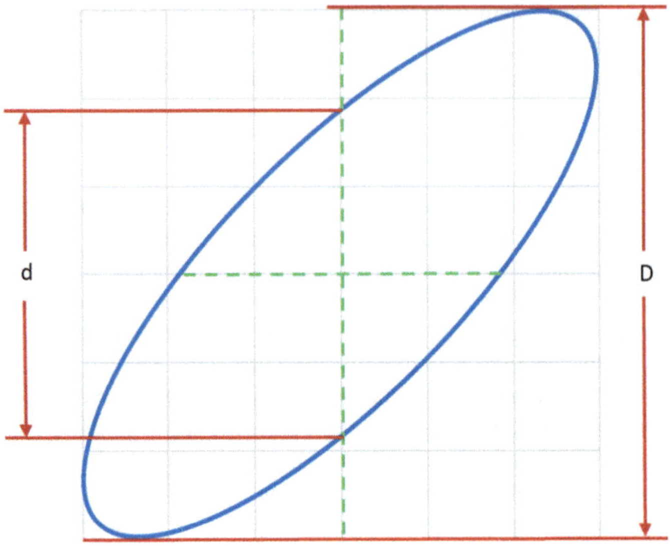

Definition of the measured quantities to derive the correlation coefficientfrom the ellipse rule

Ellipse rule: $r = \sqrt{1 - \frac{d^2}{D^2}}$

In squared form: $r^2 = 1 - \frac{d^2}{D^2}$

The geometric interpretation of the correlation coefficient r consists in the angle between two vectors. The cosine of the angle of the average deviation of standardized data vectors is the correlation coefficient. Thus, the correlation coefficient is a measure of a linear relationship. For most applications, the squared form, also known as the coefficient of determination, is preferred. Thus, the squared correlation coefficient takes only positive values between zero and one. Uncorrelation is not synonymous with lack of a certain relationship. A nonlinear relation can certainly exist.

2-3*)

The resulting system of equations can be resolved according to the Gauss-Jordan algorithm (addition of rows and extension) and the elements of the inverse matrix can be read. Another option is to calculate the determinant, whereby a matrix can always be inverted if its determinant is different from zero.

2-4*)

The features or variables, arranged as columns of numbers, can be understood as vectors in a space of corresponding dimensionality. Known for this are physical quantities like the force, which is visualized by size and direction. The geometric interpretation meets the idea. In addition, this allows a clear notation and the application of known calculation rules in connection with matrices, such as the inverse correlation matrix already introduced. Capital letters stand for vectors and matrices (X); a T as superscript stands for the transposed form, i.e., the exchange of rows and columns (X^{T}).

2-5*)

The standard deviation of the squared Mahalanobis distances of the unit group with respect to the center of the unit space (origin) results from

$$\sigma = \sqrt{\frac{1}{n}\left((D_1 - 0)^2 + (D_2 - 0)^2 + (D_3 - 0)^2 + \cdots + (D_n - 0)^2\right)}$$

$$\sigma = \sqrt{\frac{1}{n}\left(D_1^2 + D_2^2 + D_3^2 + \cdots + D_n^2\right)}$$

$$\sigma^2 = \frac{D_1^2 + D_2^2 + D_3^2 + \cdots + D_n^2}{n} = \bar{D}^2 = 1$$

2-6*)

The table displays the capital letters assigned to the clinical parameters.

Terms and units of the clinical parameters

A Hemoglobin g/dl	**B** Ck/Hematocrit (Ck = creatinine kinase) units/l	**C** Ck-MB/Hematocrit (Ck-MB = creatinine kinase-muscle-brain) µg/l	**D** Troponin/Hematocrit µg/l
E CRP/Hematocrit (CRP = C-reactive protein) mg/dl	**F** PCT/Hematocrit (PCT = procalcitonin) µg/l	**G** Interleukin6/Hematocrit ng/l	**H** TNFα/Hematocrit (TNF = tumor necrosis factor) ng/l
I TNF-Rp55/Hematocrit ng/l	**J** TNF-Rp75/Hematocrit ng/l	**K** AT III/Hematocrit (AT = angiotensin) %/%	

2-7*)

The calculation steps are made considerably easier with software support. Spreadsheet programs such as Excel already provide most of the functions.

Calculation of the correlation matrix:

In the menu: Data → Data Analysis → Correlation → Entering the fields.

The correlation coefficients are listed in the lower triangle matrix. With the functions copy, paste special, transpose the upper triangle matrix can be displayed. Summing both partial matrices results in the entire diagonal matrix, whereby the diagonal elements must be reset from 2 to 1 due to the summation.

Note: The Correlation Function is not saved, i.e., it must be called again every time if numbers are changed.

Calculation of the inverse correlation matrix:

Mark the result field to be mapped.

Call the MINVERSE function; mark the matrix to be inverted.

Execution with control + shift + enter.

Note: For all vector and matrix calculations:

- the entire area to be mapped must be marked and
- the execution always takes place with Control + Shift + Enter.

Conversion of rows, columns, fields into vectors or matrices:

Mark the result field to be mapped.

Entry in the function line: = (marked field).

Control + Shift + Enter.

The mathematical operation $D^2 = \frac{1}{k} YR^{-1}Y^T$ is performed in two steps.

Multiply vector by matrix:

Mark the result field to be mapped.

Calling the MMULT function; marking the fields to be multiplied.

Control + Shift + Enter.

Multiply vector by vector:

One can use the same function MMULT with the specifications as row vector and column vector. However, because of the column display, this does not allow simple copy and paste function line by line.

It is better to call the function SUMPRODUCT; mark both row vectors; divide by the number of variables because of the modified form of the squared Mahalanobis distance ($\bar{D}^2 = 1$).

Enter (since in this case, the result is a scalar quantity).

Foxes Team provides an Excel add-in called Matrix.xla which performs useful functions for Matrices and in Linear Algebra. The effort can be reduced enormously with extraordinarily practical function calls.

MCorr calculates the complete symmetric correlation matrix
MInv generates the inverse correlation matrix
MDet calculates the determinant of a matrix.
MT forms the transposed matrix
MProd stands for the product of any number of matrices; a single function call is sufficient for calculating the squared Mahalanobis distance: MProd [array (of data row vector); array (the inverse correlation matrix); MT (array (of data row vector))]/k
MSub Subtraction of matrices

Numerous other functions are provided in Matrix.xla. Unsuitable in the context of the algorithms used here is the function call:

MOrthoGSOrthogonalization of a matrix according to Gram–Schmidt, because the calculation is performed according to a modified procedure and the coefficients for transforming the objects not belonging to the unit group are not given.

2-8*)

For a parameter number $k < 4$, parameter screening can only be performed as an approximation. A correlation matrix exists for two or more parameters. With only one parameter the correlation with itself is equal to one, i.e., the squared Mahalanobis distance results as scalar vector product of the standardized variable.

$$D^2 = \frac{1}{k} Y \cdot 1 \cdot Y^T \quad \rightarrow \quad D^2 = Y \cdot Y^T$$

For the configurations of the orthogonal array that leave no parameters, the nonexistence of the correlation matrix is subsequently replaced by a numerical value reduced by 6 [dB] with respect to the smallest calculated *SNR* index. Thus, the orthogonal array can be evaluated for a parameter screening.

2-9*)

Interaction tables cannot be calculated for partial orthogonal arrays since the interactions are decomposed and redistributed to different degrees across several columns. Within the family of partially orthogonal arrays, there is a group in which the interaction components are distributed approximately evenly across all columns. Using such partial orthogonal arrays has the least confounding effect on each parameter column. If the greatest possible immunity to interactions is required (regardless of whether they are present or not), the following partial orthogonal arrays are recommended:

L12
L20
L28
L44
L52
L68 etc.

Note: L $2^2 \times$ prime number(≥ 3)

2-10*)

Instead of a unit group, widely used in mathematical–technical literature is the notation normal group. Consequently, nonassociated elements occur as abnormal. To ensure objectivity, the author prefers the term unit group and non-related objects or objects outside the unit group without any exception.

2-11*)

The term orthonormal base appears frequently in mathematical literature. Since any vector can be normalized to the length 1, and this does not change linear independence, any orthogonal base can be converted to an orthonormal base.

2-12*)

The identity of the result with respect to the squared Mahalanobis distance follows from the conversion into the identical Mahalanobis algorithm. The arrangement of the squared standard deviations of the orthogonalized variables as diagonals in a symmetrical diagonal matrix can be understood as a correlation matrix since all other matrix elements are equal to zero due to the orthogonality condition.

$$R = \begin{pmatrix} \sigma_1^2 & 0 & 0 & \dots & 0 \\ 0 & \sigma_2^2 & 0 & \dots & 0 \\ 0 & 0 & \sigma_3^2 & \dots & 0 \\ 0 & 0 & 0 & \dots & \sigma_k^2 \end{pmatrix}$$

Inverse matrix:

$$R^{-1} = \begin{pmatrix} \frac{1}{\sigma_1^2} & 0 & 0 & \dots & 0 \\ 0 & \frac{1}{\sigma_2^2} & 0 & \dots & 0 \\ 0 & 0 & \frac{1}{\sigma_3^2} & \dots & 0 \\ 0 & 0 & 0 & \dots & \frac{1}{\sigma_k^2} \end{pmatrix}$$

The product, consisting of the orthogonal variable vectors and the inverse correlation matrix, results in the sum of the squared components with their corresponding squared standard deviations.

$$YR^{-1}Y^T = \frac{y_1^2}{\sigma_1^2} + \frac{y_2^2}{\sigma_2^2} + \frac{y_3^2}{\sigma_3^2} + \dots + \frac{y_k^2}{\sigma_k^2}$$

The modification of the Mahalanobis algorithm divided by the number of variables, i.e., scaling the unit space to one, is identical to the squared Mahalanobis distance.

$$D_j^2 = \frac{1}{k} Y_j R^{-1} Y_j^T = \frac{1}{k}\left(\frac{y_{1j}^2}{\sigma_1^2} + \frac{y_{2j}^2}{\sigma_2^2} + \frac{y_{3j}^2}{\sigma_3^2} + \dots + \frac{y_{kj}^2}{\sigma_k^2}\right)$$

$k=$ Number of variables
$j=$ 1, 2, 3, …, n (Number of objects/individuals)

2-13*)

It cannot be excluded that two characteristics X and Y are related to a third characteristic Z, which may also be latent. If the linear influence of Z on X and Y is deducted, the correlation of the adjusted components remains, referred to as partial correlation. For the algorithms to be applied, reference is made to textbooks of multivariate statistics.

2-14*)

For a matrix of two rows and two columns each, the determinant is defined as the difference between the products of its diagonal elements.

$$A = \begin{pmatrix} a_{11} & a_{12} \\ a_{21} & a_{22} \end{pmatrix}$$

$$\mathrm{Det}A = a_{11}a_{22} - a_{12}a_{21}$$

To calculate the determinant of a 3 × 3 matrix, it is decomposed into rows or columns into the sub-determinants with alternating signs and treated in the same way.

$$A = \begin{pmatrix} a_{11} & a_{12} & a_{13} \\ a_{21} & a_{22} & a_{23} \\ a_{31} & a_{32} & a_{33} \end{pmatrix}$$

$$\mathrm{Det}A = a_{11}\begin{pmatrix} a_{22} & a_{23} \\ a_{32} & a_{33} \end{pmatrix} - a_{12}\begin{pmatrix} a_{21} & a_{23} \\ a_{31} & a_{33} \end{pmatrix} + a_{13}\begin{pmatrix} a_{21} & a_{22} \\ a_{31} & a_{32} \end{pmatrix}$$

The determinant of larger square matrices is derived from the continuous decomposition into sub-determinants. If the software Excel or similar is available, the calculation is performed with the function call MDETERM(array) in .xls or MDet(array) in .xla.

The adjoint matrix is the transpose of the cofactor matrix. Cofactors are the coefficients from the sub-determinants of the submatrices. For a square matrix of three rows and three columns each, there are nine coefficients, which are arranged in the same way as matrix elements with alternating signs.

$$A = \begin{pmatrix} a_{11} & a_{12} & a_{13} \\ a_{21} & a_{22} & a_{23} \\ a_{31} & a_{32} & a_{33} \end{pmatrix}$$

Decomposition into submatrices and cofactors

1. Cofactor is the determinant of the submatrix: $a_{11}\begin{pmatrix} a_{22} & a_{23} \\ a_{32} & a_{33} \end{pmatrix} \rightarrow$ $\mathrm{Det}\begin{pmatrix} a_{22} & a_{23} \\ a_{32} & a_{33} \end{pmatrix} = b_{11}.$

2. Cofactor is the determinant of the submatrix: $a_{12}\begin{pmatrix} a_{21} & a_{23} \\ a_{31} & a_{33} \end{pmatrix} \rightarrow$

$\mathrm{Det}\begin{pmatrix} a_{21} & a_{23} \\ a_{31} & a_{33} \end{pmatrix} = b_{12}.$

3. Cofactor is the determinant of the submatrix: $a_{13}\begin{pmatrix} a_{21} & a_{22} \\ a_{31} & a_{32} \end{pmatrix} \rightarrow$

$\mathrm{Det}\begin{pmatrix} a_{21} & a_{22} \\ a_{31} & a_{32} \end{pmatrix} = b_{13}.$

4. Cofactor is the determinant of the submatrix: $a_{21}\begin{pmatrix} a_{12} & a_{13} \\ a_{32} & a_{33} \end{pmatrix} \rightarrow$

$\mathrm{Det}\begin{pmatrix} a_{12} & a_{13} \\ a_{32} & a_{33} \end{pmatrix} = b_{21}.$

5. Cofactor is the determinant of the submatrix: $a_{22}\begin{pmatrix} a_{11} & a_{13} \\ a_{31} & a_{33} \end{pmatrix} \rightarrow$

$\mathrm{Det}\begin{pmatrix} a_{11} & a_{13} \\ a_{31} & a_{33} \end{pmatrix} = b_{22}.$

6. Cofactor is the determinant of the submatrix: $a_{23}\begin{pmatrix} a_{11} & a_{12} \\ a_{31} & a_{32} \end{pmatrix} \rightarrow$

$\mathrm{Det}\begin{pmatrix} a_{11} & a_{12} \\ a_{31} & a_{32} \end{pmatrix} = b_{23}.$

7. Cofactor is the determinant of the submatrix: $a_{31}\begin{pmatrix} a_{12} & a_{13} \\ a_{22} & a_{23} \end{pmatrix} \rightarrow$

$\mathrm{Det}\begin{pmatrix} a_{12} & a_{13} \\ a_{22} & a_{23} \end{pmatrix} = b_{31}.$

8. Cofactor is the determinant of the submatrix: $a_{32}\begin{pmatrix} a_{11} & a_{13} \\ a_{21} & a_{23} \end{pmatrix} \rightarrow$

$\mathrm{Det}\begin{pmatrix} a_{11} & a_{13} \\ a_{21} & a_{23} \end{pmatrix} = b_{32}.$

9. Cofactor is the determinant of the submatrix: $a_{33}\begin{pmatrix} a_{11} & a_{12} \\ a_{21} & a_{22} \end{pmatrix} \rightarrow$

$\mathrm{Det}\begin{pmatrix} a_{11} & a_{12} \\ a_{21} & a_{22} \end{pmatrix} = b_{33}.$

Replacement of the matrix elements by the cofactors

$$B = \begin{pmatrix} +b_{11} & -b_{12} & +b_{13} \\ -b_{21} & +b_{22} & -b_{23} \\ +b_{31} & -b_{32} & +b_{33} \end{pmatrix}$$

The adjoint matrix is the transposed cofactor matrix

$$A_{\text{adj}} = \begin{pmatrix} +b_{11} & -b_{21} & +b_{31} \\ -b_{12} & +b_{22} & -b_{32} \\ +b_{13} & -b_{23} & +b_{33} \end{pmatrix} = B^{\mathrm{T}}$$

The adjoint matrices of larger matrices are most easily calculated from the determinants of the submatrices by deleting the row and column of the element under consideration. The remaining determinant receives as sign $(-1)^{i+j}$ with $i = 1, 2, 3, \ldots, n$ (rows) and $j = 1, 2, 3, \ldots, n$ (columns).

2-15*)

The mathematical development of omitting one observation from a total amount of data is linked to the name John Wilder Tukey (1958). This comparatively simple method makes it possible to obtain very reliable estimates for statistical key figures. Its ease of use led to its frequent application and was compared in technical literature with the usefulness of a jackknife. Therefore, the term jackknife method or jackknifing has become established.

2-16*)

For $n \to k$, the correlation coefficients approximate the numerical value 1. As a result, the non-diagonal elements of the inverse correlation matrix develop toward $\pm\infty$, i.e., the squared Mahalanobis distances of the objects outside the unit group increasingly assume larger numerical values.

Part 3

3-1*)

Instead of the normalized response of the variables, their direct estimated values from the zero-point-proportional behavior can also be used. Weighted with the S/N ratio, the integral value for n signals and k variables is written as

$$\widehat{M}_i = \frac{S/N_1 \cdot \hat{\hat{M}}_{1i} + S/N_2 \cdot \hat{\hat{M}}_{2i} + S/N_3 \cdot \hat{\hat{M}}_{3i} + \cdots + S/N_k \cdot \hat{\hat{M}}_{ki}}{S/N_1 + S/N_2 + S/N_3 + \cdots + S/N_k} \qquad i = 1, 2, 3, \ldots, n$$

Just as well, the pre-factor of the normalized response with the denominator can be combined to one coefficient for each variable.

$$\widehat{M}_i = \frac{S/N_1}{\beta_1 \cdot (S/N_1 + S/N_2 + S/N_3 + \cdots + S/N_k)} x_{1i} + \frac{S/N_2}{\beta_2 \cdot (S/N_1 + S/N_2 + S/N_3 + \cdots + S/N_k)} x_{2i} + \frac{S/N_3}{\beta_3 \cdot (S/N_1 + S/N_2 + S/N_3 + \cdots + S/N_k)} x_{3i} + \cdots + \frac{S/N_k}{\beta_k \cdot (S/N_1 + S/N_2 + S/N_3 + \cdots + S/N_k)} \quad i = 1, 2, 3, \ldots, n$$

With the definition of the coefficients a_j

$$a_j = \frac{S/N_k}{\beta_k \cdot (S/N_1 + S/N_2 + S/N_3 + \cdots + S/N_k)} \quad j = 1, 2, 3, \ldots, k$$

the representation is the same as for multiple regression analyses.

$$\widehat{M}_i = a_1 y_{1i} + a_2 y_{2i} + a_3 y_{3i} + \cdots + a_k y_{ki} \quad i = 1, 2, 3, \ldots, n$$

3-2*)

Here, another look at the definition of the *SNR* index (Sect. 2.1, **1-5***, **1-9***) is worthwhile. If signal and noise are of equal size, a prognosis makes no sense, as the information disappears in the background noise. The *SNR* key figure for the case $S = N$ is

$$SNR = 0[\text{dB}]$$

If the signal can be clearly distinguished, i.e., by at least 1 order of magnitude, reliable results for predictions can be expected from experience. For the accuracy of the prognosis, this means

$$SNR \geq 10[\text{dB}]$$

This only applies to dynamic systems that consider the dimension factor S_2. *SNR* key figures of static systems (*STB*, *LTB*, *NTB*) are not invariant regarding the dimension of the response and can therefore assume any positive or negative values.

3-3*)

Instead of the *SNR* index, the squared correlation coefficient (r^2) can also be used, or comparative characteristic features appropriate to the question.

3-4*)

If the parameters interact with each other, the use of partial orthogonal arrays such as the L12, L20, L28, L44, L52, L68, etc., provides a more reliable interpretation **(2-9*)**.

3-5*)

In the 1990s, the Mahalanobis algorithm combined with G. Taguchi's data analysis was established with numerous publications. This entered the literature as the acronym MTS (Mahalanobis–Taguchi system). Modifications or alternative calculation methods for circumventing the inverse correlation matrix are called MTGS (orthogonalization approach: Mahalanobis–Taguchi–Gram–Schmidt) and MTA (approach with adjoint matrix: Mahalanobis–Taguchi adjoint). In the following years 2000–2006, G. Taguchi formulated pattern recognition and prognoses based on the algorithms developed to date. Thus, the prediction algorithm TS (Taguchi–Schmidt) and the T-method (Taguchi Method) were developed using the raw data of the variables. The application of the Mahalanobis algorithm for prognoses used the existing nomenclature. The method of variable reduction especially for pattern recognition (persons, images, objects) was given the abbreviation RT (Recognition-Taguchi). Shortly afterward, the flood of abbreviations was replaced by the categories T1, T2, and T3 introduced in Japanese literature. For a clear representation, the author allowed himself the freedom to retain as far as possible the essential and recurring definitions, but to combine similar algorithms under one umbrella term. This was inspired by the mathematical characteristics of the respective process. Apart from the abbreviations for multiple applications, the relationships are listed in the table.

Prognosis procedure	Today's categories	Historical development
Linear approach	T1	T, TS
Quadratic approach	T2	MT, MTGS, MTA
Approach with reduction of variables	T3	RT

-6*)

Known under the term "Supervised Learning" there are categories such as artificial neural network (ANN), k-nearest neighbor (kNN), decision trees, support vector machines, linear and logistic regression, and Bayesian network, which are supported by the software R or Python.

Appendix B

Orthogonal Arrays
Interaction Tables
Partial Orthogonal Arrays
Modified Orthogonal Arrays

Overview of the orthogonal arrays

Orthogonal arrays	Interaction tables	Number of levels
L4	Exists	2
L8	Exists	2 (up to 4)
L9	Exists	3
L12		2
L16	Exists	2 (up to 4 and 8)
L18		2 and 3 (up to 6)
L20		2
L25	Exists	5
L27	Exists	3 (up to 9)
L28		2
L32	Exists	2 (up to 8)
L36		2 and 3 (up to12)
L44		2
L50		2 and 5 (up to 10)
L54		2 and 3 (up to 18)
L64	Exists	
L81	Exists	3 (up to 9)
L108		3

© Springer Nature Switzerland AG 2019

H. Ruefer, *Living Without Mathematical Statistics*,
https://doi.org/10.1007/978-3-319-99632-5

Orthogonal Array L_4 (2^3)

Experiment No.	Parameter		
	1	2	3
1	1	1	1
2	1	2	2
3	2	1	2
4	2	2	1

Interaction Table L_4

Parameter	Parameter		
	1	2	3
1	(1)	3	2
2		(2)	1
3			(3)

Orthogonal Array L_8 (2^7)

Experiment No.	Parameter						
	1	2	3	4	5	6	7
1	1	1	1	1	1	1	1
2	1	1	1	2	2	2	2
3	1	2	2	1	1	2	2
4	1	2	2	2	2	1	1
5	2	1	2	1	2	1	2
6	2	1	2	2	1	2	1
7	2	2	1	1	2	2	1
8	2	2	1	2	1	1	2

Interaction Table L_8

Parameter	Parameter						
	1	2	3	4	5	6	7
1	(1)	3	2	5	4	7	6
2		(2)	1	6	7	4	5
3			(3)	7	6	5	4
4				(4)	1	2	3
5					(5)	3	2
6						(6)	1
7							(7)

Orthogonal Array L_9 (3^4)

Experiment No.	Parameter			
	1	2	3	4
1	1	1	1	1
2	1	2	2	2
3	1	3	3	3
4	2	1	2	3
5	2	2	3	1
6	2	3	1	2
7	3	1	3	2
8	3	2	1	3
9	3	3	2	1

Interaction Table L_9

Parameter	Parameter			
	1	2	3	4
1	(1)	3	2	2
		4	4	3
2		(2)	1	1
			4	3
3			(3)	1
				2
4				(4)

Partial Orthogonal Array L_{12} (2^{11})

Experiment No.	Parameter										
	1	2	3	4	5	6	7	8	9	10	11
1	1	1	1	1	1	1	1	1	1	1	1
2	1	1	1	1	1	2	2	2	2	2	2
3	1	1	2	2	2	1	1	1	2	2	2
4	1	2	1	2	2	1	2	2	1	1	2
5	1	2	2	1	2	2	1	2	1	2	1
6	1	2	2	2	1	2	2	1	2	1	1
7	2	1	2	2	1	1	2	2	1	2	1
8	2	1	2	1	2	2	2	1	1	1	2
9	2	1	1	2	2	2	1	2	2	1	1
10	2	2	2	1	1	1	1	2	2	1	2
11	2	2	1	2	1	2	1	1	1	2	2
12	2	2	1	1	2	1	2	1	2	2	1

Possible interactions are distributed across all columns.

Orthogonal Array L_{16} (2^{15})

Experiment No.	Parameter														
	1	2	3	4	5	6	7	8	9	10	11	12	13	14	15
1	1	1	1	1	1	1	1	1	1	1	1	1	1	1	1
2	1	1	1	1	1	1	1	2	2	2	2	2	2	2	2
3	1	1	1	2	2	2	2	1	1	1	1	2	2	2	2
4	1	1	1	2	2	2	2	2	2	2	2	1	1	1	1
5	1	2	2	1	1	2	2	1	1	2	2	1	1	2	2
6	1	2	2	1	1	2	2	2	2	1	1	2	2	1	1
7	1	2	2	2	2	1	1	1	1	2	2	2	2	1	1
8	1	2	2	2	2	1	1	2	2	1	1	1	1	2	2
9	2	1	2	1	2	1	2	1	2	1	2	1	2	1	2
10	2	1	2	1	2	1	2	2	1	2	1	2	1	2	1
11	2	1	2	2	1	2	1	1	2	1	2	2	1	2	1
12	2	1	2	2	1	2	1	2	1	2	1	1	2	1	2
13	2	2	1	1	2	2	1	1	2	2	1	1	2	2	1
14	2	2	1	1	2	2	1	2	1	1	2	2	1	1	2
15	2	2	1	2	1	1	2	1	2	2	1	2	1	1	2
16	2	2	1	2	1	1	2	2	1	1	2	1	2	2	1

Interaction Table L_{16}

Parameter	Parameter														
	1	2	3	4	5	6	7	8	9	10	11	12	13	14	15
1	(1)	3	2	5	4	7	6	9	8	11	10	13	12	15	14
2		(2)	1	6	7	4	5	10	11	8	9	14	15	12	13
3			(3)	7	6	5	4	11	10	9	8	15	14	13	12
4				(4)	1	2	3	12	13	14	15	8	9	10	11
5					(5)	3	2	13	12	15	14	9	8	11	10
6						(6)	1	14	15	12	13	10	11	8	9
7							(7)	15	14	13	12	11	10	9	8
8								(8)	1	2	3	4	5	6	7
9									(9)	3	2	5	4	7	6
10										(10)	1	6	7	4	5
11											(11)	7	6	5	4
12												(12)	1	2	3
13													(13)	3	2
14														(14)	1
15															(15)

Modified Orthogonal Array L_{16} $(4^1 \times 2^{12})$

Experiment No.	Parameter												
	1	2	3	4	5	6	7	8	9	10	11	12	13
1	1	1	1	1	1	1	1	1	1	1	1	1	1
2	1	1	1	1	1	2	2	2	2	2	2	2	2
3	1	2	2	2	2	1	1	1	1	2	2	2	2
4	1	2	2	2	2	2	2	2	2	1	1	1	1
5	2	1	1	2	2	1	1	2	2	1	1	2	2
6	2	1	1	2	2	2	2	1	1	2	2	1	1
7	2	2	2	1	1	1	1	2	2	2	2	1	1
8	2	2	2	1	1	2	2	1	1	1	1	2	2
9	3	1	2	1	2	1	2	1	2	1	2	1	2
10	3	1	2	1	2	2	1	2	1	2	1	2	1
11	3	2	1	2	1	1	2	1	2	2	1	2	1
12	3	2	1	2	1	2	1	2	1	1	2	1	2
13	4	1	2	2	1	1	2	2	1	1	2	2	1
14	4	1	2	2	1	2	1	1	2	2	1	1	2
15	4	2	1	1	2	1	2	2	1	2	1	1	2
16	4	2	1	1	2	2	1	1	2	1	2	2	1

Modified Orthogonal Array L_{16} $(4^2 \times 2^9)$

Experiment No.	Parameter										
	1	2	3	4	5	6	7	8	9	10	11
1	1	1	1	1	1	1	1	1	1	1	1
2	1	2	1	1	1	2	2	2	2	2	2
3	1	3	2	2	2	1	1	1	2	2	2
4	1	4	2	2	2	2	2	2	1	1	1
5	2	1	1	2	2	1	2	2	1	2	2
6	2	2	1	2	2	2	1	1	2	1	1
7	2	3	2	1	1	1	2	2	2	1	1
8	2	4	2	1	1	2	1	1	1	2	2
9	3	1	2	1	2	2	1	2	2	1	2
10	3	2	2	1	2	1	2	1	1	2	1
11	3	3	1	2	1	2	1	2	1	2	1
12	3	4	1	2	1	1	2	1	2	1	2
13	4	1	2	2	1	2	2	1	2	2	1
14	4	2	2	2	1	1	1	2	1	1	2
15	4	3	1	1	2	2	2	1	1	1	2
16	4	4	1	1	2	1	1	2	2	2	1

Modified Orthogonal Array L_{16} ($4^3 \times 2^6$)

Experiment No.	Parameter								
	1	2	3	4	5	6	7	8	9
1	1	1	1	1	1	1	1	1	1
2	1	2	2	1	1	2	2	2	2
3	1	3	3	2	2	1	1	2	2
4	1	4	4	2	2	2	2	1	1
5	2	1	2	2	2	1	2	1	2
6	2	2	1	2	2	2	1	2	1
7	2	3	4	1	1	1	2	2	1
8	2	4	3	1	1	2	1	1	2
9	3	1	3	1	2	2	2	2	1
10	3	2	4	1	2	1	1	1	2
11	3	3	1	2	1	2	2	1	2
12	3	4	2	2	1	1	1	2	1
13	4	1	4	2	1	2	1	2	2
14	4	2	3	2	1	1	2	1	1
15	4	3	2	1	2	2	1	1	1
16	4	4	1	1	2	1	2	2	2

Modified Orthogonal Array L_{16} ($4^4 \times 2^3$)

Experiment No.	Parameter						
	1	2	3	4	5	6	7
1	1	1	1	1	1	1	1
2	1	2	2	1	2	2	2
3	1	3	3	2	3	1	2
4	1	4	4	2	4	2	1
5	2	1	2	2	3	2	1
6	2	2	1	2	4	1	2
7	2	3	4	1	1	2	2
8	2	4	3	1	2	1	1
9	3	1	3	1	4	2	2
10	3	2	4	1	3	1	1
11	3	3	1	2	2	2	1
12	3	4	2	2	1	1	2
13	4	1	4	2	2	1	2
14	4	2	3	2	1	2	1
15	4	3	2	1	4	1	1
16	4	4	1	1	3	2	2

Modified Orthogonal Array L_{16} (8×2^8)

Experiment No.	Parameter								
	1	2	3	4	5	6	7	8	9
1	1	1	1	1	1	1	1	1	1
2	2	1	1	1	1	2	2	2	2
3	3	1	1	2	2	1	1	2	2
4	4	1	1	2	2	2	2	1	1
5	1	2	2	2	2	2	2	2	2
6	2	2	2	2	2	1	1	1	1
7	3	2	2	1	1	2	2	1	1
8	4	2	2	1	1	1	1	2	2
9	5	1	2	1	2	1	2	1	2
10	6	1	2	1	2	2	1	2	1
11	7	1	2	2	1	1	2	2	1
12	8	1	2	2	1	2	1	1	2
13	5	2	1	2	1	2	1	2	1
14	6	2	1	2	1	1	2	1	2
15	7	2	1	1	2	2	1	1	2
16	8	2	1	1	2	1	2	2	1

Orthogonal Array L_{16} (4^5)

Experiment No.	Parameter				
	1	2	3	4	5
1	1	1	1	1	1
2	1	2	2	2	2
3	1	3	3	3	3
4	1	4	4	4	4
5	2	1	2	3	4
6	2	2	1	4	3
7	2	3	4	1	2
8	2	4	3	2	1
9	3	1	3	4	2
10	3	2	4	3	1
11	3	3	1	2	4
12	3	4	2	1	3
13	4	1	4	2	3
14	4	2	3	1	4
15	4	3	2	4	1
16	4	4	1	3	2

Interaction Table L_{16}

Parameter	Parameter 1	2	3	4	5
1		3	2	2	2
	(1)	4	4	3	3
		5	5	5	4
2			1	1	1
		(2)	4	3	3
			5	5	4
3				1	1
			(3)	2	2
				5	4
4					1
				(4)	2
					3
5					(5)

Partial Orthogonal Array L_{18} ($2^1 \times 3^7$)

Experiment No.	Parameter							
	1	2	3	4	5	6	7	8
1	1	1	1	1	1	1	1	1
2	1	1	2	2	2	2	2	2
3	1	1	3	3	3	3	3	3
4	1	2	1	1	2	2	3	3
5	1	2	2	2	3	3	1	1
6	1	2	3	3	1	1	2	2
7	1	3	1	2	1	3	2	3
8	1	3	2	3	2	1	3	1
9	1	3	3	1	3	2	1	2
10	2	1	1	3	3	2	2	1
11	2	1	2	1	1	3	3	2
12	2	1	3	2	2	1	1	3
13	2	2	1	2	3	1	3	2
14	2	2	2	3	1	2	1	3
15	2	2	3	1	2	3	2	1
16	2	3	1	3	2	3	1	2
17	2	3	2	1	3	1	2	3
18	2	3	3	2	1	2	3	1

An interaction between parameters 1 and 2 is orthogonal to all others and can be viewed independently without abandoning other parameters. Possible interactions of parameters 3–8 are distributed over columns 3–8.

Partial Orthogonal Array L_{20} (2^{19})

Experiment No.	Parameter																		
	1	2	3	4	5	6	7	8	9	10	11	12	13	14	15	16	17	18	19
1	1	1	1	1	1	1	1	1	1	1	1	1	1	1	1	1	1	1	1
2	1	1	1	1	2	1	2	1	2	2	2	2	1	1	2	2	1	2	2
3	1	1	1	2	1	2	1	2	2	2	2	1	1	2	2	1	2	2	1
4	1	1	2	1	2	1	2	2	2	2	1	1	2	2	1	2	2	1	1
5	1	1	2	2	1	2	2	1	1	1	1	2	1	2	1	2	2	2	2
6	1	2	1	2	1	2	2	2	2	1	1	2	2	1	2	2	1	1	1
7	1	2	1	2	2	2	2	1	1	2	2	1	2	2	1	1	1	1	2
8	1	2	2	1	1	1	1	2	1	2	1	2	2	2	2	1	1	2	2
9	1	2	2	1	2	2	1	1	1	1	2	1	2	1	2	2	2	2	1
10	1	2	2	2	2	1	1	2	2	1	2	2	1	1	1	1	2	1	2
11	2	1	1	1	1	2	1	2	1	2	2	2	2	1	1	2	2	1	2
12	2	1	1	2	2	1	2	2	1	1	1	1	2	1	2	1	2	2	2
13	2	1	2	1	2	2	2	2	1	1	2	2	1	2	2	1	1	1	1
14	2	1	2	2	1	1	1	1	2	1	2	1	2	2	2	2	1	1	2
15	2	1	2	2	2	2	1	1	2	2	1	2	2	1	1	1	1	2	1
16	2	2	1	1	1	1	2	1	2	1	2	2	2	2	1	1	2	2	1
17	2	2	1	1	2	2	1	2	2	1	1	1	1	2	1	2	1	2	2
18	2	2	1	2	2	1	1	1	1	2	1	2	1	2	2	2	2	1	1
19	2	2	2	1	1	2	2	1	2	2	1	1	1	1	2	1	2	1	2
20	2	2	2	2	1	1	2	2	1	2	2	1	1	1	1	2	1	2	1

Orthogonal Array L_{27} (3^{13})

Experiment No.	Parameter												
	1	2	3	4	5	6	7	8	9	10	11	12	13
1	1	1	1	1	1	1	1	1	1	1	1	1	1
2	1	1	1	1	2	2	2	2	2	2	2	2	2
3	1	1	1	1	3	3	3	3	3	3	3	3	3
4	1	2	2	2	1	1	1	2	2	2	3	3	3
5	1	2	2	2	2	2	2	3	3	3	1	1	1
6	1	2	2	2	3	3	3	1	1	1	2	2	2
7	1	3	3	3	1	1	1	3	3	3	2	2	2

(continued)

(continued)

Experiment No.	Parameter												
	1	2	3	4	5	6	7	8	9	10	11	12	13
8	1	3	3	3	2	2	2	1	1	1	3	3	3
9	1	3	3	3	3	3	3	2	2	2	1	1	1
10	2	1	2	3	1	2	3	1	2	3	1	2	3
11	2	1	2	3	2	3	1	2	3	1	2	3	1
12	2	1	2	3	3	1	2	3	1	2	3	1	2
13	2	2	3	1	1	2	3	2	3	1	3	1	2
14	2	2	3	1	2	3	1	3	1	2	1	2	3
15	2	2	3	1	3	1	2	1	2	3	2	3	1
16	2	3	1	2	1	2	3	3	1	2	2	3	1
17	2	3	1	2	2	3	1	1	2	3	3	1	2
18	2	3	1	2	3	1	2	2	3	1	1	2	3
19	3	1	3	2	1	3	2	1	3	2	1	3	2
20	3	1	3	2	2	1	3	2	1	3	2	1	3
21	3	1	3	2	3	2	1	3	2	1	3	2	1
22	3	2	1	3	1	3	2	2	1	3	3	2	1
23	3	2	1	3	2	1	3	3	2	1	1	3	2
24	3	2	1	3	3	2	1	1	3	2	2	1	3
25	3	3	2	1	1	3	2	3	2	1	2	1	3
26	3	3	2	1	2	1	3	1	3	2	3	2	1
27	3	3	2	1	3	2	1	2	1	3	1	3	2

Interaction Table L_{27}

Parameter	1	2	3	4	5	6	7	8	9	10	11	12	13
1	(1)	3 4	2 4	2 3	6 7	5 7	5 6	9 10	8 10	8 9	12 13	11 13	11 12
2		(2)	1 4	1 3	8 11	9 12	10 13	5 11	6 12	7 13	5 8	6 9	7 10
3			(3)	1 2	9 13	10 11	8 12	7 12	5 13	6 11	6 10	7 8	5 9
4				(4)	10 12	9 13	9 11	6 13	7 11	5 12	7 9	5 10	6 8
5					(5)	1 7	1 6	2 11	3 13	4 12	2 8	4 10	3 9
6						(6)	1 5	4 13	2 12	3 11	3 10	2 9	4 8
7							(7)	3 12	4 11	2 13	4 9	3 8	2 10
8								(8)	1 10	1 9	2 5	3 7	4 6
9									(9)	1 8	4 7	2 6	3 5
10										(10)	3 6	4 5	2 7
11											(11)	1 13	1 12
12												(12)	1 11
13													(13)

Partial Orthogonal Array L_{28} (2^{27})

Experiment No.	Parameter																	
	1	2	3	4	5	6	7	8	9	10	11	12	13	14	15	16	17	18
1	1	1	1	1	1	1	1	1	1	1	1	1	1	1	1	1	1	1
2	1	1	1	1	2	2	1	1	2	2	2	2	2	2	1	1	2	1
3	1	1	1	2	2	2	2	2	2	1	1	1	2	1	2	2	2	2
4	1	1	2	1	1	2	2	1	1	2	1	2	2	1	2	2	1	1
5	1	1	2	1	2	1	1	2	1	2	2	1	1	1	2	1	2	2
6	1	1	2	2	1	1	1	2	2	1	1	2	2	2	1	1	2	2
7	1	1	2	2	1	2	2	1	2	1	2	1	1	2	1	2	1	2
8	1	2	1	1	2	1	2	1	2	1	1	2	1	2	2	1	2	2
9	1	2	1	2	1	1	1	1	1	2	2	1	2	1	2	2	2	2
10	1	2	1	2	1	2	1	2	1	2	1	1	1	2	1	2	2	1
11	1	2	1	2	2	1	2	1	1	1	2	2	2	2	2	2	1	1
12	1	2	2	1	1	2	1	2	2	1	2	2	2	1	2	1	1	1
13	1	2	2	1	2	1	2	2	2	2	1	1	1	2	1	2	1	1
14	1	2	2	2	2	2	2	2	1	2	2	2	1	1	1	1	1	2
15	2	1	1	1	1	2	2	1	2	2	2	2	1	1	1	2	2	2
16	2	1	1	1	2	2	2	2	1	2	1	1	2	2	2	1	1	2
17	2	1	1	2	1	1	2	2	2	1	2	1	1	1	2	1	1	1
18	2	1	1	2	2	1	1	2	2	2	1	2	2	1	1	2	1	1
19	2	1	2	1	2	1	1	1	1	1	2	1	2	2	1	2	1	2
20	2	1	2	2	1	1	2	1	1	2	1	2	1	2	2	1	2	1
21	2	1	2	2	2	2	1	2	1	1	2	2	1	2	2	2	2	1
22	2	2	1	1	1	1	1	2	2	2	2	2	1	2	2	2	1	2
23	2	2	1	1	1	2	2	2	1	1	2	1	2	2	1	1	2	1

(continued)

(continued)

Experiment No.	Parameter																	
	1	2	3	4	5	6	7	8	9	10	11	12	13	14	15	16	17	18
24	2	2	1	2	2	2	1	1	1	1	1	2	1	1	1	1	1	2
25	2	2	2	1	1	1	2	2	1	1	1	2	2	1	1	2	2	2
26	2	2	2	1	2	2	1	1	2	1	1	1	1	1	2	2	2	1
27	2	2	2	2	1	2	1	1	2	2	1	1	2	2	2	1	1	2
28	2	2	2	2	2	1	2	1	2	2	2	1	2	1	1	1	2	1

Partial Orthogonal Array L_{28} (2^{27}), continued

Experiment No.	Parameter								
	19	20	21	22	23	24	25	26	27
1	1	1	1	1	1	1	1	1	1
2	2	2	1	1	2	1	2	2	1
3	2	1	1	1	2	2	1	1	1
4	1	2	1	2	2	2	1	2	2
5	2	2	2	1	1	2	1	2	2
6	1	1	2	2	2	1	1	2	2
7	1	2	2	1	1	2	2	2	1
8	1	2	1	2	1	2	2	1	2
9	1	2	2	2	2	1	2	1	1
10	2	1	1	2	1	2	2	2	2
11	2	1	2	1	1	1	1	2	2
12	2	1	2	2	1	2	2	1	1
13	2	2	2	2	2	1	1	1	1
14	1	1	1	1	2	1	2	1	2
15	2	1	2	2	1	1	1	1	2
16	1	1	2	2	1	1	2	2	1
17	2	2	1	2	2	1	2	2	2
18	1	2	2	1	1	2	2	1	2
19	2	1	1	2	2	2	2	1	2
20	2	1	2	1	2	2	2	1	1
21	1	2	1	2	1	1	1	1	1
22	1	1	1	1	2	2	1	2	1
23	1	2	2	1	2	2	1	1	2
24	2	2	2	2	2	2	1	2	1
25	2	2	1	1	1	1	2	2	1
26	1	1	2	1	2	1	2	2	2
27	2	2	1	1	1	1	1	1	2
28	1	1	1	2	1	2	1	2	1

Orthogonal Array L_{32} (2^{31})

Experiment No.	Parameter																	
	1	2	3	4	5	6	7	8	9	10	11	12	13	14	15	16	17	18
1	1	1	1	1	1	1	1	1	1	1	1	1	1	1	1	1	1	1
2	1	1	1	1	1	1	1	1	1	1	1	1	1	1	1	2	2	2
3	1	1	1	1	1	1	1	2	2	2	2	2	2	2	2	1	1	1
4	1	1	1	1	1	1	1	2	2	2	2	2	2	2	2	2	2	2
5	1	1	1	2	2	2	2	1	1	1	1	2	2	2	2	1	1	1
6	1	1	1	2	2	2	2	1	1	1	1	2	2	2	2	2	2	2
7	1	1	1	2	2	2	2	2	2	2	2	1	1	1	1	1	1	1
8	1	1	1	2	2	2	2	2	2	2	2	1	1	1	1	2	2	2
9	1	2	2	1	1	2	2	1	1	2	2	1	1	2	2	1	1	2
10	1	2	2	1	1	2	2	1	1	2	2	1	1	2	2	2	2	1
11	1	2	2	1	1	2	2	2	2	1	1	2	2	1	1	1	1	2
12	1	2	2	1	1	2	2	2	2	1	1	2	2	1	1	2	2	1
13	1	2	2	2	2	1	1	1	1	2	2	2	2	1	1	1	1	2
14	1	2	2	2	2	1	1	1	1	2	2	2	2	1	1	2	2	1
15	1	2	2	2	2	1	1	2	2	1	1	1	1	2	2	1	1	2
16	1	2	2	2	2	1	1	2	2	1	1	1	1	2	2	2	2	1
17	2	1	2	1	2	1	2	1	2	1	2	1	2	1	2	1	2	1
18	2	1	2	1	2	1	2	1	2	1	2	1	2	1	2	2	1	2
19	2	1	2	1	2	1	2	2	1	2	1	2	1	2	1	1	2	1
20	2	1	2	1	2	1	2	2	1	2	1	2	1	2	1	2	1	2
21	2	1	2	2	1	2	1	1	2	1	2	2	1	2	1	1	2	1
22	2	1	2	2	1	2	1	1	2	1	2	2	1	2	1	2	1	2
23	2	1	2	2	1	2	1	2	1	2	1	1	2	1	2	1	2	1

(continued)

(continued)

Experiment No.	Parameter																	
	1	2	3	4	5	6	7	8	9	10	11	12	13	14	15	16	17	18
24	2	1	2	2	1	2	1	2	1	2	1	1	2	1	2	2	1	2
25	2	2	1	1	2	2	1	1	2	2	1	1	2	2	1	1	2	2
26	2	2	1	1	2	2	1	1	2	2	1	1	2	2	1	2	1	1
27	2	2	1	1	2	2	1	2	1	1	2	2	1	1	2	1	2	2
28	2	2	1	1	2	2	1	2	1	1	2	2	1	1	2	2	1	1
29	2	2	1	2	1	1	2	1	2	2	1	2	1	1	2	1	2	2
30	2	2	1	2	1	1	2	1	2	2	1	2	1	1	2	2	1	1
31	2	2	1	2	1	1	2	2	1	1	2	1	2	2	1	1	2	2
32	2	2	1	2	1	1	2	2	1	1	2	1	2	2	1	2	1	1

Orthogonal Array L_{32} (2^{31}), continued

Experiment No.	Parameter												
	19	20	21	22	23	24	25	26	27	28	29	30	31
1	1	1	1	1	1	1	1	1	1	1	1	1	1
2	2	2	2	2	2	2	2	2	2	2	2	2	2
3	1	1	1	1	1	2	2	2	2	2	2	2	2
4	2	2	2	2	2	1	1	1	1	1	1	1	1
5	1	2	2	2	2	1	1	1	1	2	2	2	2
6	2	1	1	1	1	2	2	2	2	1	1	1	1
7	1	2	2	2	2	2	2	2	2	1	1	1	1
8	2	1	1	1	1	1	1	1	1	2	2	2	2
9	2	1	1	2	2	1	1	2	2	1	1	2	2
10	1	2	2	1	1	2	2	1	1	2	2	1	1
11	2	1	1	2	2	2	2	1	1	2	2	1	1
12	1	2	2	1	1	1	1	2	2	1	1	2	2
13	2	2	2	1	1	1	1	2	2	2	2	1	1
14	1	1	1	2	2	2	2	1	1	1	1	2	2
15	2	2	2	1	1	2	2	1	1	1	1	2	2
16	1	1	1	2	2	1	1	2	2	2	2	1	1
17	2	1	2	1	2	1	2	1	2	1	2	1	2
18	1	2	1	2	1	2	1	2	1	2	1	2	1
19	2	1	2	1	2	2	1	2	1	2	1	2	1
20	1	2	1	2	1	1	2	1	2	1	2	1	2
21	2	2	1	2	1	1	2	1	2	2	1	2	1
22	1	1	2	1	2	2	1	2	1	1	2	1	2
23	2	2	1	2	1	2	1	2	1	1	2	1	2

(continued)

(continued)

Experiment No.	Parameter												
	19	20	21	22	23	24	25	26	27	28	29	30	31
24	1	1	2	1	2	1	2	1	2	2	1	2	1
25	1	1	2	2	1	1	2	2	1	1	2	2	1
26	2	2	1	1	2	2	1	1	2	2	1	1	2
27	1	1	2	2	1	2	1	1	2	2	1	1	2
28	2	2	1	1	2	1	2	2	1	1	2	2	1
29	1	2	1	1	2	1	2	2	1	2	1	1	2
30	2	1	2	2	1	2	1	1	2	1	2	2	1
31	1	2	1	1	2	2	1	1	2	1	2	2	1
32	2	1	2	2	1	1	2	2	1	2	1	1	2

Interaction Table L_{32}

Parameter	Parameter 1	2	3	4	5	6	7	8	9	10	11	12	13	14	15	16	17	18
1	(1)	3	2	5	4	7	6	9	8	11	10	13	12	15	14	17	16	19
2		(2)	1	6	7	4	5	10	11	8	9	14	15	12	13	18	19	16
3			(3)	7	6	5	4	11	10	9	8	15	14	13	12	19	18	17
4				(4)	1	2	3	12	13	14	15	8	9	10	11	20	21	22
5					(5)	3	2	13	12	15	14	9	8	11	10	21	20	23
6						(6)	1	14	15	12	13	10	11	8	9	22	23	20
7							(7)	15	14	13	12	11	10	9	8	23	22	21
8								(8)	1	2	3	4	5	6	7	24	25	26
9									(9)	3	2	5	4	7	6	25	24	27
10										(10)	1	6	7	4	5	26	27	24
11											(11)	7	6	5	4	27	26	25
12												(12)	1	2	3	28	29	30
13													(13)	3	2	29	28	31
14														(14)	1	30	31	28
15															(15)	31	30	29
16																(16)	1	2
17																	(17)	3
18																		(18)

Interaction Table L_{32}, continued

Parameter	Parameter												
	19	20	21	22	23	24	25	26	27	28	29	30	31
1	18	21	20	23	22	25	24	27	26	29	28	31	30
2	17	22	23	20	21	26	27	24	25	30	31	28	29
3	16	23	22	21	20	27	26	25	24	31	30	29	28
4	23	16	17	18	19	28	29	30	31	24	25	26	27
5	22	17	16	19	18	29	28	31	30	25	24	27	26
6	21	18	19	16	17	30	31	28	29	26	27	24	25
7	20	19	18	17	16	31	30	29	28	27	26	25	24
8	27	28	29	30	31	16	17	18	19	20	21	22	23
9	26	29	28	31	30	17	16	19	18	21	20	23	22
10	25	30	31	28	29	18	19	16	17	22	23	20	21
11	24	31	30	29	28	19	18	17	16	23	22	21	20
12	31	24	25	26	27	20	21	22	23	16	17	18	19
13	30	25	24	27	26	21	20	23	22	17	16	19	18
14	29	26	27	24	25	22	23	20	21	18	19	16	17
15	28	27	26	25	24	23	22	21	20	19	18	17	16
16	3	4	5	6	7	8	9	10	11	12	13	14	15
17	2	5	4	7	6	9	8	11	10	13	12	15	14
18	1	6	7	4	5	10	11	8	9	14	15	12	13
19	(19)	7	6	5	4	11	10	9	8	15	14	13	12
20		(20)	1	2	3	12	13	14	15	8	9	10	11
21			(21)	3	2	13	12	15	14	9	8	11	10
22				(22)	1	14	15	12	13	10	11	8	9
23					(23)	15	14	13	12	11	10	9	8
24						(24)	1	2	3	4	5	6	7
25							(25)	3	2	5	4	7	6
26								(26)	1	6	7	4	5
27									(27)	7	6	5	4
28										(28)	1	2	3
29											(29)	3	2
30												(30)	1
31													(31)

Modified Orthogonal Array L_{32} $(2^1 \times 4^9)$

Experiment No.	Parameter									
	1	2	3	4	5	6	7	8	9	10
1	1	1	1	1	1	1	1	1	1	1
2	1	1	2	2	2	2	2	2	2	2
3	1	1	3	3	3	3	3	3	3	3
4	1	1	4	4	4	4	4	4	4	4
5	1	2	1	1	2	2	3	3	4	4
6	1	2	2	2	1	1	4	4	3	3
7	1	2	3	3	4	4	1	1	2	2
8	1	2	4	4	3	3	2	2	1	1
9	1	3	1	2	3	4	1	2	3	4
10	1	3	2	1	4	3	2	1	4	3
11	1	3	3	4	1	2	3	4	1	2
12	1	3	4	3	2	1	4	3	2	1
13	1	4	1	2	4	3	3	4	2	1
14	1	4	2	1	3	4	4	3	1	2
15	1	4	3	4	2	1	1	2	4	3
16	1	4	4	3	1	2	2	1	3	4
17	2	1	1	4	1	4	2	3	2	3
18	2	1	2	3	2	3	1	4	1	4
19	2	1	3	2	3	2	4	1	4	1
20	2	1	4	1	4	1	3	2	3	2
21	2	2	1	4	2	3	4	1	3	2
22	2	2	2	3	1	4	3	2	4	1
23	2	2	3	2	4	1	2	3	1	4
24	2	2	4	1	3	2	1	4	2	3
25	2	3	1	3	3	1	2	4	4	2
26	2	3	2	4	4	2	1	3	3	1
27	2	3	3	1	1	3	4	2	2	4
28	2	3	4	2	2	4	3	1	1	3
29	2	4	1	3	4	2	4	2	1	3
30	2	4	2	4	3	1	3	1	2	4
31	2	4	3	1	2	4	2	4	3	1
32	2	4	4	2	1	3	1	3	4	2

An interaction between parameters 1 and 2 is orthogonal to all others and can be viewed independently without abandoning other parameters. Columns 1 and 2 can be combined into one parameter with eight levels. Possible interactions of parameters 3–10 are distributed over columns 3–10.

Partial Orthogonal Array L_{36} ($2^{11} \times 3^{12}$)

Experiment No.	Parameter																						
	1	2	3	4	5	6	7	8	9	10	11	12	13	14	15	16	17	18	19	20	21	22	23
1	1	1	1	1	1	1	1	1	1	1	1	1	1	1	1	1	1	1	1	1	1	1	1
2	1	1	1	1	1	1	1	1	1	1	1	2	2	2	2	2	2	2	2	2	2	2	2
3	1	1	1	1	1	1	1	1	1	1	1	3	3	3	3	3	3	3	3	3	3	3	3
4	1	1	1	1	1	2	2	2	2	2	2	1	1	1	1	2	2	2	2	3	3	3	3
5	1	1	1	1	1	2	2	2	2	2	2	2	2	2	2	3	3	3	3	1	1	1	1
6	1	1	1	1	1	2	2	2	2	2	2	3	3	3	3	1	1	1	1	2	2	2	2
7	1	1	2	2	2	1	1	1	2	2	2	1	1	2	3	1	2	3	3	1	2	2	3
8	1	1	2	2	2	1	1	1	2	2	2	2	2	3	1	2	3	1	1	2	3	3	1
9	1	1	2	2	2	1	1	1	2	2	2	3	3	1	2	3	1	2	2	3	1	1	2
10	1	2	1	2	2	1	2	2	1	1	2	1	1	3	2	1	3	2	3	2	1	3	2
11	1	2	1	2	2	1	2	2	1	1	2	2	2	1	3	2	1	3	1	3	2	1	3
12	1	2	1	2	2	1	2	2	1	1	2	3	3	2	1	3	2	1	2	1	3	2	1
13	1	2	2	1	2	2	1	2	1	2	1	1	2	3	1	3	2	1	3	3	2	1	2
14	1	2	2	1	2	2	1	2	1	2	1	2	3	1	2	1	3	2	1	1	3	2	3
15	1	2	2	1	2	2	1	2	1	2	1	3	1	2	3	2	1	3	2	2	1	3	1
16	1	2	2	2	1	2	2	1	2	1	1	1	2	3	2	1	1	3	2	3	3	2	1
17	1	2	2	2	1	2	2	1	2	1	1	2	3	1	3	2	2	1	3	1	1	3	2
18	1	2	2	2	1	2	2	1	2	1	1	3	1	2	1	3	3	2	1	2	2	1	3
19	2	1	2	2	1	1	2	2	1	2	1	1	2	1	3	3	3	1	2	2	1	2	3
20	2	1	2	2	1	1	2	2	1	2	1	2	3	2	1	1	1	2	3	3	2	3	1
21	2	1	2	2	1	1	2	2	1	2	1	3	1	3	2	2	2	3	1	1	3	1	2
22	2	1	2	1	2	2	2	1	1	1	2	1	2	2	3	3	1	2	1	1	3	3	2
23	2	1	2	1	2	2	2	1	1	1	2	2	3	3	1	1	2	3	2	2	1	1	3

(continued)

(continued)

Experiment No.	Parameter																						
	1	2	3	4	5	6	7	8	9	10	11	12	13	14	15	16	17	18	19	20	21	22	23
24	2	1	2	1	2	2	2	1	1	1	2	3	1	1	2	2	3	1	3	3	2	2	1
25	2	1	1	2	2	2	1	2	2	1	1	1	3	2	1	2	3	3	1	3	1	2	2
26	2	1	1	2	2	2	1	2	2	1	1	2	1	3	2	3	1	1	2	1	2	3	3
27	2	1	1	2	2	2	1	2	2	1	1	3	2	1	3	1	2	2	3	2	3	1	1
28	2	2	2	1	1	1	1	2	2	1	2	1	3	2	2	2	1	1	3	2	3	1	3
29	2	2	2	1	1	1	1	2	2	1	2	2	1	3	3	3	2	2	1	3	1	2	1
30	2	2	2	1	1	1	1	2	2	1	2	3	2	1	1	1	3	3	2	1	2	3	2
31	2	2	1	2	1	2	1	1	1	2	2	1	3	3	3	2	3	2	2	1	2	1	1
32	2	2	1	2	1	2	1	1	1	2	2	2	1	1	1	3	1	3	3	2	3	2	2
33	2	2	1	2	1	2	1	1	1	2	2	3	2	2	2	1	2	1	1	3	1	3	3
34	2	2	1	1	2	1	2	1	2	2	1	1	3	1	2	3	2	3	1	2	2	3	1
35	2	2	1	1	2	1	2	1	2	2	1	2	1	2	3	1	3	1	2	3	3	1	2
36	2	2	1	1	2	1	2	1	2	2	1	3	2	3	1	2	1	2	3	1	1	2	3

Possible interactions between any columns are distributed by all columns.

Partial Orthogonal Array L_{36} ($2^3 \times 3^{13}$)

Expt. No.	Parameter															
	1	2	3	4	5	6	7	8	9	10	11	12	13	14	15	16
1	1	1	1	1	1	1	1	1	1	1	1	1	1	1	1	1
2	1	1	1	1	2	2	2	2	2	2	2	2	2	2	2	2
3	1	1	1	1	3	3	3	3	3	3	3	3	3	3	3	3
4	1	2	2	1	1	1	1	1	2	2	2	2	3	3	3	3
5	1	2	2	1	2	2	2	2	3	3	3	3	1	1	1	1
6	1	2	2	1	3	3	3	3	1	1	1	1	2	2	2	2
7	2	1	2	1	1	1	2	3	1	2	3	3	1	2	2	3
8	2	1	2	1	2	2	3	1	2	3	1	1	2	3	3	1
9	2	1	2	1	3	3	1	2	3	1	2	2	3	1	1	2
10	2	2	1	1	1	1	3	2	1	3	2	3	2	1	3	2
11	2	2	1	1	2	2	1	3	2	1	3	1	3	2	1	3
12	2	2	1	1	3	3	2	1	3	2	1	2	1	3	2	1
13	1	1	1	2	1	2	3	1	3	2	1	3	3	2	1	2
14	1	1	1	2	2	3	1	2	1	3	2	1	1	3	2	3
15	1	1	1	2	3	1	2	3	2	1	3	2	2	1	3	1
16	1	2	2	2	1	2	3	2	1	1	3	2	3	3	2	1
17	1	2	2	2	2	3	1	3	2	2	1	3	1	1	3	2
18	1	2	2	2	3	1	2	1	3	3	2	1	2	2	1	3
19	2	1	2	2	1	2	1	3	3	3	1	2	2	1	2	3
20	2	1	2	2	2	3	2	1	1	1	2	3	3	2	3	1
21	2	1	2	2	3	1	3	2	2	2	3	1	1	3	1	2

(continued)

(continued)

Expt. No.	Parameter															
	1	2	3	4	5	6	7	8	9	10	11	12	13	14	15	16
22	2	2	1	2	1	2	2	3	3	1	2	1	1	3	3	2
23	2	2	1	2	2	3	3	1	1	2	3	2	2	1	1	3
24	2	2	1	2	3	1	1	2	2	3	1	3	3	2	2	1
25	1	1	1	3	1	3	2	1	2	3	3	1	3	1	2	2
26	1	1	1	3	2	1	3	2	3	1	1	2	1	2	3	3
27	1	1	1	3	3	2	1	3	1	2	2	3	2	3	1	1
28	1	2	2	3	1	3	2	2	2	1	1	3	2	3	1	3
29	1	2	2	3	2	1	3	3	3	2	2	1	3	1	2	1
30	1	2	2	3	3	2	1	1	1	3	3	2	1	2	3	2
31	2	1	2	3	1	3	3	3	2	3	2	2	1	2	1	1
32	2	1	2	3	2	1	1	1	3	1	3	3	2	3	2	2
33	2	1	2	3	3	2	2	2	1	2	1	1	3	1	3	3
34	2	2	1	3	1	3	1	2	3	2	3	1	2	2	2	3
35	2	2	1	3	2	1	2	3	1	3	1	2	3	3	1	2
36	2	2	1	3	3	2	3	1	2	1	2	3	1	1	2	3

Interactions between parameters 1×4, 2×4, and 3×4 are orthogonal to all others and can be viewed independently without abandoning other parameters. A triple interaction of the parameters $1 \times 2 \times 4$ can be determined by not assigning column 3. Columns 1, 2, and 4 can be combined into one parameter with 12 levels; column 3 remains free. Parameters 5–16 correspond to parameters 12–23 of the orthogonal array L_{36} $(2^{11} \times 3^{12})$.

Partial Orthogonal Array L_{44} (2^{43})

Experiment No.	Parameter																					
	1	2	3	4	5	6	7	8	9	10	11	12	13	14	15	16	17	18	19	20	21	22
1	1	1	1	1	1	1	1	1	1	1	1	1	1	1	1	1	1	1	1	1	1	1
2	1	1	1	1	1	2	2	2	1	2	1	1	1	2	2	2	2	2	1	2	2	2
3	1	1	1	1	2	2	2	1	2	1	1	1	2	2	2	2	2	1	2	2	2	1
4	1	1	1	2	1	1	1	1	1	2	2	2	1	2	1	1	1	2	2	2	2	2
5	1	1	1	2	2	2	1	2	1	1	1	2	2	2	2	2	1	2	2	2	1	1
6	1	1	1	2	2	2	2	2	1	2	2	2	1	1	2	1	2	1	1	2	2	1
7	1	1	2	1	1	1	1	1	2	2	2	1	2	1	1	1	2	2	2	2	2	1
8	1	1	2	1	2	1	1	2	2	1	2	2	1	2	1	2	2	1	1	1	2	1
9	1	1	2	2	1	2	2	1	2	1	2	2	1	1	1	2	1	1	1	1	1	2
10	1	1	2	2	2	1	2	1	1	1	2	2	2	2	2	1	2	2	2	1	1	2
11	1	1	2	2	2	2	2	1	2	2	2	1	1	2	1	2	1	1	2	2	1	2
12	1	2	1	1	1	1	1	2	2	2	1	2	1	1	1	2	2	2	2	2	1	2
13	1	2	1	1	1	2	2	2	2	2	1	2	2	2	1	1	2	1	2	1	1	2
14	1	2	1	1	2	2	1	2	2	1	2	1	2	2	1	1	1	2	1	1	1	1
15	1	2	1	2	1	1	2	2	1	2	2	1	2	1	2	2	1	1	1	2	1	1
16	1	2	1	2	2	1	1	1	2	1	1	1	1	1	2	2	2	1	2	1	1	1
17	1	2	2	1	1	1	2	1	1	1	1	1	2	2	2	1	2	1	1	1	2	2
18	1	2	2	1	2	1	2	2	1	1	1	2	1	1	1	1	1	2	2	2	1	2
19	1	2	2	1	2	2	1	2	1	2	2	1	1	1	2	1	1	1	1	1	2	2
20	1	2	2	2	1	1	2	1	2	1	1	2	2	1	2	2	1	2	1	2	2	1
21	1	2	2	2	1	2	1	1	1	2	2	2	2	2	1	2	2	2	1	1	2	1
22	1	2	2	2	2	2	1	2	2	2	1	1	2	1	2	1	1	2	2	1	2	2
23	2	1	1	1	1	1	2	2	2	1	2	1	1	1	2	2	2	2	2	1	2	2

(continued)

(continued)

Experiment No.	Parameter																					
	1	2	3	4	5	6	7	8	9	10	11	12	13	14	15	16	17	18	19	20	21	22
24	2	1	1	1	2	1	1	1	1	1	2	2	2	1	2	1	1	1	2	2	2	2
25	2	1	1	1	2	2	2	2	2	1	2	2	2	1	1	2	1	2	1	1	2	2
26	2	1	1	2	1	2	1	1	2	2	1	2	2	1	2	1	2	2	1	1	1	2
27	2	1	1	2	2	1	2	2	1	2	1	2	2	1	1	1	2	1	1	1	1	1
28	2	1	2	1	1	1	2	2	2	2	2	1	2	2	2	1	1	2	1	2	1	1
29	2	1	2	1	1	2	2	1	2	2	1	2	1	2	2	1	1	1	2	1	1	1
30	2	1	2	1	2	2	1	1	1	2	1	1	1	1	1	2	2	2	1	2	1	1
31	2	1	2	2	1	1	1	2	1	1	1	1	1	2	2	2	1	2	1	1	1	2
32	2	1	2	2	1	2	1	2	2	1	1	1	2	1	1	1	1	1	2	2	2	1
33	2	1	2	2	2	1	1	2	1	2	1	1	2	2	1	2	2	1	2	1	2	2
34	2	2	1	1	1	2	1	1	1	1	1	2	2	2	1	2	1	1	1	2	2	2
35	2	2	1	1	2	1	2	1	1	2	2	1	2	2	1	2	1	2	2	1	1	1
36	2	2	1	2	1	1	1	2	2	2	2	2	1	2	2	2	1	1	2	1	2	1
37	2	2	1	2	1	2	2	1	1	1	2	1	1	1	1	1	2	2	2	1	2	1
38	2	2	1	2	2	1	2	1	2	2	1	1	1	2	1	1	1	1	1	2	2	2
39	2	2	1	2	2	2	1	1	2	1	2	1	1	2	2	1	2	2	1	2	1	2
40	2	2	2	1	1	2	1	2	1	1	2	2	1	2	2	1	2	1	2	2	1	1
41	2	2	2	1	2	1	1	1	2	2	2	2	2	1	2	2	2	1	1	2	1	2
42	2	2	2	1	2	2	2	1	1	2	1	2	1	1	2	2	1	2	2	1	2	1
43	2	2	2	2	1	2	2	2	1	1	2	1	2	1	1	2	2	1	2	2	1	2
44	2	2	2	2	2	1	2	2	2	1	1	2	1	2	1	1	2	2	1	2	2	1

Partial Orthogonal Array L_{44} (2^{43}), continued

Experiment No.	Parameter																				
	23	24	25	26	27	28	29	30	31	32	33	34	35	36	37	38	39	40	41	42	43
1	1	1	1	1	1	1	1	1	1	1	1	1	1	1	1	1	1	1	1	1	1
2	1	1	2	1	2	1	1	2	2	1	2	2	1	2	1	2	2	1	1	1	2
3	1	2	1	2	1	1	2	2	1	2	2	1	2	1	2	2	1	1	1	2	1
4	1	2	2	2	1	1	2	1	2	1	1	2	2	1	2	2	1	2	1	2	2
5	2	1	2	1	1	2	2	1	2	2	1	2	1	2	2	1	1	1	2	1	1
6	2	2	1	2	1	2	2	1	1	1	2	1	1	1	1	1	2	2	2	1	2
7	2	2	2	1	1	2	1	2	1	1	2	2	1	2	2	1	2	1	2	2	1
8	1	1	1	1	2	2	2	1	2	1	1	1	2	2	2	2	2	1	2	2	2
9	2	2	1	2	1	1	1	2	2	2	2	2	1	2	2	2	1	1	2	1	2
10	1	2	1	1	2	2	1	2	2	1	2	1	2	2	1	1	1	2	1	1	1
11	2	1	2	1	2	2	1	1	1	2	1	1	1	1	1	2	2	2	1	2	1
12	2	2	1	1	2	1	2	1	1	2	2	1	2	2	1	2	1	2	2	1	1
13	2	1	2	2	1	2	1	2	2	1	1	1	2	1	1	1	1	1	2	2	2
14	1	2	2	2	1	2	1	1	1	2	2	2	2	2	1	2	2	2	1	1	2
15	1	1	1	2	2	2	1	2	1	1	1	2	2	2	2	2	1	2	2	2	1
16	2	2	2	2	2	1	2	2	2	1	1	2	1	2	1	1	2	2	1	2	2
17	2	2	2	1	2	2	2	1	1	2	1	2	1	1	2	2	1	2	2	1	2
18	1	1	1	2	2	2	2	2	1	2	2	2	1	1	2	1	2	1	1	2	2
19	2	1	2	1	1	1	2	2	2	2	2	1	2	2	2	1	1	2	1	2	1
20	1	1	2	1	1	1	1	1	2	2	2	1	2	1	1	1	2	2	2	2	2
21	2	1	1	2	2	1	2	2	1	2	1	2	2	1	1	1	2	1	1	1	1
22	1	2	1	2	2	1	1	1	2	1	1	1	1	1	2	2	2	1	2	1	1
23	2	1	1	2	1	2	1	1	2	2	1	2	2	1	2	1	2	2	1	1	1

(continued)

(continued)

Experiment No.	Parameter																				
	23	24	25	26	27	28	29	30	31	32	33	34	35	36	37	38	39	40	41	42	43
24	2	1	2	2	2	1	1	2	1	2	1	1	2	2	1	2	2	1	2	1	2
25	1	2	2	1	2	1	2	2	1	1	1	2	1	1	1	1	1	2	2	2	1
26	1	1	1	1	1	2	2	2	1	2	1	1	1	2	2	2	2	2	1	2	2
27	2	2	2	1	2	1	1	1	2	2	2	2	2	1	2	2	2	1	1	2	1
28	2	2	1	2	2	1	2	1	2	2	1	1	1	2	1	1	1	1	1	2	2
29	1	1	2	2	2	1	2	1	1	1	2	2	2	2	2	1	2	2	2	1	1
30	1	2	2	2	2	2	1	2	2	2	1	1	2	1	2	1	1	2	2	1	2
31	2	2	2	2	1	2	2	2	1	1	2	1	2	1	1	2	2	1	2	2	1
32	2	1	1	1	2	2	2	2	2	1	2	2	2	1	1	2	1	2	1	1	2
33	1	1	1	2	1	1	1	1	1	2	2	2	1	2	1	1	1	2	2	2	2
34	2	2	1	2	2	2	1	1	2	1	2	1	1	2	2	1	2	2	1	2	1
35	2	1	1	1	1	1	2	2	2	1	2	1	1	1	2	2	2	2	2	1	2
36	1	2	2	1	2	2	1	2	1	2	2	1	1	1	2	1	1	1	1	1	2
37	1	1	2	2	2	2	2	1	2	2	2	1	1	2	1	2	1	1	2	2	1
38	1	2	1	1	1	2	2	2	2	2	1	2	2	2	1	1	2	1	2	1	1
39	2	1	1	1	2	1	1	1	1	1	2	2	2	1	2	1	1	1	2	2	2
40	1	2	1	1	1	1	1	2	2	2	1	2	1	1	1	2	2	2	2	2	1
41	1	1	2	2	1	2	2	1	2	1	2	2	1	1	1	2	1	1	1	1	1
42	2	2	1	1	1	2	1	1	1	1	1	2	2	2	1	2	1	1	1	2	2
43	1	2	2	1	1	1	2	1	1	1	1	1	2	2	2	1	2	1	1	1	2
44	2	1	2	2	1	1	1	2	1	1	1	1	1	2	2	2	1	2	1	1	1

Partial Orthogonal Array L_{50} ($2^1 \times 5^{11}$)

Experiment No.	Parameter											
	1	2	3	4	5	6	7	8	9	10	11	12
1	1	1	1	1	1	1	1	1	1	1	1	1
2	1	1	2	2	2	2	2	2	2	2	2	2
3	1	1	3	3	3	3	3	3	3	3	3	3
4	1	1	4	4	4	4	4	4	4	4	4	4
5	1	1	5	5	5	5	5	5	5	5	5	5
6	1	2	1	2	3	4	5	1	2	3	4	5
7	1	2	2	3	4	5	1	2	3	4	5	1
8	1	2	3	4	5	1	2	3	4	5	1	2
9	1	2	4	5	1	2	3	4	5	1	2	3
10	1	2	5	1	2	3	4	5	1	2	3	4
11	1	3	1	3	5	2	4	4	1	3	5	2
12	1	3	2	4	1	3	5	5	2	4	1	3
13	1	3	3	5	2	4	1	1	3	5	2	4
14	1	3	4	1	3	5	2	2	4	1	3	5
15	1	3	5	2	4	1	3	3	5	2	4	1
16	1	4	1	4	2	5	3	5	3	1	4	2
17	1	4	2	5	3	1	4	1	4	2	5	3
18	1	4	3	1	4	2	5	2	5	3	1	4
19	1	4	4	2	5	3	1	3	1	4	2	5
20	1	4	5	3	1	4	2	4	2	5	3	1
21	1	5	1	5	4	3	2	4	3	2	1	5
22	1	5	2	1	5	4	3	5	4	3	2	1
23	1	5	3	2	1	5	4	1	5	4	3	2
24	1	5	4	3	2	1	5	2	1	5	4	3
25	1	5	5	4	3	2	1	3	2	1	5	4
26	2	1	1	1	4	5	4	3	2	5	2	3
27	2	1	2	2	5	1	5	4	3	1	3	4
28	2	1	3	3	1	2	1	5	4	2	4	5
29	2	1	4	4	2	3	2	1	5	3	5	1
30	2	1	5	5	3	4	3	2	1	4	1	2
31	2	2	1	2	1	3	3	2	4	5	5	4
32	2	2	2	3	2	4	4	3	5	1	1	5
33	2	2	3	4	3	5	5	4	1	2	2	1
34	2	2	4	5	4	1	1	5	2	3	3	2
35	2	2	5	1	5	2	2	1	3	4	4	3

continued

Experiment No.	Parameter											
	1	2	3	4	5	6	7	8	9	10	11	12
36	2	3	1	3	3	1	2	5	5	4	2	4
37	2	3	2	4	4	2	3	1	1	5	3	5
38	2	3	3	5	5	3	4	2	2	1	4	1
39	2	3	4	1	1	4	5	3	3	2	5	2
40	2	3	5	2	2	5	1	4	4	3	1	3
41	2	4	1	4	5	4	1	2	5	2	3	3
42	2	4	2	5	1	5	2	3	1	3	4	4
43	2	4	3	1	2	1	3	4	2	4	5	5
44	2	4	4	2	3	2	4	5	3	5	1	1
45	2	4	5	3	4	3	5	1	4	1	2	2
46	2	5	1	5	2	2	5	3	4	4	3	1
47	2	5	2	1	3	3	1	4	5	5	4	2
48	2	5	3	2	4	4	2	5	1	1	5	3
49	2	5	4	3	5	5	3	1	2	2	1	4
50	2	5	5	4	1	1	4	2	3	3	2	5

An interaction between parameters 1 and 2 is orthogonal to all others and can be viewed independently without abandoning other parameters. Columns 1 and 2 can be combined into one parameter with 10 levels. Possible interactions of parameters 3–12 are distributed by columns 3–12.

Partial Orthogonal Array L_{54} ($2^1 \times 3^{25}$)

Expt. No.	Parameter																									
	1	2	3	4	5	6	7	8	9	10	11	12	13	14	15	16	17	18	19	20	21	22	23	24	25	26
1	1	1	1	1	1	1	1	1	1	1	1	1	1	1	1	1	1	1	1	1	1	1	1	1	1	1
2	1	1	1	1	1	1	1	1	2	2	2	2	2	2	2	2	2	2	2	2	2	2	2	2	2	2
3	1	1	1	1	1	1	1	1	3	3	3	3	3	3	3	3	3	3	3	3	3	3	3	3	3	3
4	1	1	2	2	2	2	2	2	1	1	1	1	1	1	2	3	2	3	2	3	2	3	2	3	2	3
5	1	1	2	2	2	2	2	2	2	2	2	2	2	2	3	1	3	1	3	1	3	1	3	1	3	1
6	1	1	2	2	2	2	2	2	3	3	3	3	3	3	1	2	1	2	1	2	1	2	1	2	1	2
7	1	1	3	3	3	3	3	3	1	1	1	1	1	1	3	2	3	2	3	2	3	2	3	2	3	2
8	1	1	3	3	3	3	3	3	2	2	2	2	2	2	1	3	1	3	1	3	1	3	1	3	1	3
9	1	1	3	3	3	3	3	3	3	3	3	3	3	3	2	1	2	1	2	1	2	1	2	1	2	1
10	1	2	1	1	2	2	3	3	1	1	2	2	3	3	1	1	1	1	2	3	2	3	3	2	3	2
11	1	2	1	1	2	2	3	3	2	2	3	3	1	1	2	2	2	2	3	1	3	1	1	3	1	3
12	1	2	1	1	2	2	3	3	3	3	1	1	2	2	3	3	3	3	1	2	1	2	2	1	2	1
13	1	2	2	2	3	3	1	1	1	1	2	2	3	3	2	3	2	3	3	2	3	2	1	1	1	1
14	1	2	2	2	3	3	1	1	2	2	3	3	1	1	3	1	3	1	1	3	1	3	2	2	2	2
15	1	2	2	2	3	3	1	1	3	3	1	1	2	2	1	2	1	2	2	1	2	1	3	3	3	3
16	1	2	3	3	1	1	2	2	1	1	2	2	3	3	3	2	3	2	1	1	1	1	2	3	2	3
17	1	2	3	3	1	1	2	2	2	2	3	3	1	1	1	3	1	3	2	2	2	2	3	1	3	1
18	1	2	3	3	1	1	2	2	3	3	1	1	2	2	2	1	2	1	3	3	3	3	1	2	1	2
19	1	3	1	2	1	3	2	3	1	2	1	3	2	3	1	1	2	3	1	1	3	2	2	3	3	2
20	1	3	1	2	1	3	2	3	2	3	2	1	3	1	2	2	3	1	2	2	1	3	3	1	1	3
21	1	3	1	2	1	3	2	3	3	1	3	2	1	2	3	3	1	2	3	3	2	1	1	2	2	1
22	1	3	2	3	2	1	3	1	1	2	1	3	2	3	2	3	3	2	2	3	1	1	3	2	1	1
23	1	3	2	3	2	1	3	1	2	3	2	1	3	1	3	1	1	3	3	1	2	2	1	3	2	2

(continued)

(continued)

Expt. No.	Parameter																									
	1	2	3	4	5	6	7	8	9	10	11	12	13	14	15	16	17	18	19	20	21	22	23	24	25	26
24	1	3	2	3	2	1	3	1	3	1	3	2	1	2	1	2	2	1	1	2	3	3	2	1	3	3
25	1	3	3	1	3	2	1	2	1	2	1	3	2	3	3	2	1	1	3	2	2	3	1	1	2	3
26	1	3	3	1	3	2	1	2	2	3	2	1	3	1	1	3	2	2	1	3	3	1	2	2	3	1
27	1	3	3	1	3	2	1	2	3	1	3	2	1	2	2	1	3	3	2	1	1	2	3	3	1	2
28	2	1	1	3	3	2	2	1	1	3	3	2	2	1	1	1	3	2	3	2	2	3	2	3	1	1
29	2	1	1	3	3	2	2	1	2	1	1	3	3	2	2	2	1	3	1	3	3	1	3	1	2	2
30	2	1	1	3	3	2	2	1	3	2	2	1	1	3	3	3	2	1	2	1	1	2	1	2	3	3
31	2	1	2	1	1	3	3	2	1	3	3	2	2	1	2	3	1	1	1	1	3	2	3	2	2	3
32	2	1	2	1	1	3	3	2	2	1	1	3	3	2	3	1	2	2	2	2	1	3	1	3	3	1
33	2	1	2	1	1	3	3	2	3	2	2	1	1	3	1	2	3	3	3	3	2	1	2	1	1	2
34	2	1	3	2	2	1	1	3	1	3	3	2	2	1	3	2	2	3	2	3	1	1	1	1	3	2
35	2	1	3	2	2	1	1	3	2	1	1	3	3	2	1	3	3	1	3	1	2	2	2	2	1	3
36	2	1	3	2	2	1	1	3	3	2	2	1	1	3	2	1	1	2	1	2	3	3	3	3	2	1
37	2	2	1	2	3	1	3	2	1	2	3	1	3	2	1	1	2	3	3	2	1	1	3	2	2	3
38	2	2	1	2	3	1	3	2	2	3	1	2	1	3	2	2	3	1	1	3	2	2	1	3	3	1
39	2	2	1	2	3	1	3	2	3	1	2	3	2	1	3	3	1	2	2	1	3	3	2	1	1	2
40	2	2	2	3	1	2	1	3	1	2	3	1	3	2	2	3	3	2	1	1	2	3	1	1	3	2
41	2	2	2	3	1	2	1	3	2	3	1	2	1	3	3	1	1	3	2	2	3	1	2	2	1	3
42	2	2	2	3	1	2	1	3	3	1	2	3	2	1	1	2	2	1	3	3	1	2	3	3	2	1

Partial Orthogonal Array L_{54} ($2^1 \times 3^{25}$), continued

Experiment No.	Parameter																									
	1	2	3	4	5	6	7	8	9	10	11	12	13	14	15	16	17	18	19	20	21	22	23	24	25	26
43	2	2	3	1	2	3	2	1	1	2	3	1	3	2	3	2	1	1	2	3	3	2	2	3	1	1
44	2	2	3	1	2	3	2	1	2	3	1	2	1	3	1	3	2	2	3	1	1	3	3	1	2	2
45	2	2	3	1	2	3	2	1	3	1	2	3	2	1	2	1	3	3	1	2	2	1	1	2	3	3
46	2	3	1	3	2	3	1	2	1	3	2	3	1	2	1	1	3	2	2	3	3	2	1	1	2	3
47	2	3	1	3	2	3	1	2	2	1	3	1	2	3	2	2	1	3	3	1	1	3	2	2	3	1
48	2	3	1	3	2	3	1	2	3	2	1	2	3	1	3	3	2	1	1	2	2	1	3	3	1	2
49	2	3	2	1	3	1	2	2	1	3	2	3	1	2	2	3	1	1	3	2	1	1	2	3	3	2
50	2	3	2	1	3	1	2	3	2	1	3	1	2	3	3	1	2	2	1	3	2	2	3	1	1	3
51	2	3	2	1	3	1	2	3	3	2	1	2	3	1	1	2	3	3	2	1	3	3	1	2	2	1
52	2	3	3	2	1	2	3	1	1	3	2	3	1	2	3	2	2	3	1	1	2	3	3	2	1	1
53	2	3	3	2	1	2	3	1	2	1	3	1	2	3	1	3	3	1	2	2	3	1	1	3	2	2
54	2	3	3	2	1	2	3	1	3	2	1	2	3	1	2	1	1	2	3	3	1	2	2	1	3	3

An interaction between parameters 1 and 2 is orthogonal to all others and can be viewed independently without abandoning other parameters. Columns 1 and 2 can also be combined to a parameter with six levels. Possible interactions of parameters 1×9, 2×9, and $1 \times 2 \times 9$ can be determined by not assigning columns 10, 11, 12, 13, and 14. Parameters 1, 2, and 9 can also be combined into one parameter with 18 levels by not assigning the same columns 10–14.

Orthogonal Array L_{64} (2^{63})

Experiement No.	Parameter																									
	1	2	3	4	5	6	7	8	9	10	11	12	13	14	15	16	17	18	19	20	21	22	23	24	25	26
1	1	1	1	1	1	1	1	1	1	1	1	1	1	1	1	1	1	1	1	1	1	1	1	1	1	1
2	1	1	1	1	1	1	1	1	1	1	1	1	1	1	1	1	1	1	1	1	1	1	1	1	1	1
3	1	1	1	1	1	1	1	1	1	1	1	1	1	1	1	2	2	2	2	2	2	2	2	2	2	2
4	1	1	1	1	1	1	1	1	1	1	1	1	1	1	1	2	2	2	2	2	2	2	2	2	2	2
5	1	1	1	1	1	1	1	2	2	2	2	2	2	2	2	1	1	1	1	1	1	1	1	2	2	2
6	1	1	1	1	1	1	1	2	2	2	2	2	2	2	2	1	1	1	1	1	1	1	1	2	2	2
7	1	1	1	1	1	1	1	2	2	2	2	2	2	2	2	2	2	2	2	2	2	2	2	1	1	1
8	1	1	1	1	1	1	1	2	2	2	2	2	2	2	2	2	2	2	2	2	2	2	2	1	1	1
9	1	1	1	2	2	2	2	1	1	1	1	2	2	2	2	1	1	1	1	2	2	2	2	1	1	1
10	1	1	1	2	2	2	2	1	1	1	1	2	2	2	2	1	1	1	1	2	2	2	2	1	1	1
11	1	1	1	2	2	2	2	1	1	1	1	2	2	2	2	2	2	2	2	1	1	1	1	2	2	2
12	1	1	1	2	2	2	2	1	1	1	1	2	2	2	2	2	2	2	2	1	1	1	1	2	2	2
13	1	1	1	2	2	2	2	2	2	2	2	1	1	1	1	1	1	1	1	2	2	2	2	2	2	2
14	1	1	1	2	2	2	2	2	2	2	2	1	1	1	1	1	1	1	1	2	2	2	2	2	2	2
15	1	1	1	2	2	2	2	2	2	2	2	1	1	1	1	2	2	2	2	1	1	1	1	1	1	1
16	1	1	1	2	2	2	2	2	2	2	2	1	1	1	1	2	2	2	2	1	1	1	1	1	1	1
17	1	2	2	1	1	2	2	1	1	2	2	1	1	2	2	1	1	2	2	1	1	2	2	1	1	2
18	1	2	2	1	1	2	2	1	1	2	2	1	1	2	2	1	1	2	2	1	1	2	2	1	1	2
19	1	2	2	1	1	2	2	1	1	2	2	1	1	2	2	2	2	1	1	2	2	1	1	2	2	1
20	1	2	2	1	1	2	2	1	1	2	2	1	1	2	2	2	2	1	1	2	2	1	1	2	2	1
21	1	2	2	1	1	2	2	2	2	1	1	2	2	1	1	1	1	2	2	1	1	2	2	2	2	1
22	1	2	2	1	1	2	2	2	2	1	1	2	2	1	1	1	1	2	2	1	1	2	2	2	2	1

(continued)

(continued)

Experiment No.	Parameter																								
	1	2	3	4	5	6	7	8	9	10	11	12	13	14	15	16	17	18	19	20	21	22	23	24	25	26
23	1	2	2	1	1	2	2	2	2	1	1	2	2	1	1	2	2	1	1	2	2	1	1	1	1	2
24	1	2	2	1	1	2	2	2	2	1	1	2	2	1	1	2	2	1	1	2	2	1	1	1	1	2
25	1	2	2	2	2	1	1	1	1	2	2	2	2	1	1	1	1	2	2	2	2	1	1	1	1	2
26	1	2	2	2	2	1	1	1	1	2	2	2	2	1	1	1	1	2	2	2	2	1	1	1	1	2
27	1	2	2	2	2	1	1	1	1	2	2	2	2	1	1	2	2	1	1	1	1	2	2	2	2	1
28	1	2	2	2	2	1	1	1	1	2	2	2	2	1	1	2	2	1	1	1	1	2	2	2	2	1
29	1	2	2	2	2	1	1	2	2	1	1	1	1	2	2	1	1	2	2	2	2	1	1	2	2	1
30	1	2	2	2	2	1	1	2	2	1	1	1	1	2	2	1	1	2	2	2	2	1	1	2	2	1
31	1	2	2	2	2	1	1	2	2	1	1	1	1	2	2	2	2	1	1	1	1	2	2	1	1	2
32	1	2	2	2	2	1	1	2	2	1	1	1	1	2	2	2	2	1	1	1	1	2	2	1	1	2
33	2	1	2	1	2	1	2	1	2	1	2	1	2	1	2	1	2	1	2	1	2	1	2	1	2	1
34	2	1	2	1	2	1	2	1	2	1	2	1	2	1	2	1	2	1	2	1	2	1	2	1	2	1
35	2	1	2	1	2	1	2	1	2	1	2	1	2	1	2	2	1	2	1	2	1	2	1	2	1	2
36	2	1	2	1	2	1	2	1	2	1	2	1	2	1	2	2	1	2	1	2	1	2	1	2	1	2
37	2	1	2	1	2	1	2	2	1	2	1	2	1	2	1	1	2	1	2	1	2	1	2	2	1	2
38	2	1	2	1	2	1	2	2	1	2	1	2	1	2	1	1	2	1	2	1	2	1	2	2	1	2
39	2	1	2	1	2	1	2	2	1	2	1	2	1	2	1	2	1	2	1	2	1	2	1	1	2	1
40	2	1	2	1	2	1	2	2	1	2	1	2	1	2	1	2	1	2	1	2	1	2	1	1	2	1

Orthogonal Array L_{64} (2^{63}), continued

Experiment No.	Parameter																									
	1	2	3	4	5	6	7	8	9	10	11	12	13	14	15	16	17	18	19	20	21	22	23	24	25	26
41	2	1	2	2	1	2	1	1	2	1	2	2	1	2	1	1	2	1	2	2	1	2	1	1	2	1
42	2	1	2	2	1	2	1	1	2	1	2	2	1	2	1	1	2	1	2	2	1	2	1	1	2	1
43	2	1	2	2	1	2	1	1	2	1	2	2	1	2	1	2	1	2	1	1	2	1	2	2	1	2
44	2	1	2	2	1	2	1	1	2	1	2	2	1	2	1	2	1	2	1	1	2	1	2	2	1	2
45	2	1	2	2	1	2	1	2	1	2	1	1	2	1	2	1	2	1	2	2	1	2	1	2	1	2
46	2	1	2	2	1	2	1	2	1	2	1	1	2	1	2	1	2	1	2	2	1	2	1	2	1	2
47	2	1	2	2	1	2	1	2	1	2	1	1	2	1	2	2	1	2	1	1	2	1	2	1	2	1
48	2	1	2	2	1	2	1	2	1	2	1	1	2	1	2	2	1	2	1	1	2	1	2	1	2	1
49	2	2	1	1	2	2	1	1	2	2	1	1	2	2	1	1	2	2	1	1	2	2	1	1	2	2
50	2	2	1	1	2	2	1	1	2	2	1	1	2	2	1	1	2	2	1	1	2	2	1	1	2	2
51	2	2	1	1	2	2	1	1	2	2	1	1	2	2	1	2	1	1	2	2	1	1	2	2	1	1
52	2	2	1	1	2	2	1	1	2	2	1	1	2	2	1	2	1	1	2	2	1	1	2	2	1	1
53	2	2	1	1	2	2	1	2	1	1	2	2	1	1	2	1	2	2	1	1	2	2	1	2	1	1
54	2	2	1	1	2	2	1	2	1	1	2	2	1	1	2	1	2	2	1	1	2	2	1	2	1	1
55	2	2	1	1	2	2	1	2	1	1	2	2	1	1	2	2	1	1	2	2	1	1	2	1	2	2
56	2	2	1	1	2	2	1	2	1	1	2	2	1	1	2	2	1	1	2	2	1	1	2	1	2	2
57	2	2	1	2	1	1	2	1	2	2	1	2	1	1	2	1	2	2	1	2	1	1	2	1	2	2
58	2	2	1	2	1	1	2	1	2	2	1	2	1	1	2	1	2	2	1	2	1	1	2	1	2	2
59	2	2	1	2	1	1	2	1	2	2	1	2	1	1	2	2	1	1	2	1	2	2	1	2	1	1
60	2	2	1	2	1	1	2	1	2	2	1	2	1	1	2	2	1	1	2	1	2	2	1	2	1	1
61	2	2	1	2	1	1	2	2	1	1	2	1	2	2	1	1	2	2	1	2	1	1	2	2	1	1
62	2	2	1	2	1	1	2	2	1	1	2	1	2	2	1	1	2	2	1	2	1	1	2	2	1	1
63	2	2	1	2	1	1	2	2	1	1	2	1	2	2	1	2	1	1	2	1	2	2	1	1	2	2
64	2	2	1	2	1	1	2	2	1	1	2	1	2	2	1	2	1	1	2	1	2	2	1	1	2	2

Orthogonal Array L_{64} (2^{63}), continued

Experiment No.	Parameter																									
	27	28	29	30	31	32	33	34	35	36	37	38	39	40	41	42	43	44	45	46	47	48	49	50	51	52
1	1	1	1	1	1	1	1	1	1	1	1	1	1	1	1	1	1	1	1	1	1	1	1	1	1	1
2	1	1	1	1	1	2	2	2	2	2	2	2	2	2	2	2	2	2	2	2	2	2	2	2	2	2
3	2	2	2	2	2	1	1	1	1	1	1	1	1	1	1	1	1	1	1	1	1	2	2	2	2	2
4	2	2	2	2	2	2	2	2	2	2	2	2	2	2	2	2	2	2	2	2	2	1	1	1	1	1
5	2	2	2	2	2	1	1	1	1	1	1	1	1	2	2	2	2	2	2	2	2	1	1	1	1	1
6	2	2	2	2	2	2	2	2	2	2	2	2	2	1	1	1	1	1	1	1	1	2	2	2	2	2
7	1	1	1	1	1	1	1	1	1	1	1	1	1	2	2	2	2	2	2	2	2	2	2	2	2	2
8	1	1	1	1	1	2	2	2	2	2	2	2	2	1	1	1	1	1	1	1	1	1	1	1	1	1
9	1	2	2	2	2	1	1	1	1	2	2	2	2	1	1	1	1	2	2	2	2	1	1	1	1	2
10	1	2	2	2	2	2	2	2	2	1	1	1	1	2	2	2	2	1	1	1	1	2	2	2	2	1
11	2	1	1	1	1	1	1	1	1	2	2	2	2	1	1	1	1	2	2	2	2	2	2	2	2	1
12	2	1	1	1	1	2	2	2	2	1	1	1	1	2	2	2	2	1	1	1	1	1	1	1	1	2
13	2	1	1	1	1	1	1	1	1	2	2	2	2	2	2	2	2	1	1	1	1	1	1	1	1	2
14	2	1	1	1	1	2	2	2	2	1	1	1	1	1	1	1	1	2	2	2	2	2	2	2	2	1
15	1	2	2	2	2	1	1	1	1	2	2	2	2	2	2	2	2	1	1	1	1	2	2	2	2	1
16	1	2	2	2	2	2	2	2	2	1	1	1	1	1	1	1	1	2	2	2	2	1	1	1	1	2
17	2	1	1	2	2	1	1	2	2	1	1	2	2	1	1	2	2	1	1	2	2	1	1	2	2	1
18	2	1	1	2	2	2	2	1	1	2	2	1	1	2	2	1	1	2	2	1	1	2	2	1	1	2
19	1	2	2	1	1	1	1	2	2	1	1	2	2	1	1	2	2	1	1	2	2	2	2	1	1	2
20	1	2	2	1	1	2	2	1	1	2	2	1	1	2	2	1	1	2	2	1	1	1	1	2	2	1
21	1	2	2	1	1	1	1	2	2	1	1	2	2	2	2	1	1	2	2	1	1	1	1	2	2	1
22	1	2	2	1	1	2	2	1	1	2	2	1	1	1	1	2	2	1	1	2	2	2	2	1	1	2
23	2	1	1	2	2	1	1	2	2	1	1	2	2	2	2	1	1	2	2	1	1	2	2	1	1	2

(continued)

(continued)

Experiment No.	Parameter																									
	27	28	29	30	31	32	33	34	35	36	37	38	39	40	41	42	43	44	45	46	47	48	49	50	51	52
24	2	1	1	2	2	2	2	1	1	2	2	1	1	1	1	2	2	1	1	2	2	1	1	2	2	1
25	2	2	2	1	1	1	1	2	2	2	2	1	1	1	1	2	2	2	2	1	1	1	1	2	2	2
26	2	2	2	1	1	2	2	1	1	1	1	2	2	2	2	1	1	1	1	2	2	2	2	1	1	1
27	1	1	1	2	2	1	1	2	2	2	2	1	1	1	1	2	2	2	2	1	1	2	2	1	1	1
28	1	1	1	2	2	2	2	1	1	1	1	2	2	2	2	1	1	1	1	2	2	1	1	2	2	2
29	1	1	1	2	2	1	1	2	2	2	2	1	1	2	2	1	1	1	1	2	2	1	1	2	2	2
30	1	1	1	2	2	2	2	1	1	1	1	2	2	1	1	2	2	2	2	1	1	2	2	1	1	1
31	2	1	1	1	1	1	1	2	2	2	2	1	1	2	2	1	1	1	1	2	2	2	2	1	1	1
32	2	2	2	1	1	2	2	1	1	1	1	2	2	1	1	2	2	2	2	1	1	1	1	2	2	2
33	2	1	2	1	2	1	2	1	2	1	2	1	2	1	2	1	2	1	2	1	2	1	2	1	2	1
34	2	1	2	1	2	2	1	2	1	2	1	2	1	2	1	2	1	2	1	2	1	2	1	2	1	2
35	1	2	1	2	1	1	2	1	2	1	2	1	2	1	2	1	2	1	2	1	2	2	1	2	1	2
36	1	2	1	2	1	2	1	2	1	2	1	2	1	2	1	2	1	2	1	2	1	1	2	1	2	1
37	1	2	1	2	1	1	2	1	2	1	2	1	2	2	1	2	1	2	1	2	1	1	2	1	2	1
38	1	2	1	2	1	2	1	2	1	2	1	2	1	1	2	1	2	1	2	1	2	2	1	2	1	2
39	2	1	2	1	2	1	2	1	2	1	2	1	2	2	1	2	1	2	1	2	1	2	1	2	1	2
40	2	1	2	1	2	2	1	2	1	2	1	2	1	1	2	1	2	1	2	1	2	1	2	1	2	1

Orthogonal Array L_{64} (2^{63}), continued

Experiment No.	Parameter																									
	27	28	29	30	31	32	33	34	35	36	37	38	39	40	41	42	43	44	45	46	47	48	49	50	51	52
41	2	2	1	2	1	1	2	1	2	2	1	2	1	1	2	1	2	2	1	2	1	1	2	1	2	2
42	2	2	1	2	1	2	1	2	1	1	2	1	2	2	1	2	1	1	2	1	2	2	1	2	1	1
43	1	1	2	1	2	1	2	1	2	2	1	2	1	1	2	1	2	2	1	2	1	2	1	2	1	1
44	1	1	2	1	2	2	1	2	1	1	2	1	2	2	1	2	1	1	2	1	2	1	2	1	2	2
45	1	1	2	1	2	1	2	1	2	2	1	2	1	2	1	2	1	1	2	1	2	1	2	1	2	2
46	1	1	2	1	2	2	1	2	1	1	2	1	2	1	2	1	2	2	1	2	1	2	1	2	1	1
47	2	2	1	2	1	1	2	1	2	2	1	2	1	2	1	2	1	1	2	1	2	2	1	2	1	1
48	2	2	1	2	1	2	1	2	1	1	2	1	2	1	2	1	2	2	1	2	1	1	2	1	2	2
49	1	1	2	2	1	1	2	2	1	1	2	2	1	1	2	2	1	1	2	2	1	1	2	2	1	1
50	1	1	2	2	1	2	1	1	2	2	1	1	2	2	1	1	2	2	1	1	2	2	1	1	2	2
51	2	2	1	1	2	1	2	2	1	1	2	2	1	1	2	2	1	1	2	2	1	2	1	1	2	2
52	2	2	1	1	2	2	1	1	2	2	1	1	2	2	1	1	2	2	1	1	2	1	2	2	1	1
53	2	2	1	1	2	1	2	2	1	1	2	2	1	2	1	1	2	2	1	1	2	1	2	2	1	1
54	2	2	1	1	2	2	1	1	2	2	1	1	2	1	2	2	1	1	2	2	1	2	1	1	2	2
55	1	1	2	2	1	1	2	2	1	1	2	2	1	2	1	1	2	2	1	1	2	2	1	1	2	2
56	1	1	2	2	1	2	1	1	2	2	1	1	2	1	2	2	1	1	2	2	1	1	2	2	1	1
57	1	2	1	1	2	1	2	2	1	2	1	1	2	1	2	2	1	2	1	1	2	1	2	2	1	2
58	1	2	1	1	2	2	1	1	2	1	2	2	1	2	1	1	2	1	2	2	1	2	1	1	2	1
59	2	1	2	2	1	1	2	2	1	2	1	1	2	1	2	2	1	2	1	1	2	2	1	1	2	1
60	2	1	2	2	1	2	1	1	2	1	2	2	1	2	1	1	2	1	2	2	1	1	2	2	1	2
61	2	1	2	2	1	1	2	2	1	2	1	1	2	2	1	1	2	1	2	2	1	1	2	2	1	2
62	2	1	2	2	1	2	1	1	2	1	2	2	1	1	2	2	1	2	1	1	2	2	1	1	2	1
63	1	2	1	1	2	1	2	2	1	2	1	1	2	2	1	1	2	1	2	2	1	2	1	1	2	1
64	1	2	1	1	2	2	1	1	2	1	2	2	1	1	2	2	1	2	1	1	2	1	2	2	1	2

Orthogonal Array L_{64} (2^{63}), continued

Experiment No.	Parameter										
	53	54	55	56	57	58	59	60	61	62	63
1	1	1	1	1	1	1	1	1	1	1	1
2	2	2	2	2	2	2	2	2	2	2	2
3	2	2	2	2	2	2	2	2	2	2	2
4	1	1	1	1	1	1	1	1	1	1	1
5	1	1	1	2	2	2	2	2	2	2	2
6	2	2	2	1	1	1	1	1	1	1	1
7	2	2	2	1	1	1	1	1	1	1	1
8	1	1	1	2	2	2	2	2	2	2	2
9	2	2	2	1	1	1	1	2	2	2	2
10	1	1	1	2	2	2	2	1	1	1	1
11	1	1	1	2	2	2	2	1	1	1	1
12	2	2	2	1	1	1	1	2	2	2	2
13	2	2	2	2	2	2	2	1	1	1	1
14	1	1	1	1	1	1	1	2	2	2	2
15	1	1	1	1	1	1	1	2	2	2	2
16	2	2	2	2	2	2	2	1	1	1	1
17	1	2	2	1	1	2	2	1	1	2	2
18	2	1	1	2	2	1	1	2	2	1	1
19	2	1	1	2	2	1	1	2	2	1	1
20	1	2	2	1	1	2	2	1	1	2	2
21	1	2	2	2	2	1	1	2	2	1	1
22	2	1	1	1	1	2	2	1	1	2	2
23	2	1	1	1	1	2	2	1	1	2	2

(continued)

(continued)

Experiment No.	Parameter										
	53	54	55	56	57	58	59	60	61	62	63
24	1	2	2	2	2	1	1	2	2	1	1
25	2	1	1	1	1	2	2	2	2	1	1
26	1	2	2	2	2	1	1	1	1	2	2
27	1	2	2	2	2	1	1	1	1	2	2
28	2	1	1	1	1	2	2	2	2	1	1
29	2	1	1	2	2	1	1	1	1	2	2
30	1	2	2	1	1	2	2	2	2	1	1
31	1	2	2	1	1	2	2	2	2	1	1
32	2	1	1	2	2	1	1	1	1	2	2
33	2	1	2	1	2	1	2	1	2	1	2
34	1	2	1	2	1	2	1	2	1	2	1
35	1	2	1	2	1	2	1	2	1	2	1
36	2	1	2	1	2	1	2	1	2	1	2
37	2	1	2	2	1	2	1	2	1	2	1
38	1	2	1	1	2	1	2	1	2	1	2
39	1	2	1	1	2	1	2	1	2	1	2
40	2	1	2	2	1	2	1	2	1	2	1

Orthogonal Array L_{64} (2^{63}), continued

Experiment No.	Parameter										
	53	54	55	56	57	58	59	60	61	62	63
41	1	2	1	1	2	1	2	2	1	2	1
42	2	1	2	2	1	2	1	1	2	1	2
43	2	1	2	2	1	2	1	1	2	1	2
44	1	2	1	1	2	1	2	2	1	2	1
45	1	2	1	2	1	2	1	1	2	1	2
46	2	1	2	1	2	1	2	2	1	2	1
47	2	1	2	1	2	1	2	2	1	2	1
48	1	2	1	2	1	2	1	1	2	1	2
49	2	2	1	1	2	2	1	1	2	2	1
50	1	1	2	2	1	1	2	2	1	1	2
51	1	1	2	2	1	1	2	2	1	1	2
52	2	2	1	1	2	2	1	1	2	2	1
53	2	2	1	2	1	1	2	2	1	1	2
54	1	1	2	1	2	2	1	1	2	2	1
55	1	1	2	1	2	2	1	1	2	2	1
56	2	2	1	2	1	1	2	2	1	1	2
57	1	1	2	1	2	2	1	2	1	1	2
58	2	2	1	2	1	1	2	1	2	2	1
59	2	2	1	2	1	1	2	1	2	2	1
60	1	1	2	1	2	2	1	2	1	1	2
61	1	1	2	2	1	1	2	1	2	2	1
62	2	2	1	1	2	2	1	2	1	1	2
63	2	2	1	1	2	2	1	2	1	1	2
64	1	1	2	2	1	1	2	1	2	2	1

Interaction Table L_{64}

Para-meter	Parameter 1	2	3	4	5	6	7	8	9	10	11	12	13	14	15	16	17	18	19	20	21	22	23	24	25	26
1	(1)	3	2	5	4	7	6	9	8	11	10	13	12	15	14	17	16	19	18	21	20	23	22	25	24	27
2		(2)	1	6	7	4	5	10	11	8	9	14	15	12	13	18	19	16	17	22	23	20	21	26	27	24
3			(3)	7	6	5	4	11	10	9	8	15	14	13	12	19	18	17	16	23	22	21	20	27	26	25
4				(4)	1	2	3	12	13	14	15	8	9	10	11	20	21	22	23	16	17	18	19	28	29	30
5					(5)	3	2	13	12	15	14	9	8	11	10	21	20	23	22	17	16	19	18	29	28	31
6						(6)	1	14	15	12	13	10	11	8	9	22	23	20	21	18	19	16	17	30	31	28
7							(7)	15	14	13	12	11	10	9	8	23	22	21	20	19	18	17	16	31	30	29
8								(8)	1	2	3	4	5	6	7	24	25	26	27	28	29	30	31	16	17	18
9									(9)	3	2	5	4	7	6	25	24	27	26	29	28	31	30	17	16	19
10										(10)	1	6	7	4	5	26	27	24	25	30	31	28	29	18	19	16
11											(11)	7	6	5	4	27	26	25	24	31	30	29	28	19	18	17
12												(12)	1	2	3	28	29	30	31	24	25	26	27	20	21	22
13													(13)	3	2	29	28	31	30	25	24	27	26	21	20	23
14														(14)	1	30	31	28	29	26	27	24	25	22	23	20
15															(15)	31	30	29	28	27	26	25	24	23	22	21
16																(16)	1	2	3	4	5	6	7	8	9	10
17																	(17)	3	2	5	4	7	6	9	8	11
18																		(18)	1	6	7	4	5	10	11	8
19																			(19)	7	6	5	4	11	10	9
20																				(20)	1	2	3	12	13	14
21																					(21)	3	2	13	12	15
22																						(22)	1	14	15	12
23																							(23)	15	14	13
24																								(24)	1	2
25																									(25)	3
26																										(26)

Interaction Table L_{64}, continued

Para-meter	Parameter 27	28	29	30	31	32	33	34	35	36	37	38	39	40	41	42	43	44	45	46	47	48	49	50	51	52
1	26	29	28	31	30	33	32	35	34	37	36	39	38	41	40	43	42	45	44	47	46	49	48	51	50	53
2	25	30	31	28	29	34	35	32	33	38	39	36	37	42	43	40	41	46	47	44	45	50	51	48	49	54
3	24	31	30	29	28	35	34	33	32	39	38	37	36	43	42	41	40	47	46	45	44	51	50	49	48	55
4	31	24	25	26	27	36	37	38	39	32	33	34	35	44	45	46	47	40	41	42	43	52	53	54	55	48
5	30	25	24	27	26	37	36	39	38	33	32	35	34	45	44	47	46	41	40	43	42	53	52	55	54	49
6	29	26	27	24	25	38	39	36	37	34	35	32	33	46	47	44	45	42	43	40	41	54	55	52	53	50
7	28	27	26	25	24	39	38	37	36	35	34	33	32	47	46	45	44	43	42	41	40	55	54	53	52	51
8	19	20	21	22	23	40	41	42	43	44	45	46	47	32	33	34	35	36	37	38	39	56	57	58	59	60
9	18	21	20	23	22	41	40	43	42	45	44	47	46	33	32	35	34	37	36	39	38	57	56	59	58	61
10	17	22	23	20	21	42	43	40	41	46	47	44	45	34	35	32	33	38	39	36	37	58	59	56	57	62
11	26	23	22	21	20	43	42	41	40	47	46	45	44	35	34	33	32	39	38	37	36	59	58	57	56	63
12	23	16	17	18	19	44	45	46	47	40	41	42	43	36	37	38	39	32	33	34	35	60	61	62	63	56
13	22	17	16	19	18	45	44	47	46	41	40	43	42	37	36	39	38	33	32	35	34	61	60	63	62	57
14	21	18	19	16	17	46	47	44	45	42	43	40	41	38	39	36	37	34	35	32	33	62	63	60	61	58
15	20	19	18	17	16	47	46	45	44	43	42	41	40	39	38	37	36	35	34	33	32	63	62	61	60	59
16	11	12	13	14	15	48	49	50	51	52	53	54	55	56	57	58	59	60	61	62	63	32	33	34	35	36
17	10	13	12	15	14	49	48	51	50	53	52	55	54	57	56	59	58	61	60	63	62	33	32	35	34	37
18	9	14	15	12	13	50	51	48	49	54	55	52	53	58	59	56	57	62	63	60	61	34	35	32	33	38
19	8	15	14	13	12	51	50	49	48	55	54	53	52	59	58	57	56	63	62	61	60	35	34	33	32	39
20	15	8	9	10	11	52	53	54	55	48	49	50	51	60	61	62	63	56	57	58	59	36	37	38	39	32
21	14	9	8	11	10	53	52	55	54	49	48	51	50	61	60	63	62	57	56	59	58	37	36	39	38	33
22	13	10	11	8	9	54	55	52	53	50	51	48	49	62	63	60	61	58	59	56	57	38	39	36	37	34
23	12	11	10	9	8	55	54	53	52	51	50	49	48	63	62	61	60	59	58	57	56	39	38	37	36	35
24	3	4	5	6	7	56	57	58	59	60	61	62	63	48	49	50	51	52	53	54	55	40	41	42	43	44
25	2	5	4	7	6	57	56	59	58	61	60	63	62	49	48	51	50	53	52	55	54	41	40	43	42	45
26	1	6	7	4	5	58	59	56	57	62	63	60	61	50	51	48	49	54	55	52	53	42	43	40	41	46
27	(27)	7	6	5	4	59	58	57	56	63	62	61	60	51	50	49	48	55	54	53	52	43	42	41	40	47
28		(28)	1	2	3	60	61	62	63	56	57	58	59	52	53	54	55	48	49	50	51	44	45	46	47	40
29			(29)	3	2	61	60	63	62	57	56	59	58	53	52	55	54	49	48	51	50	45	44	47	46	41
30				(30)	1	62	63	60	61	58	59	56	57	54	55	52	53	50	51	48	49	46	47	44	45	42
31					(31)	63	62	61	60	59	58	57	56	55	54	53	52	51	50	49	48	47	46	45	44	43
32						(32)	1	2	3	4	5	6	7	8	9	10	11	12	13	14	15	16	17	18	19	20
33							(33)	3	2	5	4	7	6	9	8	11	10	13	12	15	14	17	16	19	18	21
34								(34)	1	6	7	4	5	10	11	8	9	14	15	12	13	18	19	16	17	22
35									(35)	7	6	5	4	11	10	9	8	15	14	13	12	19	18	17	16	23
36										(36)	1	2	3	12	13	14	15	8	9	10	11	20	21	22	23	16
37											(37)	3	2	13	12	15	14	9	8	11	10	21	20	23	22	17
38												(38)	1	14	15	12	13	10	11	8	9	22	23	20	21	18
39													(39)	15	14	13	12	11	10	9	8	23	22	21	20	19
40														(40)	1	2	3	4	5	6	7	24	25	26	27	28
41															(41)	3	2	5	4	7	6	25	24	27	26	29
42																(42)	1	6	7	4	5	26	27	24	25	30
43																	(43)	7	6	5	4	27	26	25	24	31
44																		(44)	1	2	3	28	29	30	31	24
45																			(45)	3	2	29	28	31	30	25
46																				(46)	1	30	31	28	29	26

Interaction Table L_{64}, continued

Parameter	Parameter										
	53	54	55	56	57	58	59	60	61	62	63
1	52	55	54	57	56	59	58	61	60	63	62
2	55	52	53	58	59	56	57	62	63	60	61
3	54	53	52	59	58	57	56	63	62	61	60
4	49	50	51	60	61	62	63	56	57	58	59
5	48	51	50	61	60	63	62	57	56	59	58
6	51	48	49	62	63	60	61	58	59	56	57
7	50	49	48	63	62	61	60	59	58	57	56
8	61	62	63	48	49	50	51	52	53	54	55
9	60	63	62	49	48	51	50	53	52	55	54
10	63	60	61	50	51	48	49	54	55	52	53
11	62	61	60	51	50	49	48	55	54	53	52
12	57	58	59	52	53	54	55	48	49	50	51
13	56	59	58	53	52	55	54	49	48	51	50
14	59	56	57	54	55	52	53	50	51	48	49
15	58	57	56	55	54	53	52	51	50	49	48
16	37	38	39	40	41	42	43	44	45	46	47
17	36	39	38	41	40	43	42	45	44	47	46
18	39	36	37	42	43	40	41	46	47	44	45
19	38	37	36	43	42	41	40	47	46	45	44
20	33	34	35	44	45	46	47	40	41	42	43
21	32	35	34	45	44	47	46	41	40	43	42
22	35	32	33	46	47	44	45	42	43	40	41
23	34	33	32	47	46	45	44	43	42	41	40

(continued)

(continued)

Parameter	Parameter										
	53	54	55	56	57	58	59	60	61	62	63
24	45	46	47	32	33	34	35	36	37	38	39
25	44	47	46	33	32	35	34	37	36	39	38
26	47	44	45	34	35	32	33	38	39	36	37
27	46	45	44	35	34	33	32	39	38	37	36
28	41	42	43	36	37	38	39	32	33	34	35
29	40	43	42	37	36	39	38	33	32	35	34
30	43	40	41	38	39	36	37	34	35	32	33
31	42	41	40	39	38	37	36	35	34	33	32
32	21	22	23	24	25	26	27	28	29	30	31
33	20	23	22	25	24	27	26	29	28	31	30
34	23	20	21	26	27	24	25	30	31	28	29
35	22	21	20	27	26	25	24	31	30	29	28
36	17	18	19	28	29	30	31	24	25	26	27
37	16	19	18	29	28	31	30	25	24	27	26
38	19	16	17	30	31	28	29	26	27	24	25
39	18	17	16	31	30	29	28	27	26	25	24
40	29	30	31	16	17	18	19	20	21	22	23
41	28	31	30	17	16	19	18	21	20	23	22
42	31	28	29	18	19	16	17	22	23	20	21
43	30	29	28	19	18	17	16	23	22	21	20
44	25	26	27	20	21	22	23	16	17	18	19
45	24	27	26	21	20	23	22	17	16	19	18
46	27	24	25	22	23	20	21	18	19	16	17
47	26	25	24	23	22	21	20	19	18	17	16

Interaction Table L_{64}, continued

Para-	Parameter										
meter	53	54	55	56	57	58	59	60	61	62	63
48	5	6	7	8	9	10	11	12	13	14	15
49	4	7	6	9	8	11	10	13	12	15	14
50	7	4	5	10	11	8	9	14	15	12	13
51	6	5	4	11	10	9	8	15	14	13	12
52	1	2	3	12	13	14	15	8	9	10	11
53	(53)	3	2	13	12	15	14	9	8	11	10
54		(54)	1	14	15	12	13	10	11	8	9
55			(55)	15	14	13	12	11	10	9	8
56				(56)	1	2	3	4	5	6	7
57					(57)	3	2	5	4	7	6
58						(58)	1	6	7	4	5
59							(59)	7	6	5	4
60								(60)	1	2	3
61									(61)	3	2
62										(62)	1

Modified Orthogonal Array L_{64} (4^{21})

Experiment No.	Parameter																				
	1	2	3	4	5	6	7	8	9	10	11	12	13	14	15	16	17	18	19	20	21
1	1	1	1	1	1	1	1	1	1	1	1	1	1	1	1	1	1	1	1	1	1
2	1	1	1	1	1	2	2	2	2	2	2	2	2	2	2	2	2	2	2	2	2
3	1	1	1	1	1	3	3	3	3	3	3	3	3	3	3	3	3	3	3	3	3
4	1	1	1	1	1	4	4	4	4	4	4	4	4	4	4	4	4	4	4	4	4
5	1	2	2	2	2	1	1	1	1	2	2	2	2	3	3	3	3	4	4	4	4
6	1	2	2	2	2	2	2	2	2	1	1	1	1	4	4	4	4	3	3	3	3
7	1	2	2	2	2	3	3	3	3	4	4	4	4	1	1	1	1	2	2	2	2
8	1	2	2	2	2	4	4	4	4	3	3	3	3	2	2	2	2	1	1	1	1
9	1	3	3	3	3	1	1	1	1	3	3	3	3	4	4	4	4	2	2	2	2
10	1	3	3	3	3	2	2	2	2	4	4	4	4	3	3	3	3	1	1	1	1
11	1	3	3	3	3	3	3	3	3	1	1	1	1	2	2	2	2	4	4	4	4
12	1	3	3	3	3	4	4	4	4	2	2	2	2	1	1	1	1	3	3	3	3
13	1	4	4	4	4	1	1	1	1	4	4	4	4	2	2	2	2	3	3	3	3
14	1	4	4	4	4	2	2	2	2	3	3	3	3	1	1	1	1	4	4	4	4
15	1	4	4	4	4	3	3	3	3	2	2	2	2	4	4	4	4	1	1	1	1
16	1	4	4	4	4	4	4	4	4	1	1	1	1	3	3	3	3	2	2	2	2
17	2	1	2	3	4	1	2	3	4	1	2	3	4	1	2	3	4	1	2	3	4
18	2	1	2	3	4	2	1	4	3	2	1	4	3	2	1	4	3	2	1	4	3
19	2	1	2	3	4	3	4	1	2	3	4	1	2	3	4	1	2	3	4	1	2
20	2	1	2	3	4	4	3	2	1	4	3	2	1	4	3	2	1	4	3	2	1
21	2	2	1	4	3	1	2	3	4	2	1	4	3	3	4	1	2	4	3	2	1
22	2	2	1	4	3	2	1	4	3	1	2	3	4	4	3	2	1	3	4	1	2
23	2	2	1	4	3	3	4	1	2	4	3	2	1	1	2	3	4	2	1	4	3

(continued)

(continued)

Experiment No.	Parameter																				
	1	2	3	4	5	6	7	8	9	10	11	12	13	14	15	16	17	18	19	20	21
24	2	2	1	4	3	4	3	2	1	3	4	1	2	2	1	4	3	1	2	3	4
25	2	3	4	1	2	1	2	3	4	3	4	1	2	4	3	2	1	2	1	4	3
26	2	3	4	1	2	2	1	4	3	4	3	2	1	3	4	1	2	1	2	3	4
27	2	3	4	1	2	3	4	1	2	1	2	3	4	2	1	4	3	4	3	2	1
28	2	3	4	1	2	4	3	2	1	2	1	4	3	1	2	3	4	3	4	1	2
29	2	4	3	2	1	1	2	3	4	4	3	2	1	2	1	4	3	3	4	1	2
30	2	4	3	2	1	2	1	4	3	3	4	1	2	1	2	3	4	4	3	2	1
31	2	4	3	2	1	3	4	1	2	2	1	4	3	4	3	2	1	1	2	3	4
32	2	4	3	2	1	4	3	2	1	1	2	3	4	3	4	1	2	2	1	4	3
33	3	1	3	4	2	1	3	4	2	1	3	4	2	1	3	4	2	1	3	4	2
34	3	1	3	4	2	2	4	3	1	2	4	3	1	2	4	3	1	2	4	3	1
35	3	1	3	4	2	3	1	2	4	3	1	2	4	3	1	2	4	3	1	2	4
36	3	1	3	4	2	4	2	1	3	4	2	1	3	4	2	1	3	4	2	1	3
37	3	2	4	3	1	1	3	4	2	2	4	3	1	3	1	2	4	4	2	1	3
38	3	2	4	3	1	2	4	3	1	1	3	4	2	4	2	1	3	3	1	2	4
39	3	2	4	3	1	3	1	2	4	4	1	1	3	1	3	4	2	2	4	3	1
40	3	2	4	3	1	4	2	1	3	3	2	2	4	2	4	3	1	1	3	4	2
41	3	3	1	2	4	1	3	4	2	3	1	2	4	4	2	1	3	2	4	3	1
42	3	3	1	2	4	2	4	3	1	4	2	1	3	3	1	2	4	1	3	4	2
43	3	3	1	2	4	3	1	2	4	1	3	4	2	2	4	3	1	4	2	1	3
44	3	3	1	2	4	4	2	1	3	2	4	3	1	1	3	4	2	3	1	2	4
45	3	4	2	1	3	1	3	4	2	4	2	1	3	2	4	3	1	3	1	2	4
46	3	4	2	1	3	2	4	3	1	3	1	2	4	1	3	4	2	4	2	1	3
47	3	4	2	1	3	3	1	2	4	2	4	3	1	4	2	1	3	1	3	4	2

Orthogonal Array L_{64} (4^{21}), continued

Experiment No.	Parameter																				
	1	2	3	4	5	6	7	8	9	10	11	12	13	14	15	16	17	18	19	20	21
48	3	4	2	1	3	4	2	1	3	1	3	4	2	3	1	2	4	2	4	3	1
49	4	1	4	2	3	1	4	2	3	1	4	2	3	1	4	2	3	1	4	2	3
50	4	1	4	2	3	2	3	1	4	2	3	1	4	2	3	1	4	2	3	1	4
51	4	1	4	2	3	3	2	4	1	3	2	4	1	3	2	4	1	3	2	4	1
52	4	1	4	2	3	4	1	3	2	4	1	3	2	4	1	3	2	4	1	3	2
53	4	2	3	1	4	1	4	2	3	2	3	1	4	3	2	4	1	4	1	3	2
54	4	2	3	1	4	2	3	1	4	1	4	2	3	4	1	3	2	3	2	4	1
55	4	2	3	1	4	3	2	4	1	4	1	3	2	1	4	2	3	2	3	1	4
56	4	2	3	1	4	4	1	3	2	3	2	4	1	2	3	1	4	1	4	2	3
57	4	3	2	4	1	1	4	2	3	3	2	4	1	4	1	3	2	2	3	1	4
58	4	3	2	4	1	2	3	1	4	4	1	3	2	3	2	4	1	1	4	2	3
59	4	3	2	4	1	3	2	4	1	1	4	2	3	2	3	1	4	4	1	3	2
60	4	3	2	4	1	4	1	3	2	2	3	1	4	1	4	2	3	3	2	4	1
61	4	4	1	3	2	1	4	2	3	4	1	3	2	2	3	1	4	3	2	4	1
62	4	4	1	3	2	2	3	1	4	3	2	4	1	1	4	2	3	4	1	3	2
63	4	4	1	3	2	3	2	4	1	2	3	1	4	4	1	3	2	1	4	2	3
64	4	4	1	3	2	4	1	3	2	1	4	2	3	3	2	4	1	2	3	1	4

Interaction Table L_{64} (4^{21})

Para-	Parameter																				
meter	1	2	3	4	5	6	7	8	9	10	11	12	13	14	15	16	17	18	19	20	21
		3	2	2	2	7	6	6	6	11	10	10	10	15	14	14	14	19	18	18	18
1	(1)	4	4	3	3	8	8	7	7	12	12	11	11	16	16	15	15	20	20	19	19
		5	5	5	4	9	9	9	8	13	13	13	12	17	17	17	16	21	21	21	20
			1	1	1	10	11	12	13	6	7	8	9	6	7	8	9	6	7	8	9
2		(2)	4	3	3	14	15	16	17	14	15	16	17	10	11	12	13	10	11	12	13
			5	5	4	18	19	20	21	18	19	20	21	18	19	20	21	14	15	16	17
				1	1	11	10	13	12	7	6	9	8	8	9	6	7	9	8	7	6
3			(3)	2	2	16	17	14	15	17	16	15	14	13	12	11	10	12	13	10	11
				5	4	21	20	19	18	20	21	18	19	19	18	21	20	15	14	17	16
					1	12	13	10	11	8	9	6	7	9	8	7	6	7	6	9	8
4				(4)	2	17	16	15	14	15	14	17	16	11	10	13	12	13	12	11	10
					3	19	18	21	20	21	20	19	18	20	21	18	19	16	17	14	15
						13	12	11	10	9	8	7	6	7	6	9	8	8	9	6	7
5					(5)	15	14	17	16	16	17	14	15	12	13	10	11	11	10	13	12
						20	21	18	19	19	18	21	20	21	20	19	18	17	16	15	14
							1	1	1	2	3	4	5	2	5	3	4	2	4	5	3
6						(6)	8	7	7	14	16	17	15	10	13	11	12	10	12	13	11
							9	9	8	18	21	19	20	18	20	21	19	14	17	15	16
								1	1	3	2	5	4	5	2	4	3	4	2	3	5
7							(7)	6	6	17	15	14	16	12	11	13	10	13	11	10	12
								9	8	20	19	21	18	21	19	18	20	16	15	17	14
									1	4	5	2	3	3	4	2	5	5	3	2	4
8								(8)	6	15	17	16	14	13	10	12	11	11	13	12	10
									7	21	18	20	19	19	21	20	18	17	14	16	15
										5	4	3	2	4	3	5	2	3	5	4	2
9									(9)	16	14	15	17	11	12	10	13	12	10	11	13
										19	20	18	21	20	18	19	21	15	16	14	17
											1	1	1	2	4	5	3	2	5	3	4
10										(10)	12	11	11	6	8	9	7	6	9	7	8
											13	13	12	18	21	19	20	14	16	17	15
												1	1	4	2	3	5	5	2	4	3
11											(11)	10	10	9	7	6	8	8	7	9	6
												13	12	20	19	21	18	17	15	14	16
													1	5	3	2	4	3	4	2	5
12												(12)	10	7	9	8	6	9	6	8	7
													11	21	18	20	19	15	17	16	14
														3	5	4	2	4	3	5	2
13													(13)	8	6	7	9	7	8	6	9
														19	20	18	21	16	14	15	17
															1	1	1	2	3	4	5
14														(14)	16	15	15	6	8	9	7
															17	17	16	10	13	11	12
																1	1	3	2	5	4
15															(15)	14	14	9	7	6	8
																17	16	12	11	13	10

Interaction Table L_{64} (4^{21}), continued

Para-	Parameter					
meter	16	17	18	19	20	21
		1	4	5	2	3
16	(16)	14	7	9	8	6
		15	13	10	12	11
			5	4	3	2
17		(17)	8	6	7	9
			11	12	10	13
				1	1	1
18			(18)	20	19	19
				21	21	20
					1	1
19				(19)	18	18
					21	20
						1
20					(20)	18
						19

Orthogonal Array L_{81} (3^{40})

Experiment No.	Parameter																			
	1	2	3	4	5	6	7	8	9	10	11	12	13	14	15	16	17	18	19	20
1	1	1	1	1	1	1	1	1	1	1	1	1	1	1	1	1	1	1	1	1
2	1	1	1	1	1	1	1	1	1	1	1	1	1	2	2	2	2	2	2	2
3	1	1	1	1	1	1	1	1	1	1	1	1	1	3	3	3	3	3	3	3
4	1	1	1	1	2	2	2	2	2	2	2	2	2	1	1	1	1	1	1	1
5	1	1	1	1	2	2	2	2	2	2	2	2	2	2	2	2	2	2	2	2
6	1	1	1	1	2	2	2	2	2	2	2	2	2	3	3	3	3	3	3	3
7	1	1	1	1	3	3	3	3	3	3	3	3	3	1	1	1	1	1	1	1
8	1	1	1	1	3	3	3	3	3	3	3	3	3	2	2	2	2	2	2	2
9	1	1	1	1	3	3	3	3	3	3	3	3	3	3	3	3	3	3	3	3
10	1	2	2	2	1	1	1	2	2	2	3	3	3	1	1	1	2	2	2	3
11	1	2	2	2	1	1	1	2	2	2	3	3	3	2	2	2	3	3	3	1
12	1	2	2	2	1	1	1	2	2	2	3	3	3	3	3	3	1	1	1	2
13	1	2	2	2	2	2	2	3	3	3	1	1	1	1	1	1	2	2	2	3
14	1	2	2	2	2	2	2	3	3	3	1	1	1	2	2	2	3	3	3	1
15	1	2	2	2	2	2	2	3	3	3	1	1	1	3	3	3	1	1	1	2
16	1	2	2	2	3	3	3	1	1	1	2	2	2	1	1	1	2	2	2	3
17	1	2	2	2	3	3	3	1	1	1	2	2	2	2	2	2	3	3	3	1
18	1	2	2	2	3	3	3	1	1	1	2	2	2	3	3	3	1	1	1	2
19	1	3	3	3	1	1	1	3	3	3	2	2	2	1	1	1	3	3	3	2
20	1	3	3	3	1	1	1	3	3	3	2	2	2	2	2	2	1	1	1	3
21	1	3	3	3	1	1	1	3	3	3	2	2	2	3	3	3	2	2	2	1
22	1	3	3	3	2	2	2	1	1	1	3	3	3	1	1	1	3	3	3	2
23	1	3	3	3	2	2	2	1	1	1	3	3	3	2	2	2	1	1	1	3

(continued)

(continued)

Experiment No.	Parameter																			
	1	2	3	4	5	6	7	8	9	10	11	12	13	14	15	16	17	18	19	20
24	1	3	3	3	2	2	2	1	1	1	3	3	3	3	3	3	2	2	2	1
25	1	3	3	3	3	3	3	2	2	2	1	1	1	1	1	1	3	3	3	2
26	1	3	3	3	3	3	3	2	2	2	1	1	1	2	2	2	1	1	1	3
27	1	3	3	3	3	3	3	2	2	2	1	1	1	3	3	3	2	2	2	1
28	2	1	2	3	1	2	3	1	2	3	1	2	3	1	2	3	1	2	3	1
29	2	1	2	3	1	2	3	1	2	3	1	2	3	2	3	1	2	3	1	2
30	2	1	2	3	1	2	3	1	2	3	1	2	3	3	1	2	3	1	2	3
31	2	1	2	3	2	3	1	2	3	1	2	3	1	1	2	3	1	2	3	1
32	2	1	2	3	2	3	1	2	3	1	2	3	1	2	3	1	2	3	1	2
33	2	1	2	3	2	3	1	2	3	1	2	3	1	3	1	2	3	1	2	3
34	2	1	2	3	3	1	2	3	1	2	3	1	2	1	2	3	1	2	3	1
35	3	1	1	2	3	1	2	3	1	2	3	3	1	2	3	1	2	3	1	2
36	2	1	2	3	3	1	2	3	1	2	3	1	2	3	1	2	3	1	2	3
37	2	2	3	1	1	2	3	2	3	1	3	1	2	1	2	3	2	3	1	3
38	2	2	3	1	1	2	3	2	3	1	3	1	2	2	3	1	3	1	2	1
39	2	2	3	1	1	2	3	2	3	1	3	1	2	3	1	2	1	2	3	2
40	2	2	3	1	2	3	1	3	1	2	1	2	3	1	2	3	2	3	1	3
41	2	2	3	1	2	3	1	3	1	2	1	2	3	2	3	1	3	1	2	1
42	2	2	3	1	2	3	1	3	1	2	1	2	3	3	1	2	1	2	3	2
43	2	2	3	1	3	1	2	1	2	3	2	3	1	1	2	3	2	3	1	3
44	2	2	3	1	3	1	2	1	2	3	2	3	1	2	3	1	3	1	2	1
45	2	2	3	1	3	1	2	1	2	3	2	3	1	3	1	2	1	2	3	2

Orthogonal Array L_{81} (3^{40}), continued

Experiment No.	Parameter																			
	1	2	3	4	5	6	7	8	9	10	11	12	13	14	15	16	17	18	19	20
46	2	3	1	2	1	2	3	3	1	2	2	3	1	1	2	3	3	1	2	2
47	2	3	1	2	1	2	3	3	1	2	2	3	1	2	3	1	1	2	3	3
48	2	3	1	2	1	2	3	3	1	2	2	3	1	3	1	2	2	3	1	1
49	2	3	1	2	2	3	1	1	2	3	3	1	2	1	2	3	3	1	2	2
50	2	3	1	2	2	3	1	1	2	3	3	1	2	2	3	1	1	2	3	3
51	2	3	1	2	2	3	1	1	2	3	3	1	2	3	1	2	2	3	1	1
52	2	3	1	2	3	1	2	2	3	1	1	2	3	1	2	3	3	1	2	2
53	2	3	1	2	3	1	2	2	3	1	1	2	3	2	3	1	1	2	3	3
54	2	3	1	2	3	1	2	2	3	1	1	2	3	3	1	2	2	3	1	1
55	3	1	3	2	1	3	2	1	3	2	1	3	2	1	3	2	1	3	2	1
56	3	1	3	2	1	3	2	1	3	2	1	3	2	2	1	3	2	1	3	2
57	3	1	3	2	1	3	2	1	3	2	1	3	2	3	2	1	3	2	1	3
58	3	1	3	2	2	1	3	2	1	3	2	1	3	1	3	2	1	3	2	1
59	3	1	3	2	2	1	3	2	1	3	2	1	3	2	1	3	2	1	3	2
60	3	1	3	2	2	1	3	2	1	3	2	1	3	3	2	1	3	2	1	3
61	3	2	3	2	1	3	2	1	3	2	1	2	1	3	2	1	3	2	1	3
62	3	1	3	2	3	2	1	3	2	1	3	2	1	2	1	3	2	1	3	2
63	3	1	3	2	3	2	1	3	2	1	3	2	1	3	2	1	3	2	1	3
64	3	2	1	3	1	3	2	2	1	3	3	2	1	1	3	2	2	1	3	3
65	3	2	1	3	1	3	2	2	1	3	3	2	1	2	1	3	3	2	1	1
66	3	2	1	3	1	3	2	2	1	3	3	2	1	3	2	1	1	3	2	2
67	3	2	1	3	2	1	3	3	2	1	1	3	2	1	3	2	2	1	3	3
68	3	2	1	3	2	1	3	3	2	1	1	3	2	2	1	3	3	2	1	1

(continued)

(continued)

Experiment No.	Parameter																			
	1	2	3	4	5	6	7	8	9	10	11	12	13	14	15	16	17	18	19	20
69	3	2	1	3	2	1	3	3	2	1	1	3	2	3	2	1	1	3	2	2
70	3	2	1	3	3	2	1	1	3	2	2	1	3	1	3	2	2	1	3	3
71	3	2	1	3	3	2	1	1	3	2	2	1	3	2	1	3	3	2	1	1
72	3	2	1	3	3	2	1	1	3	2	2	1	3	3	2	1	1	3	2	2
73	3	3	2	1	1	3	2	3	2	1	2	1	3	1	3	2	3	2	1	2
74	3	3	2	1	1	3	2	3	2	1	2	1	3	2	1	3	1	3	2	3
75	3	3	2	1	1	3	2	3	2	1	2	1	3	3	2	1	2	1	3	1
76	3	3	2	1	2	1	3	1	3	2	3	2	1	1	3	2	3	2	1	2
77	3	3	2	1	2	1	3	1	3	2	3	2	1	2	1	3	1	3	2	3
78	3	3	2	1	2	1	3	1	3	2	3	2	1	3	2	1	2	1	3	1
79	3	3	2	1	3	2	1	2	1	3	1	3	2	1	3	2	3	2	1	2
80	3	3	2	1	3	2	1	2	1	3	1	3	2	2	1	3	1	3	2	3
81	3	3	2	1	3	2	1	2	1	3	1	3	2	3	2	1	2	1	3	1

Orthogonal Array L_{81} (3^{40}), continued

Experiment No.	Parameter																			
	21	22	23	24	25	26	27	28	29	30	31	32	33	34	35	36	37	38	39	40
1	1	1	1	1	1	1	1	1	1	1	1	1	1	1	1	1	1	1	1	1
2	2	2	2	2	2	2	2	2	2	2	2	2	2	2	2	2	2	2	2	2
3	3	3	3	3	3	3	3	3	3	3	3	3	3	3	3	3	3	3	3	3
4	1	1	2	2	2	2	2	2	2	2	2	3	3	3	3	3	3	3	3	3
5	2	2	3	3	3	3	3	3	3	3	3	1	1	1	1	1	1	1	1	1
6	3	3	1	1	1	1	1	1	1	1	1	2	2	2	2	2	2	2	2	2
7	1	1	3	3	3	3	3	3	3	3	3	2	2	2	2	2	2	2	2	2
8	2	2	1	1	1	1	1	1	1	1	1	3	3	3	3	3	3	3	3	3
9	3	3	2	2	2	2	2	2	2	2	2	1	1	1	1	1	1	1	1	1
10	3	3	1	1	1	2	2	2	3	3	3	1	1	1	2	2	2	3	3	3
11	1	1	2	2	2	3	3	3	1	1	1	2	2	2	3	3	3	1	1	1
12	2	2	3	3	3	1	1	1	2	2	2	3	3	3	1	1	1	2	2	2
13	3	3	2	2	2	3	3	3	1	1	1	3	3	3	1	1	1	2	2	2
14	1	1	3	3	3	1	1	1	2	2	2	1	1	1	2	2	2	3	3	3
15	2	2	1	1	1	2	2	2	3	3	3	2	2	2	3	3	3	1	1	1
16	3	3	3	3	3	1	1	1	2	2	2	2	2	2	3	3	3	1	1	1
17	1	1	1	1	1	2	2	2	3	3	3	3	3	3	1	1	1	2	2	2
18	2	2	2	2	2	3	3	3	1	1	1	1	1	1	2	2	2	3	3	3
19	2	2	1	1	1	3	3	3	2	2	2	1	1	1	3	3	3	2	2	2
20	3	3	2	2	2	1	1	1	3	3	3	2	2	2	1	1	1	3	3	3
21	1	1	3	3	3	2	2	2	1	1	1	3	3	3	2	2	2	1	1	1
22	2	2	2	2	2	1	1	1	3	3	3	3	3	3	2	2	2	1	1	1
23	3	3	3	3	3	2	2	2	1	1	1	1	1	1	3	3	3	2	2	2

(continued)

(continued)

Experiment No.	Parameter																			
	21	22	23	24	25	26	27	28	29	30	31	32	33	34	35	36	37	38	39	40
24	1	1	1	1	1	3	3	3	2	2	2	2	2	2	1	1	1	3	3	3
25	2	2	3	3	3	2	2	2	1	1	1	2	2	2	1	1	1	3	3	3
26	3	3	1	1	1	3	3	3	2	2	2	3	3	3	2	2	2	1	1	1
27	1	1	2	2	2	1	1	1	3	3	3	1	1	1	3	3	3	2	2	2
28	2	3	1	2	3	1	2	3	1	2	3	1	2	3	1	2	3	1	2	3
29	3	1	2	3	1	2	3	1	2	3	1	2	3	1	2	3	1	2	3	1
30	1	2	3	1	2	3	1	2	3	1	2	3	1	2	3	1	2	3	1	2
31	2	3	2	3	1	2	3	1	2	3	1	3	1	2	3	1	2	3	1	2
32	3	1	3	1	2	3	1	2	3	1	2	1	2	3	1	2	3	1	2	3
33	1	2	1	2	3	1	2	3	1	2	3	2	3	1	2	3	1	2	3	1
34	2	3	3	1	2	3	1	2	3	1	2	2	3	1	2	3	1	2	3	1
35	1	2	2	3	1	2	3	1	2	3	1	1	2	3	1	2	3	1	2	3
36	1	2	2	3	1	2	3	1	2	3	1	1	2	3	1	2	3	1	2	3
37	1	2	1	2	3	2	3	1	3	1	2	1	2	3	2	3	1	3	1	2
38	2	3	2	3	1	3	1	2	1	2	3	2	3	1	3	1	2	1	2	3
39	3	1	3	1	2	1	2	3	2	3	1	3	1	2	1	2	3	2	3	1
40	1	2	2	3	1	3	1	2	1	2	3	3	1	2	1	2	3	2	3	1
41	2	3	3	1	2	1	2	3	2	3	1	1	2	3	2	3	1	3	1	2
42	3	1	1	2	3	2	3	1	3	1	2	2	3	1	3	1	2	1	2	3
43	1	2	3	1	2	1	2	3	2	3	1	2	3	1	3	1	2	1	2	3
44	2	3	1	2	3	2	3	1	3	1	2	3	1	2	1	2	3	2	3	1
45	3	1	2	3	1	3	1	2	1	2	3	1	2	3	2	3	1	3	1	2

Orthogonal Array L_{81} (3^{40}), continued

Experiment No.	Parameter																			
	21	22	23	24	25	26	27	28	29	30	31	32	33	34	35	36	37	38	39	40
46	3	1	1	2	3	3	1	2	2	3	1	1	2	3	3	1	2	2	3	1
47	1	2	2	3	1	1	2	3	3	1	2	2	3	1	1	2	3	3	1	2
48	2	3	3	1	2	2	3	1	1	2	3	3	1	2	2	3	1	1	2	3
49	3	1	2	3	1	1	2	3	3	1	2	3	1	2	2	3	1	1	2	3
50	1	2	3	1	2	2	3	1	1	2	3	1	2	3	3	1	2	2	3	1
51	2	3	1	2	3	3	1	2	2	3	1	2	3	1	1	2	3	3	1	2
52	3	1	3	1	2	2	3	1	1	2	3	2	3	1	1	2	3	3	1	2
53	1	2	1	2	3	3	1	2	2	3	1	3	1	2	2	3	1	1	2	3
54	2	3	2	3	1	1	2	3	3	1	2	1	2	3	3	1	2	2	3	1
55	3	2	1	3	2	1	3	2	1	3	2	1	3	2	1	3	2	1	3	2
56	1	3	2	1	3	2	1	3	2	1	3	2	1	3	2	1	3	2	1	3
57	2	1	3	2	1	3	2	1	3	2	1	3	2	1	3	2	1	3	2	1
58	3	2	2	1	3	2	1	3	2	1	3	3	2	1	3	2	1	3	2	1
59	1	3	3	2	1	3	2	1	3	2	1	1	3	2	1	3	2	1	3	2
60	2	1	1	3	2	1	3	2	1	3	2	2	1	3	2	1	3	2	1	3
61	2	1	2	1	3	2	1	3	2	1	3	1	3	2	1	3	2	1	3	2
62	1	3	1	3	2	1	3	2	1	3	2	3	2	1	3	2	1	3	2	1
63	2	1	2	1	3	2	1	3	2	1	3	1	3	2	1	3	2	1	3	2
64	2	1	1	3	2	2	1	3	3	2	1	1	3	2	2	1	3	3	2	1
65	3	2	2	1	3	3	2	1	1	3	2	2	1	3	3	2	1	1	3	2
66	1	3	3	2	1	1	3	2	2	1	3	3	2	1	1	3	2	2	1	3
67	2	1	2	1	3	3	2	1	1	3	2	3	2	1	1	3	2	2	1	3
68	3	2	3	2	1	1	3	2	2	1	3	1	3	2	2	1	3	3	2	1

(continued)

(continued)

Experiment No.	Parameter																			
	21	22	23	24	25	26	27	28	29	30	31	32	33	34	35	36	37	38	39	40
69	1	3	1	3	2	2	1	3	3	2	1	2	1	3	3	2	1	1	3	2
70	2	1	3	2	1	1	3	2	2	1	3	2	1	3	3	2	1	1	3	2
71	3	2	1	3	2	2	1	3	3	2	1	3	2	1	1	3	2	2	1	3
72	1	3	2	1	3	3	2	1	1	3	2	1	3	2	2	1	3	3	2	1
73	1	3	1	3	2	3	2	1	2	1	3	1	3	2	3	2	1	2	1	3
74	2	1	2	1	3	1	3	2	3	2	1	2	1	3	1	3	2	3	2	1
75	3	2	3	2	1	2	1	3	1	3	2	3	2	1	2	1	3	1	3	2
76	1	3	2	1	3	1	3	2	3	2	1	3	2	1	2	1	3	1	3	2
77	2	1	3	2	1	2	1	3	1	3	2	1	3	2	3	2	1	2	1	3
78	3	2	1	3	2	3	2	1	2	1	3	2	1	3	1	3	2	3	2	1
79	1	3	3	2	1	2	1	3	1	3	2	2	1	3	1	3	2	3	2	1
80	2	1	1	3	2	3	2	1	2	1	3	3	2	1	2	1	3	1	3	2
81	3	2	2	1	3	1	3	2	3	2	1	1	3	2	3	2	1	2	1	3

Interaction Table L_{81}

Para-	Parameter																			
meter	1	2	3	4	5	6	7	8	9	10	11	12	13	14	15	16	17	18	19	20
1	(1)	3	2	2	6	5	5	9	8	8	12	11	11	15	14	14	18	17	17	21
		4	4	3	7	7	6	10	10	9	13	13	12	16	16	15	19	19	18	22
2		(2)	1	1	8	9	10	5	6	7	5	6	7	17	18	19	14	15	16	14
			4	3	11	12	13	11	12	13	8	9	10	20	21	22	20	21	22	17
3			(3)	1	9	10	8	7	5	6	6	7	5	18	19	17	16	14	15	15
				2	13	11	12	12	13	11	10	8	9	22	20	21	21	22	20	19
4				(4)	10	8	9	6	7	5	7	5	6	19	17	18	15	16	14	16
					12	13	11	13	11	12	9	10	8	21	22	20	22	20	21	18
5					(5)	1	1	2	3	4	2	4	3	23	24	25	26	27	28	29
						7	6	11	13	12	8	10	9	32	33	34	35	36	37	38
6						(6)	1	4	2	3	3	2	4	24	25	23	27	28	26	30
							5	13	12	11	10	9	8	34	32	33	37	35	36	40
7							(7)	3	4	2	4	3	2	25	23	24	28	26	27	31
								12	11	13	9	8	10	33	34	32	36	37	35	39
8								(8)	1	1	2	3	4	26	27	28	29	30	31	23
									10	9	5	7	6	38	39	40	32	33	34	35
9									(9)	1	4	2	3	27	28	26	30	31	29	24
										8	7	6	5	40	38	39	34	32	33	37
10										(10)	3	4	2	28	26	27	31	29	30	25
											6	5	7	39	40	38	33	34	32	36
11											(11)	1	1	29	30	31	23	24	25	26
												13	12	35	36	37	38	39	40	32
12												(12)	1	30	31	29	24	25	23	27
													11	37	35	36	40	38	39	34
13													(13)	31	29	30	25	23	24	28
														36	37	35	39	40	38	33
14														(14)	1	1	2	3	4	2
															16	15	20	22	21	17
15															(15)	1	4	2	3	3
																14	22	21	20	19
16																(16)	3	4	2	4
																	21	20	22	18
17																	(17)	1	1	2
																		19	18	14
18																		(18)	1	4
																			17	16
19																			(19)	3
																				15

Interaction Table L_{81}, continued

Parameter	Parameter																			
	21	22	23	24	25	26	27	28	29	30	31	32	33	34	35	36	37	38	39	40
1	20	20	24	23	23	27	26	26	30	29	29	33	32	32	36	35	35	39	38	38
	22	21	25	25	24	28	28	27	31	31	30	34	34	33	37	37	36	40	40	39
2	15	16	26	27	28	23	24	25	23	24	25	35	36	37	32	33	34	32	33	34
	18	19	29	30	31	29	30	31	26	27	28	38	39	40	38	39	40	35	36	37
3	16	14	27	28	26	25	23	24	24	25	23	36	37	35	34	32	33	33	34	32
	17	18	31	29	30	30	31	29	28	26	27	40	38	39	39	40	38	37	35	36
4	14	15	28	26	27	24	25	23	25	23	24	37	35	36	33	34	32	34	32	33
	19	17	30	31	29	31	29	30	27	28	26	39	40	38	40	38	39	36	37	35
5	30	31	14	15	16	17	18	19	20	21	22	14	15	16	17	18	19	20	21	22
	39	40	32	33	34	35	36	37	38	39	40	23	24	25	26	27	28	29	30	31
6	31	29	16	14	15	19	17	18	22	20	21	15	16	14	18	19	17	21	22	20
	38	39	33	34	32	36	37	35	39	40	38	25	23	24	28	26	27	31	29	30
7	29	30	15	16	14	18	19	17	21	22	20	16	14	15	19	17	18	22	20	21
	40	38	34	32	33	37	35	36	40	38	39	24	25	23	27	28	26	30	31	29
8	24	25	20	21	22	14	15	16	17	18	19	17	18	19	20	21	22	14	15	16
	36	37	35	36	37	38	39	40	32	33	34	29	30	31	23	24	25	26	27	28
9	25	23	22	20	21	16	14	15	19	17	18	18	19	17	21	22	20	15	16	14
	35	36	36	37	35	39	40	38	33	34	32	31	29	30	25	23	24	28	26	27
10	23	24	21	22	20	15	16	14	18	19	17	19	17	18	22	20	21	16	14	15
	37	35	37	35	36	40	38	39	34	32	33	30	31	29	24	25	23	27	28	26
11	27	28	17	18	19	20	21	22	14	15	16	20	21	22	14	15	16	17	18	19
	33	34	38	39	40	32	33	34	35	36	37	26	27	28	29	30	31	23	24	25
12	28	26	19	17	18	22	20	21	16	14	15	21	22	20	15	16	14	18	19	17

(continued)

(continued)

Parameter	Parameter																			
	21	22	23	24	25	26	27	28	29	30	31	32	33	34	35	36	37	38	39	40
	32	33	39	40	38	33	34	32	36	37	35	28	26	27	31	29	30	25	23	24
13	26	27	18	19	17	21	22	20	15	16	14	22	20	21	16	14	15	19	17	18
	34	32	40	38	39	34	32	33	37	35	36	27	28	26	30	31	29	24	25	23
14	4	3	5	6	7	8	9	10	11	12	13	5	7	6	11	13	12	8	10	9
	19	18	32	34	33	38	40	39	35	37	36	23	25	24	29	31	30	26	28	27
15	2	4	7	5	6	10	8	9	13	11	12	6	5	7	12	11	13	9	8	10
	18	17	34	33	32	40	39	38	37	36	35	25	24	23	31	30	29	28	27	26
16	3	2	6	7	5	9	10	8	12	13	11	7	6	5	13	12	11	10	9	8
	17	19	33	32	34	39	38	40	36	35	37	24	23	25	30	29	31	27	26	28
17	3	4	11	12	13	5	6	7	8	9	10	8	10	9	5	7	6	11	13	12
	16	15	38	40	39	35	37	36	32	34	33	29	31	30	26	28	27	23	25	24
18	2	3	13	11	12	7	5	6	10	8	9	9	8	10	6	5	7	12	11	13
	15	14	40	39	38	37	36	35	34	33	32	31	30	29	28	27	26	25	24	24
19	4	2	12	13	11	6	7	5	9	10	8	10	9	8	7	6	5	13	12	11
	14	16	39	38	40	36	35	37	33	32	34	30	29	31	27	26	28	24	23	25

Interaction Table L_{81}, continued

Para-meter	20	21	22	23	24	25	26	27	28	29	30	31	32	33	34	35	36	37	38	39	40
20	(20)	1	1	8	9	10	11	12	13	5	6	7	11	13	12	8	10	9	5	7	6
		22	21	35	37	36	32	34	33	38	40	39	26	28	27	23	25	24	29	31	30
21		(21)	1	10	8	9	13	11	12	7	5	6	12	11	13	9	8	10	6	5	7
			20	37	36	35	34	33	32	40	39	38	28	27	26	25	24	23	31	30	29
22			(22)	9	10	8	12	13	11	6	7	5	13	12	11	10	9	8	7	6	5
				36	35	37	33	32	34	39	38	40	27	26	28	24	23	25	30	29	31
23				(23)	1	1	2	3	4	2	4	3	5	6	7	8	9	10	11	12	13
					25	24	29	31	30	26	28	27	14	16	15	20	22	21	17	19	18
24					(24)	1	4	2	3	3	2	4	7	5	6	10	8	9	13	11	12
						23	31	30	29	28	27	26	16	15	14	22	21	20	19	18	17
25						(25)	3	4	2	4	3	2	6	7	5	9	10	8	12	13	11
							30	29	31	27	26	28	15	14	16	21	20	22	18	17	19
26							(26)	1	1	2	3	4	11	12	13	5	6	7	8	9	10
								28	27	23	25	24	20	22	21	17	19	18	14	16	15
27								(27)	1	4	2	3	13	11	12	7	5	6	10	8	9
									23	25	24	23	22	21	20	19	18	17	16	15	14
28									(28)	3	4	2	12	13	11	6	7	5	9	10	8
										24	23	25	21	20	22	18	17	19	15	14	16
29										(29)	1	1	8	9	10	11	12	13	5	6	7
											31	30	17	19	18	14	16	15	20	22	21
30											(30)	1	10	8	9	13	11	12	7	5	6
												29	19	18	17	16	15	14	22	21	20
31												(31)	9	10	8	12	13	11	6	7	5
													18	17	19	15	14	16	21	20	22
32													(32)	1	1	2	3	4	2	4	3
														34	33	38	40	39	35	37	36
33														(33)	1	4	2	3	3	2	4
															32	40	39	38	37	36	35
34															(34)	3	4	2	4	3	2
																39	38	40	36	35	37
35																(35)	1	1	2	3	4
																	37	36	32	34	33
36																	(36)	1	4	2	3
																		35	34	33	32
37																		(37)	3	4	2
																			33	32	34
38																			(38)	1	1
																				40	39
39																				(39)	1
																					38

Modified Orthogonal Array L_{81} (9^{10})

Experiment No.	Parameter									
	1	2	3	4	5	6	7	8	9	10
1	1	1	1	1	1	1	1	1	1	1
2	1	2	3	4	5	6	7	8	9	2
3	1	3	2	7	9	8	4	6	5	3
4	1	4	7	8	2	5	6	9	3	4
5	1	5	9	2	6	7	3	4	8	5
6	1	6	8	5	7	3	9	2	4	6
7	1	7	4	6	3	9	8	5	2	7
8	1	8	6	9	4	2	5	3	7	8
9	1	9	5	3	8	4	2	7	6	9
10	2	1	3	8	7	9	5	4	6	3
11	2	2	2	2	2	2	2	2	2	1
12	2	3	1	5	6	4	8	9	7	2
13	2	4	9	6	8	1	7	3	5	6
14	2	5	8	9	3	6	4	7	1	4
15	2	6	7	3	4	8	1	5	9	5
16	2	7	6	1	9	5	3	8	4	9
17	2	8	5	4	1	7	9	6	3	7
18	2	9	4	7	5	3	6	1	8	8
19	3	1	2	6	4	5	9	7	8	2
20	3	2	1	9	8	7	6	5	4	3
21	3	3	3	3	3	3	3	3	3	1
22	3	4	8	1	5	9	2	6	7	5
23	3	5	7	4	9	2	8	1	6	6
24	3	6	9	7	1	4	5	8	2	4
25	3	7	5	8	6	1	4	2	9	8
26	3	8	4	2	7	6	1	9	5	9
27	3	9	6	5	2	8	7	4	1	7
28	4	1	7	9	6	3	2	8	5	7
29	4	2	9	3	7	5	8	6	1	8
30	4	3	8	6	2	7	5	1	9	9
31	4	4	4	4	4	4	4	4	4	1
32	4	5	6	7	8	9	1	2	3	2
33	4	6	5	1	3	2	7	9	8	3
34	4	7	1	2	5	8	9	3	6	4
35	4	8	3	5	9	1	6	7	2	5
36	4	9	2	8	1	6	3	5	7	6
37	5	1	9	4	3	8	6	2	7	9
38	5	2	8	7	4	1	3	9	6	7
39	5	3	7	1	8	6	9	4	2	8

(continued)

(continued)

Experiment No.	Parameter									
	1	2	3	4	5	6	7	8	9	10
40	5	4	6	2	1	3	8	7	9	3
41	5	5	5	5	5	5	5	5	5	1
42	5	6	4	8	9	7	2	3	1	2
43	5	7	3	9	2	4	1	6	8	6
44	5	8	2	3	6	9	7	1	4	4
45	5	9	1	6	7	2	4	8	3	5

Modified Orthogonal Array L_{81} (9^{10}), continued

Experiment No.	Parameter									
	1	2	3	4	5	6	7	8	9	10
46	6	1	8	2	9	4	7	5	3	8
47	6	2	7	5	1	9	4	3	8	9
48	6	3	9	8	5	2	1	7	4	7
49	6	4	5	9	7	8	3	1	2	2
50	6	5	4	3	2	1	9	8	7	3
51	6	6	6	6	6	6	6	6	6	1
52	6	7	2	4	8	3	5	9	1	5
53	6	8	1	7	3	5	2	4	9	6
54	6	9	3	1	4	7	8	2	5	4
55	7	1	4	5	8	2	3	6	9	4
56	7	2	6	8	3	4	9	1	5	5
57	7	3	5	2	4	9	6	8	1	6
58	7	4	1	3	9	6	5	2	8	7
59	7	5	3	6	1	8	2	9	4	8
60	7	6	2	9	5	1	8	4	3	9
61	7	7	7	7	7	7	7	7	7	1
62	7	8	9	1	2	3	4	5	6	2
63	7	9	8	4	6	5	1	3	2	3

(continued)

(continued)

Experiment No.	Parameter									
	1	2	3	4	5	6	7	8	9	10
64	8	1	6	3	5	7	4	9	2	6
65	8	2	5	6	9	3	1	4	7	4
66	8	3	4	9	1	5	7	2	6	5
67	8	4	3	7	6	2	9	5	1	9
68	8	5	2	1	7	4	6	3	9	7
69	8	6	1	4	2	9	3	7	5	8
70	8	7	9	5	4	6	2	1	3	3
71	8	8	8	8	8	8	8	8	8	1
72	8	9	7	2	3	1	5	6	4	2
73	9	1	5	7	2	6	8	3	4	5
74	9	2	4	1	6	8	5	7	3	6
75	9	3	6	4	7	1	2	5	8	4
76	9	4	2	5	3	7	1	8	6	8
77	9	5	1	8	4	3	7	6	2	9
78	9	6	3	2	8	5	4	1	7	7
79	9	7	8	3	1	2	6	4	5	2
80	9	8	7	6	5	4	3	2	1	3
81	9	9	9	9	9	9	9	9	9	1

Partial Orthogonal Array L_{108} (3^{49})

Experiment No.	Parameter																								
	1	2	3	4	5	6	7	8	9	10	11	12	13	14	15	16	17	18	19	20	21	22	23	24	25
1	1	1	1	1	1	1	1	1	1	1	1	1	1	1	1	1	1	1	1	1	1	1	1	1	1
2	1	1	1	1	1	1	1	1	1	1	1	1	1	2	2	2	2	2	2	2	2	2	2	2	2
3	1	1	1	1	1	1	1	1	1	1	1	1	1	3	3	3	3	3	3	3	3	3	3	3	3
4	1	2	2	2	2	2	2	2	2	2	2	2	2	1	1	1	1	1	1	1	1	1	1	1	1
5	1	2	2	2	2	2	2	2	2	2	2	2	2	2	2	2	2	2	2	2	2	2	2	2	2
6	1	2	2	2	2	2	2	2	2	2	2	2	2	3	3	3	3	3	3	3	3	3	3	3	3
7	1	3	3	3	3	3	3	3	3	3	3	3	3	1	1	1	1	1	1	2	2	2	2	2	2
8	1	3	3	3	3	3	3	3	3	3	3	3	3	2	2	2	2	2	2	3	3	3	3	3	3
9	1	3	3	3	3	3	3	3	3	3	3	3	3	3	3	3	3	3	3	1	1	1	1	1	1
10	1	1	1	1	1	2	2	2	2	3	3	3	3	1	1	1	1	1	1	3	3	3	3	3	3
11	1	1	1	1	1	2	2	2	3	3	3	3	3	2	2	2	2	2	2	1	1	1	1	1	1
12	1	1	1	1	1	2	2	2	2	3	3	3	3	3	3	3	3	3	3	2	2	2	2	2	2
13	1	2	2	2	2	3	3	3	3	1	1	1	1	1	1	1	1	1	1	2	2	2	2	2	2
14	1	2	2	2	2	3	3	3	3	1	1	1	1	2	2	2	2	2	2	3	3	3	3	3	3
15	1	2	2	2	2	3	3	3	3	1	1	1	1	3	3	3	3	3	3	1	1	1	1	1	1
16	1	3	3	3	3	1	1	1	1	2	2	2	2	1	1	1	1	1	1	3	3	3	3	3	3
17	1	3	3	3	3	1	1	1	1	2	2	2	2	2	2	2	2	2	2	1	1	1	1	1	1
18	1	3	3	3	3	1	1	1	1	2	2	2	2	3	3	3	3	3	3	2	2	2	2	2	2
19	1	1	1	2	3	1	2	3	3	1	2	2	3	1	1	2	2	3	3	1	1	2	2	3	3
20	1	1	1	2	3	1	2	3	3	1	2	2	3	2	2	3	3	1	1	2	2	3	3	1	1
21	1	1	1	2	3	1	2	3	3	1	2	2	3	3	3	1	1	2	2	3	3	1	1	2	2
22	1	2	2	3	1	2	3	1	1	2	3	3	1	1	1	2	2	3	3	1	1	2	2	3	3
23	1	2	2	3	1	2	3	1	1	2	3	3	1	2	2	3	3	1	1	2	2	3	3	1	1

(continued)

(continued)

Experiment No.	Parameter																								
	1	2	3	4	5	6	7	8	9	10	11	12	13	14	15	16	17	18	19	20	21	22	23	24	25
24	1	2	2	3	1	2	3	1	1	2	3	3	1	3	3	1	1	2	2	3	3	1	1	2	2
25	1	3	3	1	2	3	1	2	2	3	1	1	2	1	1	2	2	3	3	2	2	3	3	1	1
26	1	3	3	1	2	3	1	2	2	3	1	1	2	2	2	3	3	1	1	3	3	1	1	2	2
27	1	3	3	1	2	3	1	2	2	3	1	1	2	3	3	1	1	2	2	1	1	2	2	3	3
28	1	1	1	3	2	1	3	2	3	2	1	3	2	1	1	2	2	3	3	3	3	1	1	2	2
29	1	1	1	3	2	1	3	2	3	2	1	3	2	2	2	3	3	1	1	1	1	2	2	3	3
30	1	1	1	3	2	1	3	2	3	2	1	3	2	3	3	1	1	2	2	2	2	3	3	1	1
31	1	2	2	1	3	2	1	3	1	3	2	1	3	1	1	2	2	3	3	2	2	3	3	1	1
32	1	2	2	1	3	2	1	3	1	3	2	1	3	2	2	3	3	1	1	3	3	1	1	2	2
33	1	2	2	1	3	2	1	3	1	3	2	1	3	3	3	1	1	2	2	1	1	2	2	3	3
34	1	3	3	2	1	3	2	1	2	1	3	2	1	1	1	2	2	3	3	3	3	1	1	2	2
35	1	3	3	2	1	3	2	1	2	1	3	2	1	2	2	3	3	1	1	1	1	2	2	3	3
36	1	3	3	2	1	3	2	1	2	1	3	2	1	3	3	1	1	2	2	2	2	3	3	1	1
37	2	1	2	3	1	3	2	1	3	3	2	1	2	1	2	1	3	2	3	1	2	1	3	2	3
38	2	1	2	3	1	3	2	1	3	3	2	1	2	2	3	2	1	3	1	2	3	2	1	3	1
39	2	1	2	3	1	3	2	1	3	3	2	1	2	3	1	3	2	1	2	3	1	3	2	1	2
40	2	2	3	1	2	1	3	2	1	1	3	2	3	1	2	1	3	2	3	1	2	1	3	2	3
41	2	2	3	1	2	1	3	2	1	1	3	2	3	2	3	2	1	3	1	2	3	2	1	3	1
42	2	2	3	1	2	1	3	2	1	1	3	2	3	3	1	3	2	1	2	3	1	3	2	1	2
43	2	3	1	2	3	2	1	3	2	2	1	3	1	1	2	1	3	2	3	2	3	2	1	3	1
44	2	3	1	2	3	2	1	3	2	2	1	3	1	2	3	2	1	3	1	3	1	3	2	1	2
45	2	3	1	2	3	2	1	3	2	2	1	3	1	3	1	3	2	1	2	1	2	1	3	2	3
46	2	1	2	3	2	1	1	3	2	3	3	2	1	1	2	1	3	2	3	3	1	3	2	1	2

Partial Orthogonal Array L_{108} (3^{49}), continued

Experiment No.	Parameter																							
	1	2	3	4	5	6	7	8	9	10	11	12	13	14	15	16	17	18	19	20	21	22	23	24	25
47	2	1	2	3	2	1	1	3	2	3	3	2	1	2	3	2	1	3	1	1	2	1	3	2	3
48	2	1	2	3	2	1	1	3	2	3	3	2	1	3	1	3	2	1	2	2	3	2	1	3	1
49	2	2	3	1	3	2	2	1	3	1	1	3	2	1	2	1	3	2	3	2	3	2	1	3	1
50	2	2	3	1	3	2	2	1	3	1	1	3	2	2	3	2	1	3	1	3	1	3	2	1	2
51	2	2	3	1	3	2	2	1	3	1	1	3	2	3	1	3	2	1	2	1	2	1	3	2	3
52	2	3	1	2	1	3	3	2	1	2	2	1	3	1	2	1	3	2	3	3	1	3	2	1	2
53	2	3	1	2	1	3	3	2	1	2	2	1	3	2	3	2	1	3	1	1	2	1	3	2	3
54	2	3	1	2	1	3	3	2	1	2	2	1	3	3	1	3	2	1	2	2	3	2	1	3	1
55	2	1	2	1	3	3	3	1	2	2	1	2	3	1	3	3	2	2	1	1	3	3	2	2	1
56	2	1	2	1	3	3	3	1	2	2	1	2	3	2	1	1	3	3	2	2	1	1	3	3	2
57	2	1	2	1	3	3	3	1	2	2	1	2	3	3	2	2	1	1	3	3	2	2	1	1	3
58	2	2	3	2	1	1	1	2	3	3	2	3	1	1	3	3	2	2	1	1	3	3	2	2	1
59	2	2	3	2	1	1	1	2	3	3	2	3	1	2	1	1	3	3	2	2	1	1	3	3	2
60	2	2	3	2	1	1	1	2	3	3	2	3	1	3	2	2	1	1	3	3	2	2	1	1	3
61	2	3	1	3	2	2	2	3	1	1	3	1	2	1	3	3	2	2	1	2	1	1	3	3	2
62	2	3	1	3	2	2	2	3	1	1	3	1	2	2	1	1	3	3	2	3	2	2	1	1	3
63	2	3	1	3	2	2	2	3	1	1	3	1	2	3	2	2	1	1	3	1	3	3	2	2	1
64	2	1	2	2	3	3	1	2	1	1	3	3	2	1	3	3	2	2	1	3	2	2	1	1	3
65	2	1	2	2	3	3	1	2	1	1	3	3	2	2	1	1	3	3	2	1	3	3	2	2	1
66	2	1	2	2	3	3	1	2	1	1	3	3	2	3	2	2	1	1	3	2	1	1	3	3	2
67	2	2	3	3	1	1	2	3	2	2	1	1	3	1	3	3	2	2	1	2	1	1	3	3	2
68	2	2	3	3	1	1	2	3	2	2	1	1	3	2	1	1	3	3	2	3	2	2	1	1	3
69	2	2	3	3	1	1	2	3	2	2	1	1	3	3	2	2	1	1	3	1	3	3	2	2	1

(continued)

(continued)

Experiment No.	Parameter																								
	1	2	3	4	5	6	7	8	9	10	11	12	13	14	15	16	17	18	19	20	21	22	23	24	25
70	2	3	1	1	2	2	3	1	3	3	2	2	1	1	3	3	2	2	1	3	2	2	1	1	3
71	2	3	1	1	2	2	3	1	3	3	2	2	1	2	1	1	3	3	2	1	3	3	2	2	1
72	2	3	1	1	2	2	3	1	3	3	2	2	1	3	2	2	1	1	3	2	1	1	3	3	2
73	3	1	3	2	1	2	3	3	1	3	1	2	2	1	2	3	1	3	2	1	2	3	1	3	2
74	3	1	3	2	1	2	3	3	1	3	1	2	2	2	3	1	2	1	3	2	3	1	2	1	3
75	3	1	3	2	1	2	3	3	1	3	1	2	2	3	1	2	3	2	1	3	1	2	3	2	1
76	3	2	1	3	2	3	1	1	2	1	2	3	3	1	2	3	1	3	2	1	2	3	1	3	2
77	3	2	1	3	2	3	1	1	2	1	2	3	3	2	3	1	2	1	3	2	3	1	2	1	3
78	3	2	1	3	2	3	1	1	2	1	2	3	3	3	1	2	3	2	1	3	1	2	3	2	1
79	3	3	2	1	3	1	2	2	3	2	3	1	1	1	2	3	1	3	2	2	3	1	2	1	3
80	3	3	2	1	3	1	2	2	3	2	3	1	1	2	3	1	2	1	3	3	1	2	3	2	1
81	3	3	2	1	3	1	2	2	3	2	3	1	1	3	1	2	3	2	1	1	2	3	1	3	2
82	3	1	3	2	2	2	1	1	3	2	3	1	3	1	2	3	1	3	2	3	1	2	3	2	1
83	3	1	3	2	2	2	1	1	3	2	3	1	3	2	3	1	2	1	3	1	2	3	1	3	2
84	3	1	3	2	2	2	1	1	3	2	3	1	3	3	1	2	3	2	1	2	3	1	2	1	3
85	3	2	1	3	3	3	2	2	1	3	1	2	1	1	2	3	1	3	2	2	3	1	2	1	3
86	3	2	1	3	3	3	2	2	1	3	1	2	1	2	3	1	2	1	3	3	1	2	3	2	1
87	3	2	1	3	3	3	2	2	1	3	1	2	1	3	1	2	3	2	1	1	2	3	1	3	2
88	3	3	2	1	1	1	3	3	2	1	2	3	2	1	2	3	1	3	2	3	1	2	3	2	1
89	3	3	2	1	1	1	3	3	2	1	2	3	2	2	3	1	2	1	3	1	2	3	1	3	2
90	3	3	2	1	1	1	3	3	2	1	2	3	2	3	1	2	3	2	1	2	3	1	2	1	3
91	3	1	3	3	3	2	3	2	2	1	2	1	1	1	3	2	3	1	2	1	3	2	3	1	2
92	3	1	3	3	3	2	3	2	2	1	2	1	1	2	1	3	1	2	3	2	1	3	1	2	3

Partial Orthogonal Array L_{108} (3^{49}), continued

Experiment No.	Parameter																								
	1	2	3	4	5	6	7	8	9	10	11	12	13	14	15	16	17	18	19	20	21	22	23	24	25
93	3	1	3	3	3	2	3	2	2	1	2	1	1	3	2	1	2	3	1	3	2	1	2	3	1
94	3	2	1	1	1	3	1	3	3	2	3	2	2	1	3	2	3	1	2	1	3	2	3	1	2
95	3	2	1	1	1	3	1	3	3	2	3	2	2	2	1	3	1	2	3	2	1	3	1	2	3
96	3	2	1	1	1	3	1	3	3	2	3	2	2	3	2	1	2	3	1	3	2	1	2	3	1
97	3	3	2	2	2	1	2	1	1	3	1	3	3	1	3	2	3	1	2	2	1	3	1	2	3
98	3	3	2	2	2	1	2	1	1	3	1	3	3	2	1	3	1	2	3	3	2	1	2	3	1
99	3	3	2	2	2	1	2	1	1	3	1	3	3	3	2	1	2	3	1	1	3	2	3	1	2
100	3	1	3	1	2	3	2	3	1	2	2	3	1	1	3	2	3	1	2	3	2	1	2	3	1
101	3	1	3	1	2	3	2	3	1	2	2	3	1	2	1	3	1	2	3	1	3	2	3	1	2
102	3	1	3	1	2	3	2	3	1	2	2	3	1	3	2	1	2	3	1	2	1	3	1	2	3
103	3	2	1	2	3	1	3	1	2	3	3	1	2	1	3	2	3	1	2	2	1	3	1	2	3
104	3	2	1	2	3	1	3	1	2	3	3	1	2	2	1	3	1	2	3	3	2	1	2	3	1
105	3	2	1	2	3	1	3	1	2	3	3	1	2	3	2	1	2	3	1	1	3	2	3	1	2
106	3	3	2	3	1	2	1	2	3	1	1	2	3	1	3	2	3	1	2	3	2	1	2	3	1
107	3	3	2	3	1	2	1	2	3	1	1	2	3	2	1	3	1	2	3	1	3	2	3	1	2
108	3	3	2	3	1	2	1	2	3	1	1	2	3	3	2	1	2	3	1	2	1	3	1	2	3

Partial Orthogonal Array L_{108} (3^{49}), continued

Experiment No.	Parameter																							
	26	27	28	29	30	31	32	33	34	35	36	37	38	39	40	41	42	43	44	45	46	47	48	49
1	1	1	1	1	1	1	1	1	1	1	1	1	1	1	1	1	1	1	1	1	1	1	1	1
2	2	2	2	2	2	2	2	2	2	2	2	2	2	2	2	2	2	2	2	2	2	2	2	2
3	3	3	3	3	3	3	3	3	3	3	3	3	3	3	3	3	3	3	3	3	3	3	3	3
4	2	2	2	2	2	2	2	2	2	2	2	2	3	3	3	3	3	3	3	3	3	3	3	3
5	3	3	3	3	3	3	3	3	3	3	3	3	1	1	1	1	1	1	1	1	1	1	1	1
6	1	1	1	1	1	1	1	1	1	1	1	1	2	2	2	2	2	2	2	2	2	2	2	2
7	1	1	1	1	1	1	3	3	3	3	3	3	2	2	2	2	2	2	3	3	3	3	3	3
8	2	2	2	2	2	2	1	1	1	1	1	1	3	3	3	3	3	3	1	1	1	1	1	1
9	3	3	3	3	3	3	2	2	2	2	2	2	1	1	1	1	1	1	2	2	2	2	2	2
10	3	3	3	3	3	3	2	2	2	2	2	2	2	2	2	2	2	2	1	1	1	1	1	1
11	1	1	1	1	1	1	3	3	3	3	3	3	3	3	3	3	3	3	2	2	2	2	2	2
12	2	2	2	2	2	2	1	1	1	1	1	1	1	1	1	1	1	1	3	3	3	3	3	3
13	3	3	3	3	3	3	1	1	1	1	1	1	3	3	3	3	3	3	2	2	2	2	2	2
14	1	1	1	1	1	1	2	2	2	2	2	2	1	1	1	1	1	1	3	3	3	3	3	3
15	2	2	2	2	2	2	3	3	3	3	3	3	2	2	2	2	2	2	1	1	1	1	1	1
16	2	2	2	2	2	2	3	3	3	3	3	3	1	1	1	1	1	1	2	2	2	2	2	2
17	3	3	3	3	3	3	1	1	1	1	1	1	2	2	2	2	2	2	3	3	3	3	3	3
18	1	1	1	1	1	1	2	2	2	2	2	2	3	3	3	3	3	3	1	1	1	1	1	1
19	1	1	2	2	3	3	1	1	2	2	3	3	1	1	2	2	3	3	1	1	2	2	3	3
20	2	2	3	3	1	1	2	2	3	3	1	1	2	2	3	3	1	1	2	2	3	3	1	1
21	3	3	1	1	2	2	3	3	1	1	2	2	3	3	1	1	2	2	3	3	1	1	2	2
22	2	2	3	3	1	1	2	2	3	3	1	1	3	3	1	1	2	2	3	3	1	1	2	2
23	3	3	1	1	2	2	3	3	1	1	2	2	1	1	2	2	3	3	1	1	2	2	3	3

(continued)

(continued)

Experiment No.	Parameter																							
	26	27	28	29	30	31	32	33	34	35	36	37	38	39	40	41	42	43	44	45	46	47	48	49
24	1	1	2	2	3	3	1	1	2	2	3	3	2	2	3	3	1	1	2	2	3	3	1	1
25	1	1	2	2	3	3	3	3	1	1	2	2	2	2	3	3	1	1	3	3	1	1	2	2
26	2	2	3	3	1	1	1	1	2	2	3	3	3	3	1	1	2	2	1	1	2	2	3	3
27	3	3	1	1	2	2	2	2	3	3	1	1	1	1	2	2	3	3	2	2	3	3	1	1
28	3	3	1	1	2	2	2	2	3	3	1	1	2	2	3	3	1	1	1	1	2	2	3	3
29	1	1	2	2	3	3	3	3	1	1	2	2	3	3	1	1	2	2	2	2	3	3	1	1
30	2	2	3	3	1	1	1	1	2	2	2	3	1	1	2	2	3	3	3	3	1	1	2	2
31	3	3	1	1	2	2	1	1	2	2	3	3	3	3	1	1	2	2	2	2	3	3	1	1
32	1	1	2	2	3	3	2	2	3	3	1	1	1	1	2	2	3	3	3	3	1	1	2	2
33	2	2	3	3	1	1	3	3	1	1	2	2	2	2	3	3	1	1	1	1	2	2	3	3
34	2	2	3	3	1	1	3	3	1	1	2	2	1	1	2	2	3	3	2	2	3	3	1	1
35	3	3	1	1	2	2	1	1	2	2	3	3	2	2	3	3	1	1	3	3	1	1	2	2
36	1	1	2	2	3	3	2	2	3	3	1	1	3	3	1	1	2	2	1	1	2	2	3	3
37	1	2	1	3	2	3	1	2	1	3	2	3	1	2	1	3	2	3	1	2	1	3	2	3
38	2	3	2	1	3	1	2	3	2	1	3	1	2	3	2	1	3	1	2	3	2	1	3	1
39	3	1	3	2	1	2	3	1	3	2	1	2	3	1	3	2	1	2	3	1	3	2	1	2
40	2	3	2	1	3	1	2	3	2	1	3	1	3	1	3	2	1	2	3	1	3	2	1	2
41	3	1	3	2	1	2	3	1	3	2	1	2	1	2	1	3	2	3	1	2	1	3	2	3
42	1	2	1	3	2	3	1	2	1	3	2	3	2	3	2	1	3	1	2	3	2	1	3	1
43	1	2	1	3	2	3	3	1	3	2	1	2	2	3	2	1	3	1	3	1	3	2	1	2
44	2	3	2	1	3	1	1	2	1	3	2	3	3	1	3	2	1	2	1	2	1	3	2	3
45	3	1	3	2	1	2	2	3	2	1	3	1	1	2	1	3	2	3	2	3	2	1	3	1
46	3	1	3	2	1	2	2	3	2	1	3	1	2	3	2	1	3	1	1	2	1	3	2	3

Partial Orthogaonal Array L_{108} (3^{49}), continued

Experiment No.	Parameter																							
	26	27	28	29	30	31	32	33	34	35	36	37	38	39	40	41	42	43	44	45	46	47	48	49
47	1	2	1	3	2	3	3	1	3	2	1	2	3	1	3	2	1	2	2	3	2	1	3	1
48	2	3	2	1	3	1	1	2	1	3	2	3	1	2	1	3	2	3	3	1	3	2	1	2
49	3	1	3	2	1	2	1	2	1	3	2	3	3	1	3	2	1	2	2	3	2	1	3	1
50	1	2	1	3	2	3	2	3	2	1	3	1	1	2	1	3	2	3	3	1	3	2	1	2
51	2	3	2	1	3	1	3	1	3	2	1	2	2	3	2	1	3	1	1	2	1	3	2	3
52	2	3	2	1	3	1	3	1	3	2	1	2	1	2	1	3	2	3	2	3	2	1	3	1
53	3	1	3	2	1	2	1	2	1	3	2	3	2	3	2	1	3	1	3	1	3	2	1	2
54	1	2	1	3	2	3	2	3	2	1	3	1	3	1	3	2	1	2	1	2	1	3	2	3
55	1	3	3	2	2	1	1	3	3	2	2	1	1	3	3	2	2	1	1	3	3	2	2	1
56	2	1	1	3	3	2	2	1	1	3	3	2	2	1	1	3	3	2	2	1	1	3	3	2
57	3	2	2	1	1	3	3	2	2	1	1	3	3	2	2	1	1	3	3	2	2	1	1	3
58	2	1	1	3	3	2	2	1	1	3	3	2	3	2	2	1	1	3	3	2	2	1	1	3
59	3	2	2	1	1	3	3	2	2	1	1	3	1	3	3	2	2	1	1	3	3	2	2	1
60	1	3	3	2	2	1	1	3	3	2	2	1	2	1	1	3	3	2	2	1	1	3	3	2
61	1	3	3	2	2	1	3	2	2	1	1	3	2	1	1	3	3	2	3	2	2	1	1	3
62	2	1	1	3	3	2	1	3	3	2	2	1	3	2	2	1	1	3	1	3	3	2	2	1
63	3	2	2	1	1	3	2	1	1	3	3	2	1	3	3	2	2	1	2	1	1	3	3	2
64	3	2	2	1	1	3	2	1	1	3	3	2	2	1	1	3	3	2	1	3	3	2	2	1
65	1	3	3	2	2	1	3	2	2	1	1	3	3	2	2	1	1	3	2	1	1	3	3	2
66	2	1	1	3	3	2	1	3	3	2	2	1	1	3	3	2	2	1	3	2	2	1	1	3
67	3	2	2	1	1	3	1	3	3	2	2	1	3	2	2	1	1	3	2	1	1	3	3	2
68	1	3	3	2	2	1	2	1	1	3	3	2	1	3	3	2	2	1	3	2	2	1	1	3
69	2	1	1	3	3	2	3	2	2	1	1	3	2	1	1	3	3	2	1	3	3	2	2	1

(continued)

(continued)

Experiment No.	Parameter																							
	26	27	28	29	30	31	32	33	34	35	36	37	38	39	40	41	42	43	44	45	46	47	48	49
70	2	1	1	3	3	2	3	2	2	1	1	3	1	3	3	2	2	1	2	1	1	3	3	2
71	3	2	2	1	1	3	1	3	3	2	2	1	2	1	1	3	3	2	3	2	2	1	1	3
72	1	3	3	2	2	1	2	1	1	3	3	2	3	2	2	1	1	3	1	3	3	2	2	1
73	1	2	3	1	3	2	1	2	3	1	3	2	1	2	3	1	3	2	1	2	3	1	3	2
74	2	3	1	2	1	3	2	3	1	2	1	3	2	3	1	2	1	3	2	3	1	2	1	3
75	3	1	2	3	2	1	3	1	2	3	2	1	3	1	2	3	2	1	3	1	2	3	2	1
76	2	3	1	2	1	3	2	3	1	2	1	3	3	1	2	3	2	1	3	1	2	3	2	1
77	3	1	2	3	2	1	3	1	2	3	2	1	1	2	3	1	3	2	1	2	3	1	3	2
78	1	2	3	1	3	2	1	2	3	1	3	2	2	3	1	2	1	3	2	3	1	2	1	3
79	1	2	3	1	3	2	3	1	2	3	2	1	2	3	1	2	1	3	3	1	2	3	2	1
80	2	3	1	2	1	3	1	2	3	1	3	2	3	1	2	3	2	1	1	2	3	1	3	2
81	3	1	2	3	2	1	2	3	1	2	1	3	1	2	3	1	3	2	2	3	1	2	1	3
82	3	1	2	3	2	1	2	3	1	2	1	3	2	3	1	2	1	3	1	2	3	1	3	2
83	1	2	3	1	3	2	3	1	2	3	2	1	3	1	2	3	2	1	2	3	1	2	1	3
84	2	3	1	2	1	3	1	2	3	1	3	2	1	2	3	1	3	2	3	1	2	3	2	1
85	3	1	2	3	2	1	1	2	3	1	3	2	3	1	2	3	2	1	2	3	1	2	1	3
86	1	2	3	1	3	2	2	3	1	2	1	3	1	2	3	1	3	2	3	1	2	3	2	1
87	2	3	1	2	1	3	3	1	2	3	2	1	2	3	1	2	1	3	1	2	3	1	3	2
88	2	3	1	2	1	3	3	1	2	3	2	1	1	2	3	1	3	2	2	3	1	2	1	3
89	3	1	2	3	2	1	1	2	3	1	3	2	2	3	1	2	1	3	3	1	2	3	2	1
90	1	2	3	1	3	2	2	3	1	2	1	3	3	1	2	3	2	1	1	2	3	1	3	2
91	1	3	2	3	1	2	1	3	2	3	1	2	1	3	2	3	1	2	1	3	2	3	1	2
92	2	1	3	1	2	3	2	1	3	1	2	3	2	1	3	1	2	3	2	1	3	1	2	3

Partial Orthogonal Array L_{108} (3^{49}), continued

Experiment No.	Parameter																							
	26	27	28	29	30	31	32	33	34	35	36	37	38	39	40	41	42	43	44	45	46	47	48	49
93	3	2	1	2	3	1	3	2	1	2	3	1	3	2	1	2	3	1	3	2	1	2	3	1
94	2	1	3	1	2	3	2	1	3	1	2	3	3	2	1	2	3	1	3	2	1	2	3	1
95	3	2	1	2	3	1	3	2	1	2	3	1	1	3	2	3	1	2	1	3	2	3	1	2
96	1	3	2	3	1	2	1	3	2	3	1	2	2	1	3	1	2	3	2	1	3	1	2	3
97	1	3	2	3	1	2	3	2	1	2	3	1	2	1	3	1	2	3	3	2	1	2	3	1
98	2	1	3	1	2	3	1	3	2	3	1	2	3	2	1	2	3	1	1	3	2	3	1	2
99	3	2	1	2	3	1	2	1	3	1	2	3	1	3	2	3	1	2	2	1	3	1	2	3
100	3	2	1	2	3	1	2	1	3	1	2	3	2	1	3	1	2	3	1	3	2	3	1	2
101	1	3	2	3	1	2	3	2	1	2	3	1	3	2	1	2	3	1	2	1	3	1	2	3
102	2	1	3	1	2	3	1	3	2	3	1	2	1	3	2	3	1	2	3	2	1	2	3	1
103	3	2	1	2	3	1	1	3	2	3	1	2	3	2	1	2	3	1	2	1	3	1	2	3
104	1	3	2	3	1	2	2	1	3	1	2	3	1	3	2	3	1	2	3	2	1	2	3	1
105	2	1	3	1	2	3	3	2	1	2	3	1	2	1	3	1	2	3	1	3	2	3	1	2
106	2	1	3	1	2	3	3	2	1	2	3	1	1	3	2	3	1	2	2	1	3	1	2	3
107	3	2	1	2	3	1	1	3	2	3	1	2	2	1	3	1	2	3	3	2	1	2	3	1
108	1	3	2	3	1	2	2	1	3	1	2	3	3	2	1	2	3	1	1	3	2	3	1	2

References

Part 1

Accompanying literature in Japanese journals

Main article by Dr. Genichi Taguchi and other authors:

Journal of Quality Engineering Forum **1**(1), (1993)—**6**(6), (1998)
Journal of Quality Engineering Society **7**(1), (1999)—**14**(3), (2006)
Standardization and Quality Control (1993–2006)

Articles by numerous authors

Journal of Quality Engineering Society **14**(4), (2006)—today
Standardization and Quality Control (2007)—today

Accompanying English literature

The ASI Journal **1**(1), (1988), ASI Press, (American Supplier Institute)
The ASI Journal **2**(1, 2), (1989), ASI Press, (American Supplier Institute)
The ASI Journal **3**(1, 2), (1990), ASI Press, (American Supplier Institute)

Books

Genichi Taguchi

System of Experimental Design, Volume I, Volume II
UNIPUB, Kraus International Publications, White Plains, New York (1987)

Genichi Taguchi, S. Konishi

Taguchi Methods: Orthogonal Arrays and Linear Graphs
ASI Press, (American Supplier Institute), (1987)

© Springer Nature Switzerland AG 2019

H. Ruefer, *Living Without Mathematical Statistics*,
https://doi.org/10.1007/978-3-319-99632-5

Genichi Taguchi and Yuin Wu

Quality Engineering Series: Taguchi Methods

Volume 1: Research and Development
Volume 2: On-line Production
Volume 3: Signal-to-Noise Ratio for Quality Evaluation
Volume 4: Design of Experiments
Volume 6: Case Studies from the U.S. and Europe

ASI Press, (American Supplier Institute) (1989–1994)

Genichi Taguchi

Taguchi on Robust Technology Development
ASME, Press, (The American Society of Mechanical Engineering), (1993)

Genichi Taguchi, Subir Chowdhury, and Yuin Wu

Taguchi's Quality Engineering Handbook
John Wiley & Sons Inc. (2005)

Genichi Taguchi, Rajesh Jugulum, and Shin Taguchi

Computer-Based Robust Engineering
American Society for Quality
Quality Press, Milwaukee (2004)

Yuin Wu and Alan Wu

Taguchi Methods for Robust Design
ASME, Press, (The American Society of Mechanical Engineering), (2000)

Teruo Mori

The new Experimental Design
ASI Press, (American Supplier Institute) (1990)

Teruo Mori

Taguchi Techniques for Image and Pattern Developing Technology
Prentice-Hall Inc., Englewood Cliffs, New Jersey (1995)

William Y. Fowlkes, Clyde M. Creveling

Engineering Methods for Robust Product Design
Addison-Wesley Publishing Company (1995)

Clyde M. Creveling

Tolerance Design, a Handbook for Developing Optimal Specifications
Addison-Wesley Publishing Company (1997)

Madhav Phadke

Quality Engineering using Robust Design
Prentice Hall (1989)

Sung H. Park

Robust Design and Analysis for Quality Engineering
Chapman & Hall (1996)

Literature on the text

1-1) Don P. Clausing

Total Quality Development (pages 204 cont.)
ASME Press, New York (1994)

1-2) Ira N. Levine

Physical Chemistry
McGraw Hill, (2001)

1-3) Tetsuo Hosokawa, Yoshitsugu Kouno, Masahiro Toyoda

Dispersion evaluation method for tolerance design using direct product table of orthogonal array
Journal of Quality Engineering Society **20**(4), 26–35 (2012) (in Japanese language)

1-4) Kei Takada, Masaru Uchikawa, Kazuhiro Kajimoto, and Jun-ichi Deguchi

Efficient debugging of a software using an orthogonal array
Quality Engineering **8**(1), 60–69 (2000)

1-5) Jonathan M. Samet

Concepts of Time in Clinical Research
Annals of Internal Medicine 132 (1), 37–44 (January 2000)

1-6) Altman, G., Douglas, Bland, J., Martin

Time to event (survival) data
British Medical Journal (BMJ) 317, 468–469 (August 1998)

Part 2

Accompanying English literature

Genichi Taguchi, Rajesh Jugulum

The Mahalanobis—Taguchi Strategy: A pattern technology system
John Wiley & Sons, New York (2002)

Literature on the text

2-1) Prasanta Chandra Mahalanobis

On the generalized distance in statistics
Proceedings of the National Institute of Science in India, Calcutta, **2**, 49–55 (1936)

2-2) Genichi Taguchi, Subir Chowdhury, Yuin Wu

The Mahalanobis—Taguchi System
McGraw—Hill, New York, NY, 10121–2298 (2001)

2-3) Shoichi Teshima, Tomonori Bando, Dan Jin

A research of defect detection using the Mahalanobis-Taguchi System Method
Journal of Quality Engineering Forum **6**(6), 74–80 (1998)

2-4) Genichi Taguchi

Application of Mahalanobis Distance for Medical Treatment
Journal of Quality Engineering Forum **2**(6), 2–6 (1994)
(Original in Japanese language; revised and translated in English in **(2-2)**)

2-5) Tatsuji Kanetaka

Diagnosis of a Special Health Check using Mahalanobis' Generalized Distance
The ASI Journal **3**(1), 57–74 (1990)
American Supplier Institute (ASI) Press
(Translated from the Japanese Journal Standardization and Quality Control, **40** (10), 57–64 (1987))

2-6) Sarah Parcak

Satellite Remote Sensing for Archaeology
Taylor & Francis LTD (2009)

2-7) Genichi Taguchi

Method for Pattern Recognition
United States Patent: US 5,684,892 A (4. November 1997)

2-8) Takashi Kamoshita, Kazuto Tabata, Harutoshi Okano, Kazuhito Takahashi, Hiroshi Yano

Optimization of a multi-dimensional information system using Mahalanobis Distance
(The case of a fire alarm system)
Quality Engineering Forum **3**(6), 91–99 (1998)

2-9) Kazutaka Mizoguchi, Akio Kawai, Kazutaka Hamada, Takashi Kamoshita

A research on a sensing system using Mahalanobis Distance for Preventing of driving accidents
Journal of Quality Engineering Society **3**(7), 79–86 (1999), (in Japanese language)

2-10) Matthias S. Goepfert, Alwin E. Goetz, and Herbert Ruefer

Surveillance of patients after open heart surgery by Mahalanobis—Taguchi—System
American Supplier Institute—Client Conference Proceedings 184–195 (2008), Detroit, Michigan, USA

2-11) Sakano Susumu

US Dollar Bill Inspection
Quality Engineering Forum Symposium, Proceedings (1998), (in Japanese language)
Editor QES, Tokyo, Japan

2-12) Shames, D.S., Minna, J.D., Gazdar, A.I.

DNA methylation in health, disease, and cancer
Current Molecular Medicine **7**(1), 85–102 (2007)

2-13) Global seismic activity data are processed and archived at

Geoforschungszentrum (GFZ), Potsdam, Germany
Data and additional information are accessible on the Internet
www.gfz-potsdam.de (first access Sept. 7, 2011)

2-14) Sloan Digital Sky Survey: www.sdss.org (first access July 1, 2014)

Part 3

Accompanying English literature and exercises with real data

Shoichi Teshima, Yoshiko Hasegawa, Kazuo Tatebayashi

Quality Recognition and Prediction: Smarter Pattern Technology with the Mahalanobis—Taguchi System
Momentum Press, LLC, New York (2012)

Genichi Taguchi, Rajesh Jugulum, and Shin Taguchi

Multivariate Data Analysis Method and Uses Thereof
United States Patent: US 7,043,401 B2 (9. Mai 2006)

For exercises, even with very high demands, data sets are available on the Internet, e.g.

www.kaggle.com
www.kdnuggetts.com

Literature on the text

3-1) Theodore Modis

Predictions. Society's Telltale Signature Reveals the Past and Forecasts the Future
Simon & Schuster, New York (1992)

3-2) Shoichi Teshima and Tatsuji Kanetaka

Comparison between Variation Pressure Method and MT-Method and their Diagnosis—The Medical Check Data as an example
Journal of Quality Engineering (QES) **20 (No. 1)**, 74–79 (2012)
(Abstract in English)

Index

© Springer Nature Switzerland AG 2019
H. Ruefer, *Living Without Mathematical Statistics*,
https://doi.org/10.1007/978-3-319-99632-5

MIX
Papier aus verantwortungsvollen Quellen
Paper from responsible sources
FSC® C105338

If you have any concerns about our products,
you can contact us on
ProductSafety@springernature.com

In case Publisher is established outside the EU,
the EU authorized representative is:
Springer Nature Customer Service Center GmbH
Europaplatz 3, 69115 Heidelberg, Germany

Printed by Libri Plureos GmbH
in Hamburg, Germany